COORDINATING AUTHOR

JILL PRICE

CONTRIBUTING AUTHORS

JILL PRICE

TRISH DOUGLAS

DENISE MILES

CATHY BEDSON

KINGSLEY HEAD

JANE WILSON

CLEO WESTHORPE

jacaranda
A Wiley Brand

Second edition published 2018 by
John Wiley & Sons Australia, Ltd
42 McDougall Street, Milton, Qld 4064

First edition published 2013

Typeset in 11/14 pt Times LT Std Roman

National Library of Australia
Cataloguing-in-publication data

Author:	Price, Jill, author.
Title:	Jacaranda Geography Alive 10 for the Australian Curriculum / Jill Price, Trish Douglas, Denise Miles, Cathy Bedson, Kingsley Head, Jane Wilson, Cleo Westhorpe
Edition:	Second edition.
ISBN:	978-0-730-34785-9 (paperback)
Notes:	Includes index.
Target Audience:	For secondary school age.
Subjects:	Geography — Textbooks. Geography — Study and teaching (Secondary)
Other Creators/ Contributors:	Douglas, Trish, author. Miles, Denise, author. Bedson, Cathy, author. Head, Kingsley, author. Wilson, Jane, author. Westhorpe, Cleo, author.

Front cover image: Jess Kraft / Shutterstock

Cartography by Spatial Vision, Melbourne and MAPgraphics Pty Ltd, Brisbane

Illustrated by various artists and the Wiley Art Studio.

Typeset in India by diacriTech

Printed in Singapore by
Markono Print Media Pte Ltd

10 9 8 7 6 5 4 3 2

This textbook contains images of Indigenous people who are, or may be, deceased. The publisher appreciates that this inclusion may distress some Indigenous communities. These images have been included so that the young multicultural audience for this book can better appreciate specific aspects of Indigenous history and experience.

It is recommended that teachers should first preview resources on Indigenous topics in relation to their suitability for the class level or situation. It is also suggested that Indigenous parents or community members be invited to help assess the resources to be shown to Indigenous children. At all times the guidelines laid down by the relevant educational authorities should be followed.

CONTENTS

4 Inland water — dammed, diverted and drained 69

5 Managing change in coastal environments 89

6 Marine environments — are we trashing our oceans? 109

7 Sustaining urban environments 132

8 Geographical inquiry: Developing an environmental management plan 156

UNIT 2 GEOGRAPHIES OF HUMAN WELLBEING 159

9 What makes a good life? 161

10 Human wellbeing and change 184

11 Is life the same everywhere? 207

12 Trapped by conflict 230

13 Fieldwork inquiry: Comparing wellbeing in the local area 261

HOW TO USE

the *Jacaranda Geography Alive* resource suite

For more effective learning, the *Jacaranda Geography Alive* series is now available on the learnON platform. The features described here show how you can use *Jacaranda Geography Alive* to optimise your learning experience.

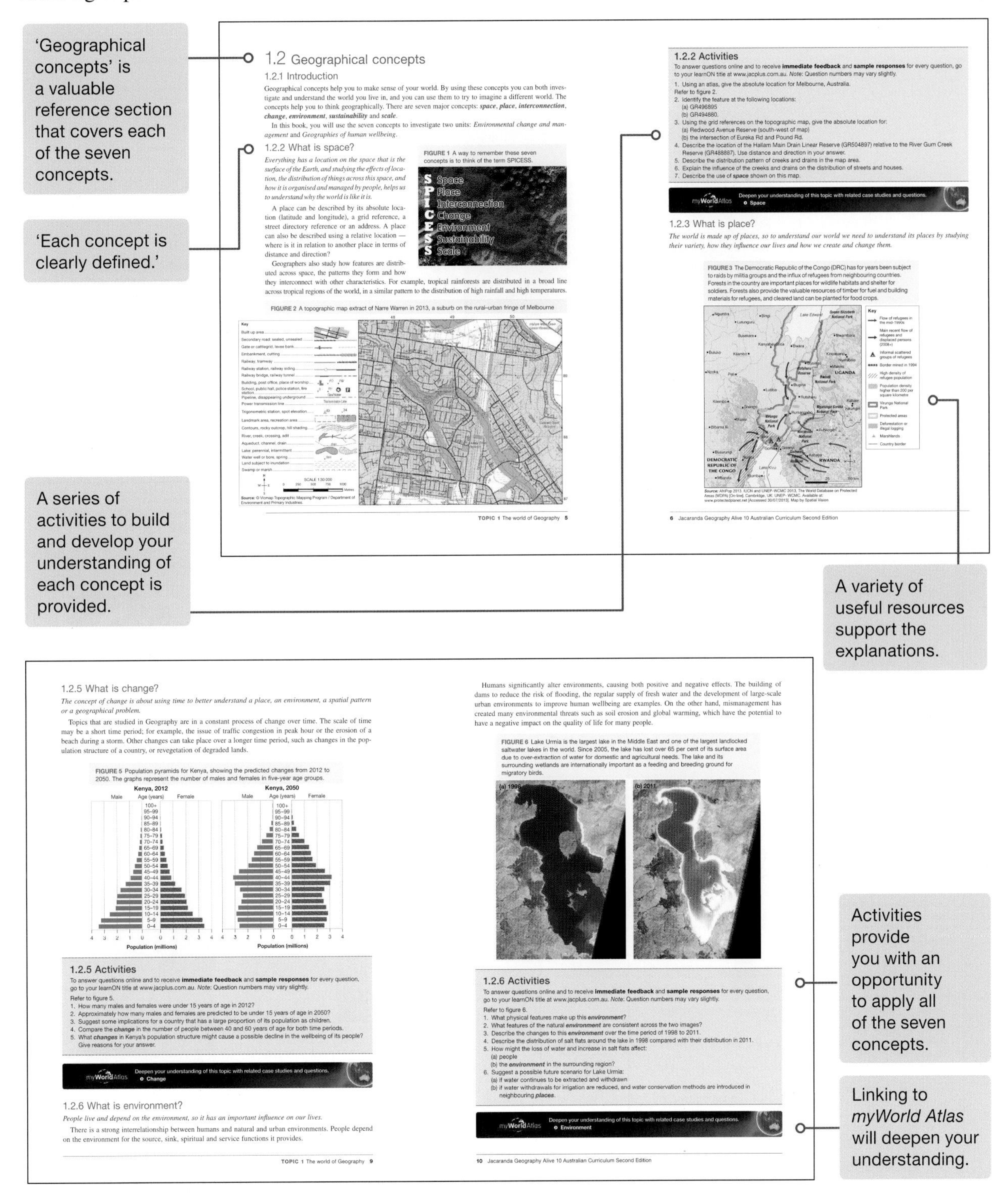

Questions raise issues, link the unit to your life, and prompt you to think about what you already know and feel about the unit.

A sequence for your inquiry.

A thought-provoking topic opener sets the scene for your inquiry.

Evocative and informative images stimulate interest and discussion.

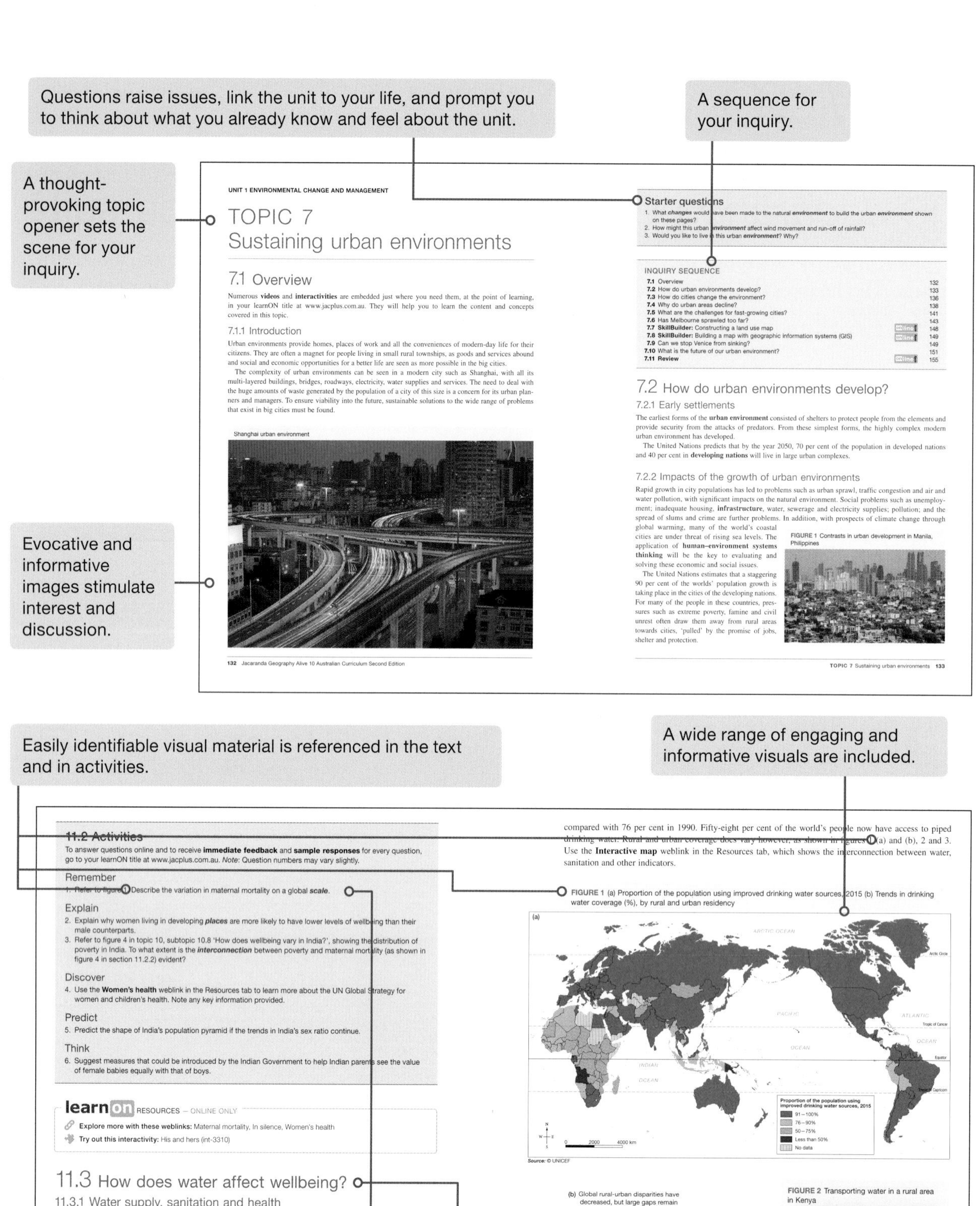

Easily identifiable visual material is referenced in the text and in activities.

A wide range of engaging and informative visuals are included.

Italicised key concepts are applied to the activities.

Each section begins with a clearly identifiable subtopic number and inquiry question.

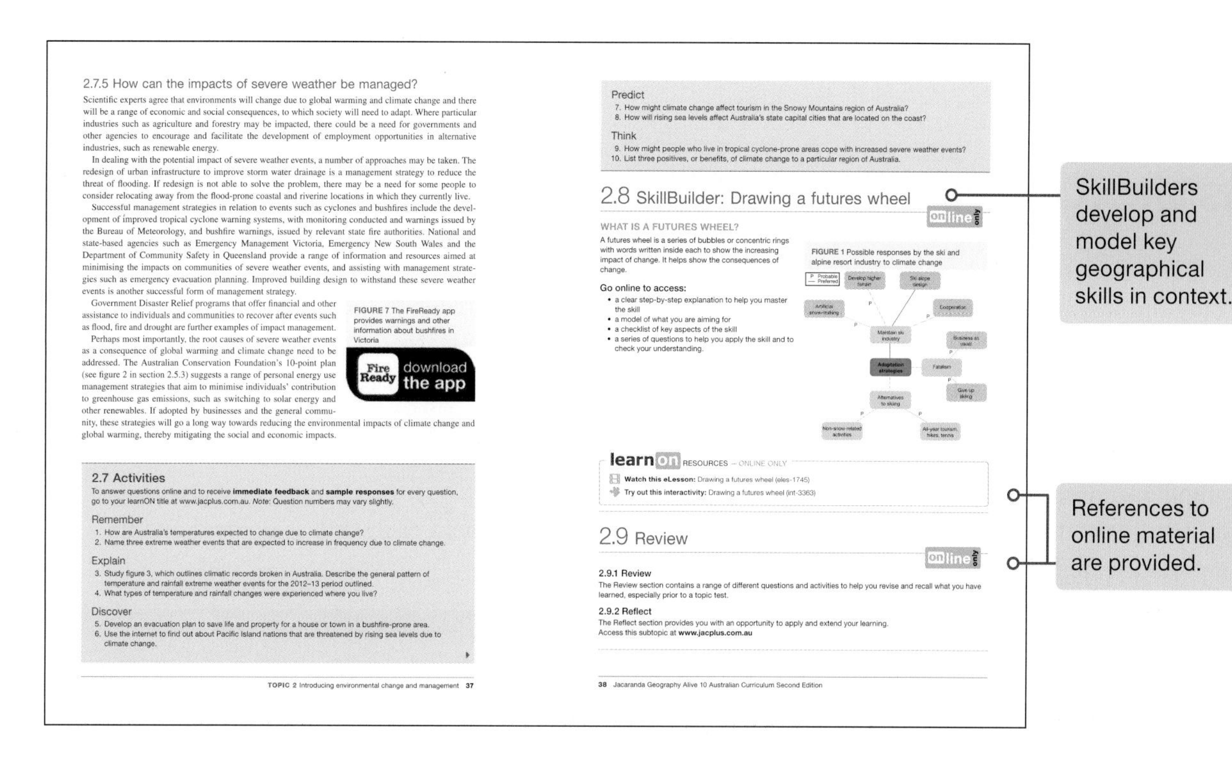

2.7.5 How can the impacts of severe weather be managed?

Scientific experts agree that environments will change due to global warming and climate change and there will be a range of economic and social consequences, to which society will need to adapt. Where particular industries such as agriculture and forestry may be impacted, there could be a need for governments and other agencies to encourage and facilitate the development of employment opportunities in alternative industries, such as renewable energy.

In dealing with the potential impact of severe weather events, a number of approaches may be taken. The redesign of urban infrastructure to improve storm water drainage is a management strategy to reduce the threat of flooding. If redesign is not able to solve the problem, there may be a need for some people to consider relocating away from the flood-prone coastal and riverine locations in which they currently live.

Successful management strategies in relation to events such as cyclones and bushfires include the development of improved tropical cyclone warning systems, with monitoring conducted and warnings issued by the Bureau of Meteorology, and bushfire warnings, issued by relevant state fire authorities. National and state-based agencies such as Emergency Management Victoria, Emergency New South Wales and the Department of Community Safety in Queensland provide a range of information and resources aimed at minimising the impacts on communities of severe weather events, and assisting with management strategies such as emergency evacuation planning. Improved building design to withstand these severe weather events is another successful form of management strategy.

Government Disaster Relief programs that offer financial and other assistance to individuals and communities to recover after events such as flood, fire and drought are further examples of impact management.

Perhaps most importantly, the root causes of severe weather events as a consequence of global warming and climate change need to be addressed. The Australian Conservation Foundation's 10-point plan (see figure 2 in section 2.5.3) suggests a range of personal energy use management strategies that aim to minimise individuals' contribution to greenhouse gas emissions, such as switching to solar energy and other renewables. If adopted by businesses and the general community, these strategies will go a long way towards reducing the environmental impacts of climate change and global warming, thereby mitigating the social and economic impacts.

FIGURE 7 The FireReady app provides warnings and other information about bushfires in Victoria

Fire Ready download the app

2.7 Activities

To answer questions online and to receive **immediate feedback** and **sample responses** for every question, go to your learnON title at www.jacplus.com.au. *Note:* Question numbers may vary slightly.

Remember
1. How are Australia's temperatures expected to change due to climate change?
2. Name three extreme weather events that are expected to increase in frequency due to climate change.

Explain
3. Study figure 3, which outlines climatic records broken in Australia. Describe the general pattern of temperature and rainfall extreme weather events for the 2012–13 period outlined.
4. What types of temperature and rainfall changes were experienced where you live?

Discover
5. Develop an evacuation plan to save life and property for a house or town in a bushfire-prone area.
6. Use the internet to find out about Pacific Island nations that are threatened by rising sea levels due to climate change.

TOPIC 2 Introducing environmental change and management 37

Predict
7. How might climate change affect tourism in the Snowy Mountains region of Australia?
8. How will rising sea levels affect Australia's state capital cities that are located on the coast?

Think
9. How might people who live in tropical cyclone-prone areas cope with increased severe weather events?
10. List three positives, or benefits, of climate change to a particular region of Australia.

2.8 SkillBuilder: Drawing a futures wheel

online only

WHAT IS A FUTURES WHEEL?

A futures wheel is a series of bubbles or concentric rings with words written inside each to show the increasing impact of change. It helps show the consequences of change.

Go online to access:
- a clear step-by-step explanation to help you master the skill
- a model of what you are aiming for
- a checklist of key aspects of the skill
- a series of questions to help you apply the skill and to check your understanding.

FIGURE 1 Possible responses by the ski and alpine resort industry to climate change

learnON RESOURCES – ONLINE ONLY

Watch this eLesson: Drawing a futures wheel (eles-1745)

Try out this interactivity: Drawing a futures wheel (int-3363)

2.9 Review

online only

2.9.1 Review

The Review section contains a range of different questions and activities to help you revise and recall what you have learned, especially prior to a topic test.

2.9.2 Reflect

The Reflect section provides you with an opportunity to apply and extend your learning.
Access this subtopic at **www.jacplus.com.au**

38 Jacaranda Geography Alive 10 Australian Curriculum Second Edition

SkillBuilders develop and model key geographical skills in context.

References to online material are provided.

The Fieldwork inquiry and Geographical inquiry provide you with an opportunity to develop your inquiry skills in the field and through research.

UNIT 2 GEOGRAPHIES OF HUMAN WELLBEING

TOPIC 13 Fieldwork inquiry: Comparing wellbeing in the local area

13.1 Overview

Numerous **videos** and **interactivities** are embedded just where you need them, at the point of learning, in your learnON title at www.jacplus.com.au. They will help you to learn the content and concepts covered in this topic.

13.1.1 Scenario and your task

You may have noticed that there are distinct variations across space in any city, suburb or regional community in terms of human wellbeing. Your council has asked for locals to inform them about differences in wellbeing they notice within their local areas, and what could or should be done about these in the future. Investigation of the topic will require you to undertake some fieldwork in order to make first-hand observations in the field, collect, process and analyse data.

Your task

Your task is to produce a fieldwork report you could present to your local council that outlines variations within your local area, reasons for the differences and strategies to improve the situation in the future. The aim of the fieldwork is for you to explore some of these variations by comparing two places at the local scale. The key inquiry questions the council wants to know answers to are:

- How does wellbeing vary between area X and area Y in the local area?
- What factors might explain the variations in wellbeing?
- How can wellbeing be improved in the local area?

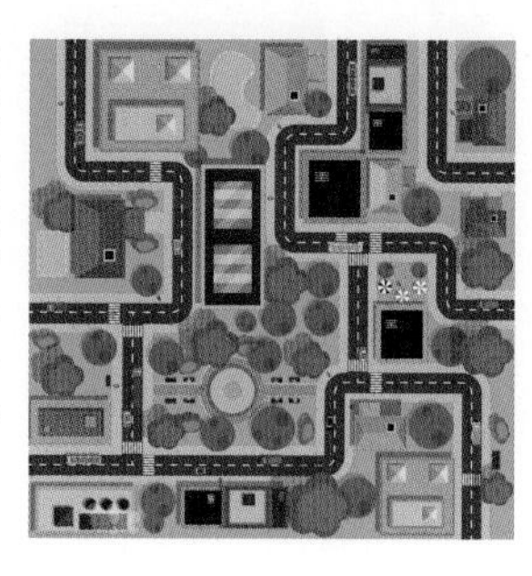

262 Jacaranda Geography Alive 10 Australian Curriculum Second Edition

13.2 Process

13.2.1 Process

- Go to **www.jacplus.com.au** to access and watch the introductory video lesson for this fieldwork enquiry.
- **Planning:**
  1. As a class, discuss the types of indicators you would use as a basis for comparing wellbeing in your local area; for example, surveys. Decide on teams and allocate tasks, or different streets, to each team member. Download the fieldwork planning document from the Resources tab. Use the task list to help you plan your fieldwork.
  2. You will need to determine the features of the houses and streets that you wish to gain data about. How will you record this data on the day (per house block or per street block)? Think carefully and plan your data record sheet so it is easy to use and also easy to summarise. Download the sample street analysis template from the Resources tab to help you plan and to record your housing data.
  3. If you are planning to survey people, you will need to plan and prepare survey questions. Download the community sample survey template from the Resources tab to help you plan and to record your data.

13.2.2 Collecting and recording data

1. Prior to going on your field trip, prepare a simple map to show the location of your fieldwork site(s) relative to key features such as your school or city centre. You will need a separate location map if your second site is not in the same area.
2. Prepare a more detailed map of your fieldwork site(s). Use a street directory, Google Earth or local council map as a guide. Include streets, street names, schools, preschools, shops and shopping centres, parks, public transport and other community facilities. Complete your map with BOLTSS.
3. During the field trip you may be required to survey houses whereby you record key features, take photographs (*Hint:* Keep a record of the location of photographs taken) and survey local residents in public places such as parks and shopping centres.
4. Download the fieldwork report document from the Resources tab to help you prepare your report.

13.2.3 Analysing your information and data

- An important skill is the ability to analyse the information you have collected on your field trip and any other supplementary data, in order to write the findings of your inquiry into a fieldwork report. A key part of your report is to determine any patterns or trends revealed in the data. At the same time, try to identify any anomalies (variations) from the patterns or trends. Download the analysis document from the Resources tab to help you further analyse the data you have collected.

TOPIC 13 Fieldwork inquiry: Comparing wellbeing in the local area 263

Inside your *Jacaranda Geography Alive learnON*

Jacaranda Geography Alive learnON is an immersive digital learning platform that enables real-time learning through peer-to-peer connections, complete visibility and immediate feedback. It includes:

- a wide variety of embedded videos and interactivities to engage the learner and bring ideas to life
- the **Capabilities** of the Australian Curriculum, available in and throughout the course in activities and **Discussion** widgets
- links to the *myWorld Atlas* for media-rich case studies
- sample responses and immediate feedback for every question
- **SkillBuilders** that present a step-by-step approach to each skill, where each skill is defined and its importance clearly explained
- collaborative activities
- and much more.

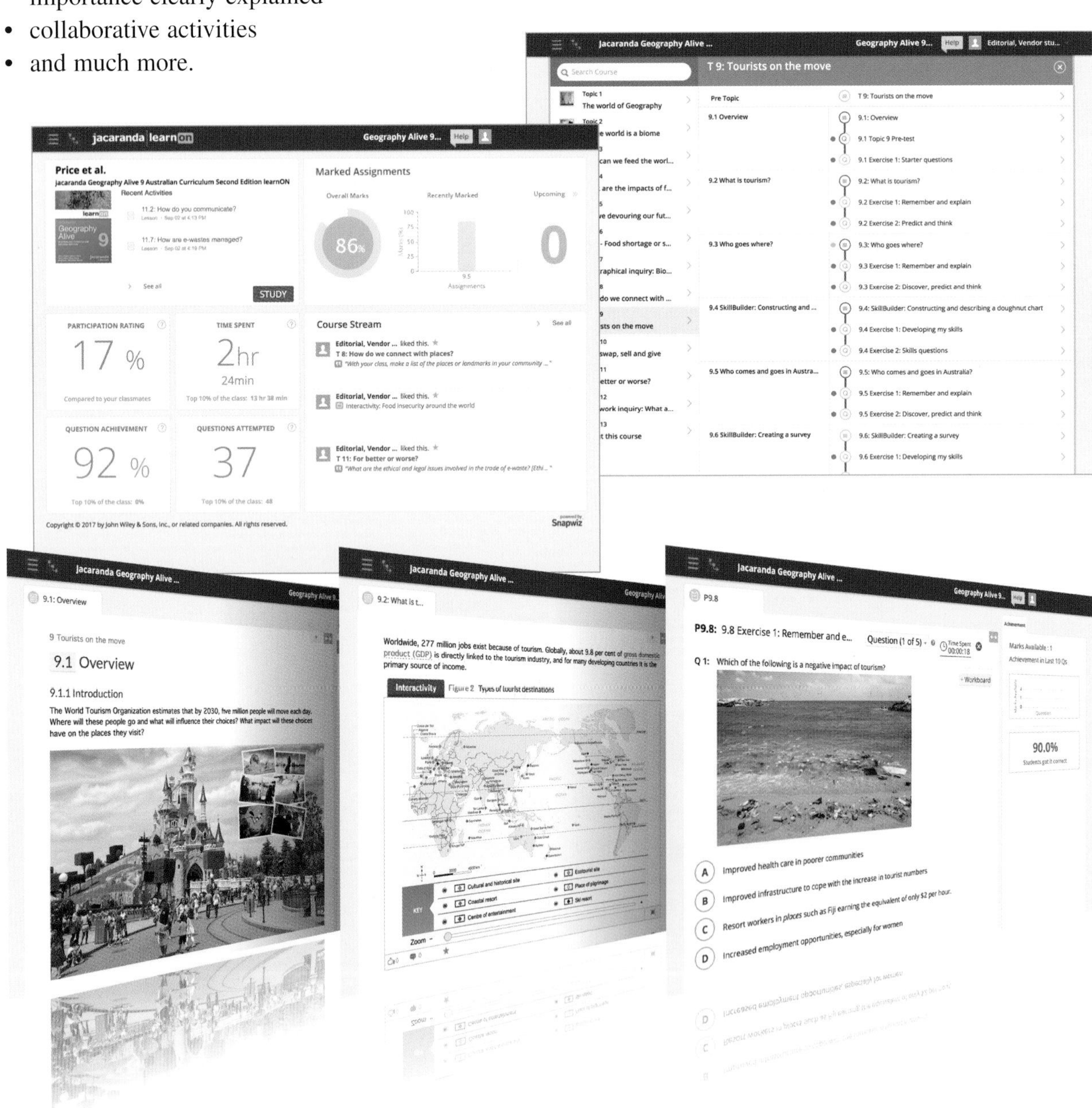

ACKNOWLEDGEMENTS

The authors and publisher would like to thank the following copyright holders, organisations and individuals for their assistance and for permission to reproduce copyright material in this book.

Images

• AAP Newswire: **36** (top right)/David Mariuz; **92**/AP Photo/Mohammed Seeneen; **126**/AP Photo/Gerald Herbert; **150** (bottom)/AP Photo/Luigi Costantini; **245** (bottom)/Alexander Kots / Komsomolskaya Pravda via AP; **248** (middle)/Thanassis Stavrakis; **251** (c)/EPA / Darko Dozet • AIHW: **217** (left), **217** (right) • Alamy Australia Pty Ltd: **55**/Green Eyes; **63**/Pete Titmuss; **67** (left)/J C Clamp; **67** (right)/Paul Dymond; **79**/SCPhotos; **101**/imageBROKER; **114**/F.Bettex - Mysterra.org; **139** (top)/Roy Garner; **140** (a)/Paul Mayall Australia; **142** (right)/National Geographic Creative; **203** (top left)/Bill Bachman; **207** (b)/redbrickstock.com; **219** (bottom)/Dinodia Photos; **220**/Susanna Bennett; **245** (middle)/epa european press photo agency b.v.; **246**/epa european pressphoto agency b.v.; **251** (a)/dpa picture alliance; **251** (b)/epa european pressphoto agency b.v. • Ashden: **77** • Australian Bureau of Statistic: **203** (top right) • Australian Bureau of Statistics: **222**/Closing the Gap Prime Ministers Report 2016: https://www.dpmc.gov.au/indigenous-affairs/publication/closing-gap-prime-ministers-report-2016 • Climate Council: **35** • Copyright Clearance Centre: **115**/© The American Association for the Advancement of Science; **245** (top)/© Spatial Vision • Core Logic RP Data : **218** (bottom a), **218** (bottom b) • Country Womens Association of Australia: **218** (top) • Creative Commons: **24**/© Bureau of Meterology; **27**, **31**/© Commonwealth of Australia 2010; **34**, **127**/© Bureau of Meteorology; **56** (a)/Allan Fox & DSEWPAC Australia © Commonwealth of Australia 2013.; **60** (left)/ Landcare Research / Hawkes Bay regional council; **60** (right)/Tim Doherty; **87**/Department of Environment, Land, Water and Planning © State of Victoria.; **97**/© Commonwealth of Australia, Geoscience Australia 1982. Topographic map of Daintree National Park — Mossman, QLD. 1:100 000 Series R631, Sheet 7965, Edition 1 1- AAS. 1982. Redrawn by MAPgraphics; **137**/Imre Solt; **173**/© AusAID Photolibrary; **203** (middle)/Unpublished ABS data and Treasury projections.; **210** (left)/Published and issued by Office of the Registrar General, India, Ministry of Home Affairs http://www.censusindia.gov.in/vital_statistics/SRS_Bulletins/MMR_Bulletin-2010-12.pdf; **233** (middle)/DoD photo by Spc. DeYonte Mosley, U.S. Army/Released; **250**/European Migrant Crisis 2015 © Spatial Vision; **253**/© 2012 World Vision Australia • CSIRO Land and Water: **48** (a)/John Coppi; **48** (b)/Greg Heath; **48** (c); **48** (d)/Willem van Aken • Defence Digital Media-Imagery: **258**/ Warrant Officer Class Two Gary Ramage WO2 / Department of Defence • Department of Environment, Land, Water & Planning : **5**, **146**, **147**/TO BE PRINTED WITH MAP: Copyright © The State of Victoria, Department of Environment, Land, Water and Planning, 2016. This publication may be of assistance to you, but the State of Victoria and its employees do not guarantee that the publication is without flaw of any kind or is wholly appropriate for your particular purposes and therefore disclaims all liability for any error, loss or other consequence which may arise from you relying on any information in this publication. • Dept of Agriculture and Food: **55**/© Western Australian Agriculture Authority Department of Primary Industries and Regional Development, WA • Dept of Primary Industries Vic: **49** (bottom)/Photo by Clem Sturmfels © State of Victoria, Department of Primary Industries. Victorian Resources Online www.dpi.vic.gov.au/vro. Reproduced with permission. • Earth Policy Institute: **25** • Emergency Management Australia: **37** • Getty Images: **48** (e)/Science Photo Library/MICHAEL MARTEN; **89**/Vince Streano; **91** (top)/Peter Harrison; **99**/Ashley Cooper; **104**/STR/AFP; **106**/The Asahi Shimbun; **139** (bottom)/In Pictures Ltd./Corbis; **142** (left)/Matt Mawson; **142** (bottom)/Sam Panthaky; **150** (top)/ Jonathan Blair; **264**/2xSamara.com / Shutterstock.com • Getty Images Australia: **36** (top left)/Saeed Khan / AFP; **140** (e)/ Jodi Cobb/National Geographic; **207** (a)/Peter Harrison; **217** (middle)/Auscape / Universal Images Group; **230**/Abdullah Doma/AFP; **236**/MOHAMMED ABED/AFP; **238** (bottom)/Odd Andersen/AFP; **241** (bottom)/Per-Anders Pettersson; **248** (top)/BULENT KILIC/AFP; **251** (e)/Denis Charlet/AFP; **254** (top)/David Hancock/AFP • GhostNets Australia: **122**; **123**/ Aurukun ghost net art workshop 2009/Photo by Sue Ryan • Global Footprint Network: **22** • Institute of Economics and Welfare: **258** (bottom) • Internal Displacement Monitoring Centre : **243** • John Rasic - PERMISSION GRANTED ONE BLANKET REQUEST FOR ALL IMAGES : **50** (top)/c John Ivo Rasic <NEEDS TO APPEAR IN BOOK> • MAPgraphics: **45**, **51**, **98** • Matter Of Trust: **129**/Amanda Bacon matteroftrust.org • Medecins Sans Frontieres Aus: **251** (d); **257**/© Anne Yzebe • NASA: **10**, **81** (a), **81** (b)/NASA Earth Observatory image created by Jesse Allen and Robert Simmon, using Landsat data provided by the United States Geological Survey.; **71**/NASA Earth Observatory; **187**/Goddard Space Flight Center Scientific Visualization Studio • National Geographic: **59** • Newspix: **58**/Mike Keating; **221**/Brett Hartwig • NSW Land and Property: **95**/c LPI - NSW Department of Finance and Services [2013] Panorama Avenue, Bathurst 2795 www.lpi..nsw.gov.au • Ocean Conservancy: **114**, **116** (left) • Outback stores: **223** • Oz Aerial Photography: **96**/© 2012 Oz Aerial Photos Australia Photo by Daryl Jones • Pauline English: **42** • Peter Coyne: **56** (b) • Photodisc: **32** (a), **32** (b) • Population Reference Bureau: **188**, **191** (bottom), **192**/© United Nations Publications : http://www.prb.org/Publications/Datasheets/2015/2015-world-population-data-sheet/world-map.aspx#map/world/births/bpk • Project Ploughshares: **232**, **233** (top left), **233** (top right)/Project Ploughshares and the Armed Conflicts Report • Public Domain: **31**/Energy Information

Administration EIA; **136** (bottom), **152**, **175-177**, **175-177**, **214** (right); **168**; **179** (top)/Freedom House; **189** (left), **193**; **209** (bottom right)/ http://apps.who.int/iris/bitstream/10665/43185/1/924156315X_eng.pdf; **226** (bottom)/© 2016 The National Campaign to Prevent Teen and Unplanned Pregnancy.; **227**/ http://www.census.gov/hhes/www/poverty/data/incpovhlth/2014/index.html; **244**/UN High Commissioner for Refugees UNHCR, Syria Situation Map As of 29 February 2016, 29 February 2016, available at: http://www.refworld.org/docid/56de7fdd4.html [accessed 22 June 2016] • Reefbase: **124** • Shutterstock: **1** (a), **1** (c)/goodluz; **1** (b)/Free Wind 2014; **5**/Christian Draghici; **12**, **50** (bottom)/Caleb Holder; **13**, **219** (middle)/JHMimaging; **17**/Tony Campbell; **19** (b)/Vladimir Wrangel; **32** (c)/Sky Light Pictures; **32** (d)/N.Minton; **36** (bottom), **91** (bottom)/Pete Niesen; **39**/Dirk Ercken; **46**/mark higgins; **48** (f)/Alberto Loyo; **49** (top)/HHelene; **49** (middle)/Mark Winfrey; **49** (middle)/Neil Bradfield; **52**/David Salcedo; **54** (a)/mrfotos; **57** (a)/Grezova Olga; **57** (b)/Susan Flashman; **66**/Sam DCruz; **69**/Mark Schwettmann; **72**/Warren Price Photography; **86**/Sherrianne Talon; **100**/Mohamed Shareef; **109**/Mikadun; **113**/Antonio V. Oquias; **132**/chungking; **133**/donsimon; **136** (top)/Nataliya Hora; **137**/FCG; **138**/Vladimir Korostyshevskiy; **140** (b)/Ivonne Wierink; **140** (c)/SvedOliver; **140** (d)/Zack Frank; **144**/View Factor Images; **154**/Paolo Bona; **156**/goodluz; **158** (left)/Johnny Lye; **158** (right)/John Sartin; **161**/Travel Stock; **172** (top)/GNEs; **184**/Milles Studio; **186** (middle)/Aaron Gekoski; **189** (right)/Andrei Shumskiy; **194** (bottom)/wizdata1; **196** (top)/meunierd; **197** (top)/mykeyruna; **199**/Matyas Rehak; **202** (right)/Brisbane; **209** (bottom left)/Lucian Coman; **210** (right)/paul prescott; **211** (right)/zeber; **213** (bottom right)/africa924; **214** (left)/Neale Cousland; **242**/Sadik Gulec; **261** (a), **261** (f)/Ryan Rodrick Beiler; **261** (b)/Alexandre Rotenberg; **261** (c)/Malcolm Chapman; **261** (d)/mikhail; **261** (e), **261** (g)/ART production; **261** (h)/thomas koch; **262**/robuart; **263**/Johnny Habell • Snowy Monaro Regional Council : **54** (b) • Spatial Vision: **6**/AfriPop 2013. IUCN and UNEP-WCMC 2013, The World Database on Protected Areas WDPA [On-line]. Cambridge, UK: UNEP- WCMC. Available at: www.protectedplanet.net [Accessed 30/07/2013]. Made with Natural Earth Map by Spatial Vision.; **11**, **200** (bottom), **201** (right), **211** (left)/Government of India, Ministry of Home Affairs, Office of Registrar General Made with Natural Earth. Map by Spatial Vision; **19** (a)/Made with Natural Earth.Map by Spatial Vision.; **46**/© Commonwealth of Australia Geoscience Australia 2013. © Commonwealth of Australia Department of Sustainability, Environment, Water, Population and Communities 2013. Map by Spatial Vision.; **53**, **56**/© Commonwealth of Australia Geoscience Australia 2013. © Commonwealth of Australia Department of Sustainability, Environment, Water, Population and Communities 2013. Map by Spatial Vision; **73**/Made with Natural Earth. University of New Hampshire UNH/Global Runoff Data Centre GRDC http://www.grdc.sr.unh.edu/ Map by Spatial Vision; **76**, **88**; **78**/World Climate - http://www.worldclim.org/ Made with Natural Earth. Map by Spatial Vision; **82**/United Nations Environment Programme Made with Natural Earth. Vector Map Level 0 Digital Chart of the World Map by Spatial Vision; **84**/BGR & UNESCO 2008: Groundwater Resources of the World 1 : 25 000 000. Hannover, Paris. Made with Natural Earth. Map by Spatial Vision; **99**/Made with Natural Earth. Map by Spatial Vision.; **111**/Made with Natural Earth. Map by Spatial Vision; **116**/Greenpeace International Made with Natural Earth. Map by Spatial Vision; **122**/© Commonwealth of Australia Geoscience Australia 2013. Ghost Nets Australia, http://www.ghostnets.com.au/index.html Made with Natural Earth. Map by Spatial Vision; **130**/National Oceanic and Atmospheric Administration, Office of Ocean Exploration and Research, U.S. Department of Commerce. Adapted by Spatial Vision.; **144**/.idplacemaker © The State of Victoria, Department of Environment and Primary Industries 2013 © Commonwealth of Australia Geoscience Australia 2013. Map by Spatial Vision.; **145**/© Commonwealth of Australia Geoscience Australia 2013. © The State of Victoria, Department of Environment and Primary Industries 2013 Map by Spatial Vision; **149**/Made with Natural Earth. © OpenStreetMap contributors Map by Spatial Vision.; **152** (a)/United Nations, Department of Economic and Social Affairs, Population Division 2012. World Urbanization Prospects: The 2011 Revision Made with Natural Earth Map by Spatial Vision.; **153** (b)/United Nations, Department of Economic and Social Affairs, Population Division 2012. World Urbanization Prospects: The 2011 Revision Made with Natural Earth. Map by Spatial Vision.; **166**/United Nations Development Report Made with Natural Earth. Map by Spatial Vision; **167**/Source: The World Bank: Poverty headcount ratio at $1.25 a day PPP % of population: World Development Indicators. Map by Spatial Vision; **170**/Abdallah S, Michaelson J, Shah S, Stoll L, Marks N 2012 The Happy Planet Index: 2012 Report. A global index of sustainable well-being nef: London Made with Natural Earth. Map by Spatial vision.; **181**/Maplecroft Made with Natural Earth. Map by Spatial Vision; **186** (top)/NASA Socioeconomic Data and Applications Center SEDAC. http://sedac.ciesin.columbia.edu/data/collection/gpw-v3 Made with Natural Earth. Map by Spatial Vision; **201** (left)/Percentage of Population Below Poverty Line URP Consumption, ChartsBin.com, viewed 27th August, 2013, <http://chartsbin.com/view/2797> Made with Natural Earth. Map by Spatial Vision; **202** (left)/© Australian Bureau of Statistics / Spatial Vision; **216**/Copyright Commonwealth of Australia , Australian Bureau of Statistics http://www.abs.gov.au/AUSSTATS/abs@.nsf/DetailsPage/1270.0.55.005July%20 2011?OpenDocument © Commonwealth of Australia Geoscience Australia 2013. Map by Spatial Vision; **224**/© Commonwealth of Australia Geoscience Australia 2013. © Commonwealth of Australia Australian Bureau of Statistics 2013. Map by Spatial Vision; **234**/Made with Natural Earth. Map by Spatial Vision.; **238** (top)/EPOS - International Mediating and Negotiating Operational Agency, http://eposweb.org/ Lowy Institute, http://lowyinstitute.org/ Made with Natural Earth Map by Spatial Vision • Surfrider Foundation: **119** • Susan Middleton: **116** (right) • UNHCR: **240**, **241** (a), **241** (b) • UNICEF: **213** (middle), **213** (bottom left)/© UNICEF; **226** (top a), **226** (top b)/UNICEF Office of Research 2013. 'Child Well-being in Rich Countries: A comparative overview', Innocenti Report Card 11, UNICEF Office of Research,

Florence • United Nations Publications: **134**, **191** (middle)/© United Nations Publications : http://www.prb.org/Publications/Datasheets/2015/2015-world-population-data-sheet/world-map.aspx#map/world/births/bpk; **165**/© UNDP; **235**/"The Gaza Strip: The Humanitarian Impact of the Blockade July 2015 © 2016 United Nations. Reprinted with the permission of the United Nations." • US Census Bureau: **9** (a), **9** (b), **196** (middle left a), **196** (middle b), **197** (middle a), **197** (middle b), **200** (top a), **200** (top b) • Walther-Maria Scheid: **125**/Walther-Maria Scheid, Berlin, Germany for World Ocean Review 2010 • World Bank: **174**, **174** • World Health Organisation: **209** (top)/WHO DOCUMENT REFERENCE: WHO/RHR/15.23* Map of Maternal mortality ratio MMR, maternal deaths per 100 000 live births, 2015 page 3 of this document http://data.unicef.org/corecode/uploads/document6/uploaded_pdfs/corecode/MMR_executive_summary_final_mid-res_243.pdf • Wrights Media: **249**/Business Insider, Inc. © Spatial Vision

Text

• © Australian Curriculum, Assessment and Reporting Authority (ACARA) 2010 to present, unless otherwise indicated. This material was downloaded from the Australian Curriculum website (www.australiancurriculum.edu.au) (Website) (accessed September 5, 2017) and was not modified. The material is licensed under CC BY 4.0 (https://creativecommons.org/licenses/by/4.0). Version updates are tracked on the 'Curriculum version history' page (www.australiancurriculum.edu.au/Home/CurriculumHistory) of the Australian Curriculum website. • Australian Bureau Meteorol.: **35** • Australian Conservation Foundation: **29** • Creative Commons: **28**/The Garnaut Climate Change Review 2008, published by the Commonwealth of Australia • IPCC: **32-33**/Based on Table SPM.3 from Climate Change 2007: Mitigation of Climate Change. Working Group III Contribution to the Fourth Assessment Report of the Intergovernmental Panel on Climate Change [B. Metz, O.R. Davidson, P.R. Bosch, R. Dave, L.A. Meyer eds.]. Cambridge University Press, Cambridge, United Kingdom and New York, NY, USA. • Ocean Conservancy: **117** • Public Domain: **141**/Government of India Census • Surfrider Foundation: **119** • US Census Bureau: **195**/PRB, Data Sheet, 2012, 2011 US Census Bureau

Every effort has been made to trace the ownership of copyright material. Information that will enable the publisher to rectify any error or omission in subsequent reprints will be welcome. In such cases, please contact the Permissions Section of John Wiley & Sons Australia, Ltd.

TOPIC 1
The world of Geography

1.1 Overview

Numerous **videos** and **interactivities** are embedded just where you need them, at the point of learning, in your learnON title at www.jacplus.com.au. They will help you to learn the content and concepts covered in this topic.

1.1.1 Work and careers in Geography

As the world's population increases and the impacts of environmental changes affect living conditions, people and organisations will need to adapt and develop strategies to manage and sustain fragile environments and resources. Land degradation, marine pollution and feeding the future world populations are just three environmental challenges that will be the focus for many occupations in the future. Which careers will be helpful in managing environmental change?

TABLE 1 Careers that will manage environmental change

Conservationists 	Conservationists will work to find solutions to land degradation. They will work for governments, in national parks and on policy development, and with local communities on environmental protection projects.
Oceanographers 	Work for oceanographers will mainly involve research and monitoring of the marine environment. They may work for governments, providing data and advice on pollution levels, or they may work for private or not-for-profit organisations, helping to suggest and implement plans for cleaning up the oceans.
Agricultural scientists 	Agricultural scientists will be employed by the government, and agricultural and horticultural producers. They will work with farmer groups and agribusiness to carry out research, and with mining companies, working on regeneration projects.

Profile of a geographer

Geographers have a love of learning. They are the explorers of the modern world. Geographers are lifelong learners; they expand their knowledge to adapt their skills to the tasks required.

Expansion of knowledge requires a willingness to learn. How many of these skills and attributes have you developed?

- Willingness to learn
- Curiosity and adaptability

- Active listening
- Good communication
- Critical thinking
- Time management
- Problem solving

You can develop your skills and work attributes by undertaking work experience or volunteering activities while you are still at school.

1.1.2 The importance of work experience

The activities you undertake and skills that you develop in Geography will be useful to you in many aspects of your life and your career. In building and managing your career options, it is also helpful to have an understanding of the interconnections between various careers. One way of building your knowledge of these interconnections, particularly in relation to Geography and the career paths that lead from it, is to undertake work experience in the field. Work experience can help you to understand the tasks involved in various roles and the training required to specialise in a particular area. You can gain first-hand experience through observation of and participation in the day-to-day tasks of workplaces.

Volunteering

Volunteering in your community is a great way to find out about different work environments and the things that impact on the delivery of the services or programs. Volunteering your time to support local communities and businesses also demonstrates your willingness to learn and support others and it can provide a great boost to your self-confidence, as well as important skill development.

Thinking of volunteering? Why not consider …

FIGURE 1 Some options for volunteering

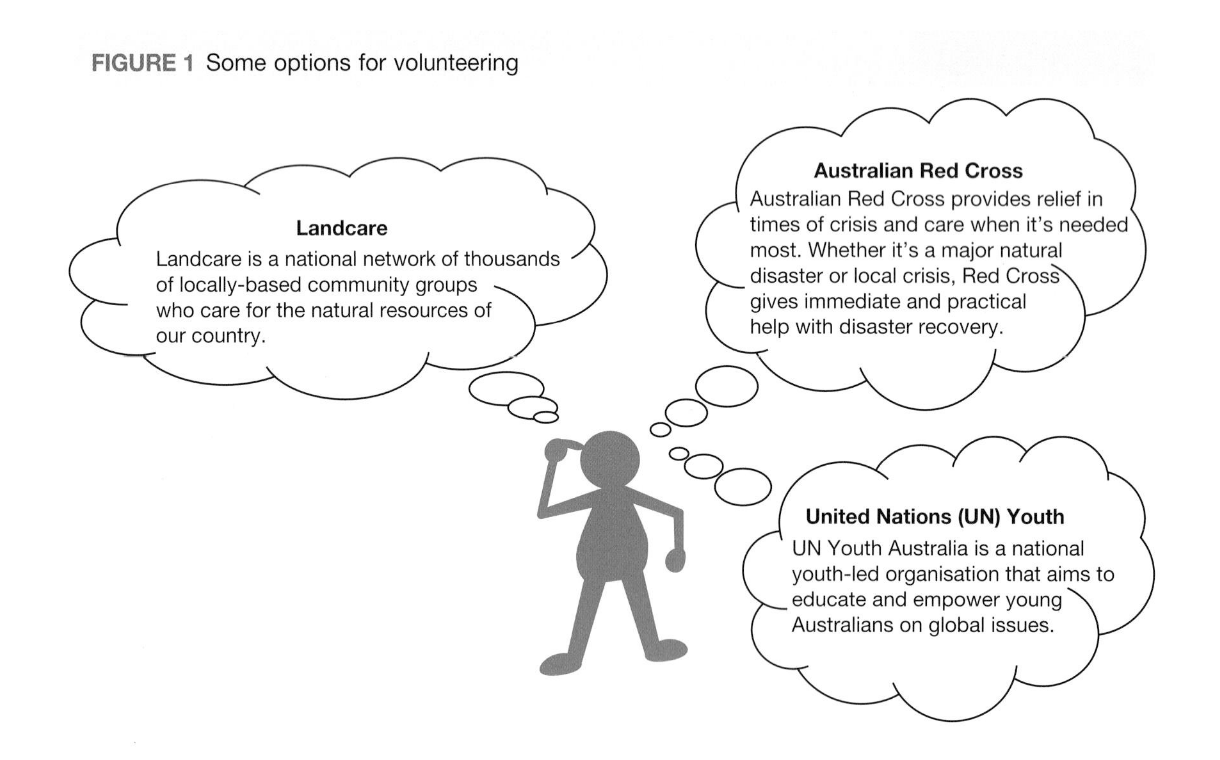

Landcare
Landcare is a national network of thousands of locally-based community groups who care for the natural resources of our country.
Australian Red Cross
Australian Red Cross provides relief in times of crisis and care when it's needed most. Whether it's a major natural disaster or local crisis, Red Cross gives immediate and practical help with disaster recovery.
United Nations (UN) Youth
UN Youth Australia is a national youth-led organisation that aims to educate and empower young Australians on global issues.

Learning directly from industry experts through volunteering can help you to consolidate your interests while also picking up valuable core skills for work (refer to table 2). These core skills are considered the most important component of a career portfolio. The study of Geography also assists in the development of these skills.

How many of the core skills for work have you developed? Use the figure 2 chart to help you think about your own skills. You may find you have particular strengths and other areas you need to improve upon. If you do this periodically, you can monitor your progress in this area. Figure 3 is an example of a completed chart.

TABLE 2 The core skills for work

Communication	Ability to use effective listening and speaking skills
Teamwork	Ability to connect and work with others
Learning	Ability to recognise and utilise diverse perspectives
Planning and organisation	Ability to develop plans and see things through to completion
Self-management	Ability to make decisions
Problem solving	Ability to identify and solve problems
Initiative and enterprise	Ability to create and innovate through new ideas
Use of technology	Ability to work in a digital world

FIGURE 2 Evaluating my core skills for work

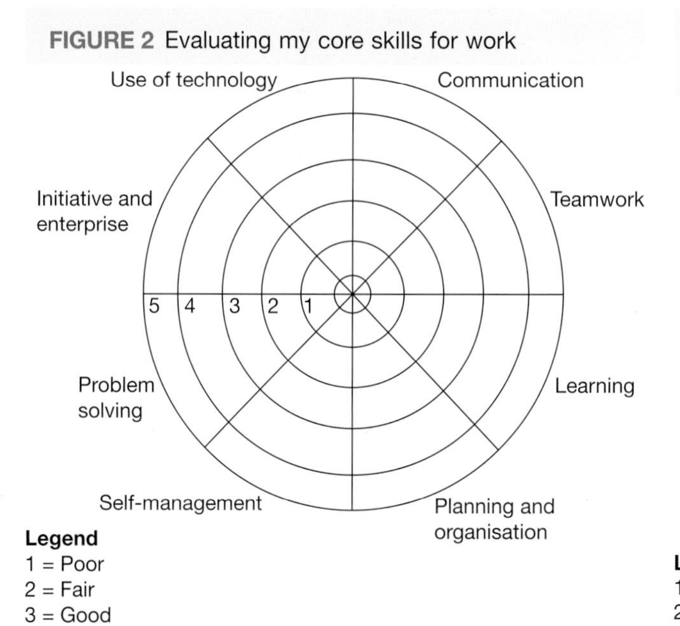

FIGURE 3 Core skills for work – Ashley Green, Semester 1 2017

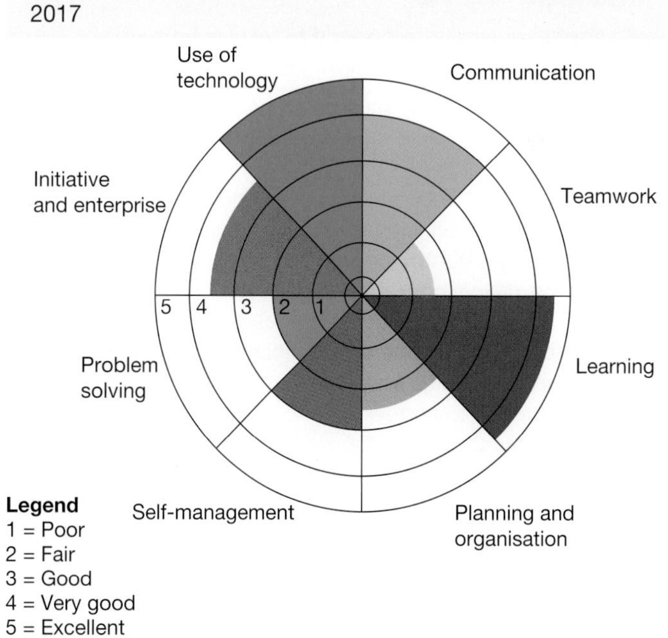

1.1.3 Future careers and Geography

Studies in Geography, along with other Social Science subjects and evidence of your work experience or volunteering, can demonstrate your adaptability, creativity and enterprise skills for future work.

In the future, the type of work that will be available will change in response to the impact of climate change, population growth and decline, and technological innovation. The rapid expansion of world economies will mean that industries will adapt their workforces. Migration and a borderless world will mean that individuals will become global citizens working in large teams around the world. Many of the occupations of this century are yet to be created, while others have been imagined and offer a glimpse into the future.

FIGURE 4 **Agroecologist** – a career of the future

SEEKING AN AGROECOLOGIST...

Agroecologists help restore ecological balance while feeding and fuelling the planet. Agroecologists work with farmers to design and manage agricultural ecosystems whose parts (plants, water, nutrients and insects) work together to create an effective and sustainable means of producing the food and environmentally friendly biofuel crops of the future.

Agroecologists also work with Ecosystem Managers to reintroduce native species and biodiversity to repair the damage done by the ecosystem-disruptive farming techniques of the past.

Job requirements/skills

You will need an undergraduate degree in Agroecology, in which you'll have learned how plants, soil, insects, animals, nutrients, water and weather interact with one another to create the living systems in which crop-based foods are grown. You'll also have learned about the technologies and methods involved in growing food in a sustainable way.

To be successful in this role, you will need to be responsive to change, demonstrate adaptability by working as part of a global team, and be creative and enterprising in all elements of the business to ensure that business growth is sustainable.

1.1 Activities

To answer questions online and to receive **immediate feedback** and **sample responses** for every question, go to your learnON title at www.jacplus.com.au. *Note*: Question numbers may vary slightly.

1. Part-time, casual or vacation work are all useful ways to build your core skills for work. Use the **ACTU Worksite** weblink in the Resources tab to locate information on work experience, volunteering and being ready for your first job.
2. Geographers work in primary, secondary and tertiary industries.
   (a) Provide a definition for each of these industries.
   (b) What is a quaternary industry?
   (c) Give an example of a quaternary industry career that may use geographical skills (Hint: spatial technologies)
3. Over the coming decades, new careers in geography will emerge. Ecosystem auditors, localisers and rewilders will become commonplace in the future. Exploring these careers today can provide an insight into the type of studies and further training you will need to undertake to ensure that you are ready for the workforce of tomorrow.
   (a) Use the **Careers 2030** weblink in the Resources tab to learn about the work of an Ecosystem auditor, a localiser or a rewilder.
   (b) Develop a career profile for this emerging career. Include the following details in your profile:
      (i) a definition for this occupation
      (ii) the core skills needed in this field
      (iii) the study or training required to successfully carry out the tasks of the role
      (iv) the industries that will employ these occupations.

RESOURCES – ONLINE ONLY

Explore more with these weblinks: ACTU Worksite, Careers 2030

1.2 Geographical concepts

1.2.1 Introduction

Geographical concepts help you to make sense of your world. By using these concepts you can both investigate and understand the world you live in, and you can use them to try to imagine a different world. The concepts help you to think geographically. There are seven major concepts: ***space***, ***place***, ***interconnection***, ***change***, ***environment***, ***sustainability*** and ***scale***.

In this book, you will use the seven concepts to investigate two units: *Environmental change and management* and *Geographies of human wellbeing*.

1.2.2 What is space?

Everything has a location on the space that is the surface of the Earth, and studying the effects of location, the distribution of things across this space, and how it is organised and managed by people, helps us to understand why the world is like it is.

FIGURE 1 A way to remember these seven concepts is to think of the term SPICESS.

A place can be described by its absolute location (latitude and longitude), a grid reference, a street directory reference or an address. A place can also be described using a relative location — where is it in relation to another place in terms of distance and direction?

Geographers also study how features are distributed across space, the patterns they form and how they interconnect with other characteristics. For example, tropical rainforests are distributed in a broad line across tropical regions of the world, in a similar pattern to the distribution of high rainfall and high temperatures.

FIGURE 2 A topographic map extract of Narre Warren in 2013, a suburb on the rural–urban fringe of Melbourne

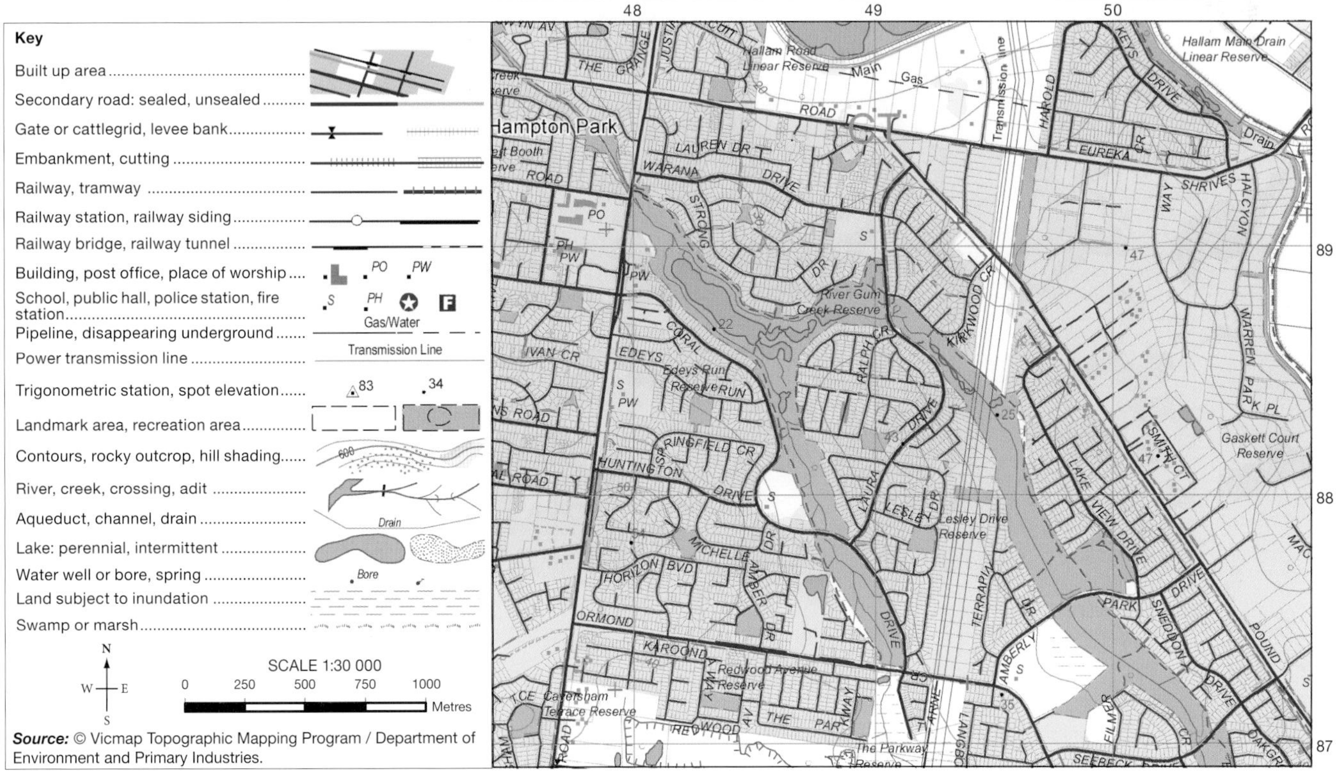

Source: © Vicmap Topographic Mapping Program / Department of Environment and Primary Industries.

1.2.2 Activities

To answer questions online and to receive **immediate feedback** and **sample responses** for every question, go to your learnON title at www.jacplus.com.au. *Note*: Question numbers may vary slightly.

1. Using an atlas, give the absolute location for Melbourne, Australia.

Refer to figure 2.

2. Identify the feature at the following locations:
   (a) GR496895
   (b) GR494880.
3. Using the grid references on the topographic map, give the absolute location for:
   (a) Redwood Avenue Reserve (south-west of map)
   (b) the intersection of Eureka Rd and Pound Rd.
4. Describe the location of the Hallam Main Drain Linear Reserve (GR504897) relative to the River Gum Creek Reserve (GR488887). Use distance and direction in your answer.
5. Describe the distribution pattern of creeks and drains in the map area.
6. Explain the influence of the creeks and drains on the distribution of streets and houses.
7. Describe the use of ***space*** shown on this map.

Deepen your understanding of this topic with related case studies and questions.
Space

1.2.3 What is place?

The world is made up of places, so to understand our world we need to understand its places by studying their variety, how they influence our lives and how we create and change them.

FIGURE 3 The Democratic Republic of the Congo (DRC) has for years been subject to raids by militia groups and the influx of refugees from neighbouring countries. Forests in the country are important places for wildlife habitats and shelter for soldiers. Forests also provide the valuable resources of timber for fuel and building materials for refugees, and cleared land can be planted for food crops.

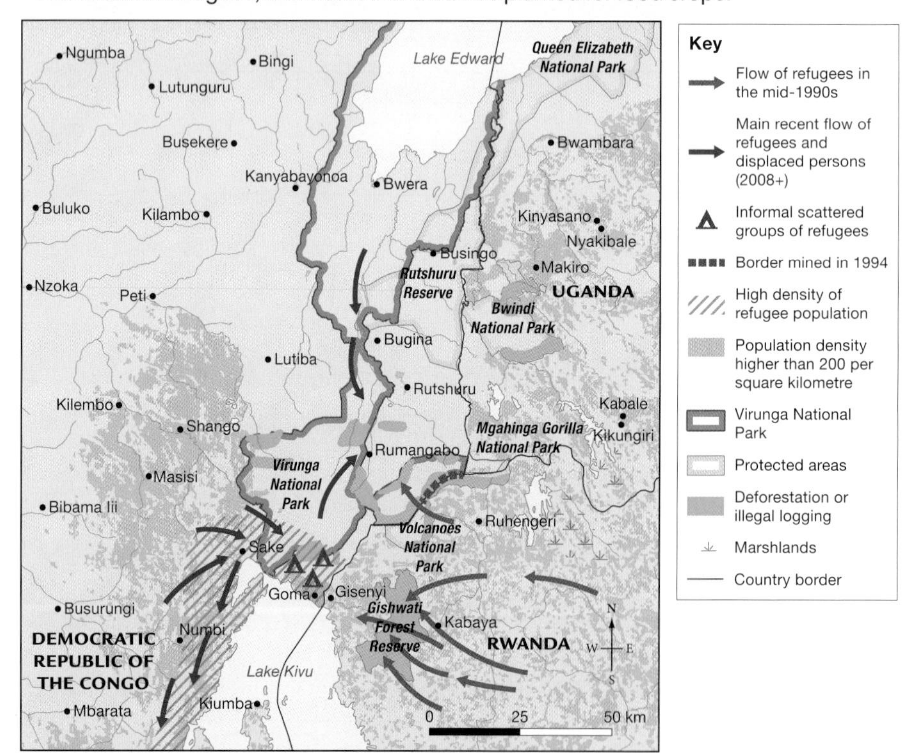

Source: AfriPop 2013. IUCN and UNEP-WCMC 2013, The World Database on Protected Areas (WDPA) [On-line]. Cambridge, UK: UNEP- WCMC. Available at: www.protectedplanet.net [Accessed 30/07/2013]. Map by Spatial Vision

Places may be natural (such as an undisturbed wetland) or highly modified (like a large urban conurbation). Places provide us with the services and facilities we need in our everyday life. The physical and human characteristics of places, their location and environmental quality can influence the quality of life and wellbeing of people living there.

1.2.3 Activities

To answer questions online and to receive **immediate feedback** and **sample responses** for every question, go to your learnON title at www.jacplus.com.au. *Note*: Question numbers may vary slightly.

Refer to figure 3.

1. Identify and name a ***place*** on the map that would be considered a natural ***environment***, and a ***place*** that is a human-developed ***environment***.
2. How many ***places*** in the region have been subject to deforestation or illegal logging?
3. What services and functions do the forests provide people and wildlife?
4. In what ways do the people of Rwanda ***interconnect*** with the DRC?
5. How would forest clearing change the importance of a ***place*** for:
   (a) soldiers
   (b) refugees
   (c) mountain gorillas?
6. Using direction and ***place*** names, compare the pattern of refugee movements in the 1990s with the more recent movements around southern DRC.

Deepen your understanding of this topic with related case studies and questions.
Place

1.2.4 What is interconnection?

People and things are connected to other people and things in their own and other places, and understanding these connections helps us to understand how and why places are changing.

The interconnection between people and environments in one place can lead to changes in another location. The damming of a river upstream can significantly alter the river environment downstream and affect the people who depend on it. Similarly, the economic development of a place can influence its population characteristics; for example, an isolated mining town will tend to attract a large percentage of young males, while a coastal town with a mild climate will attract retirees who will require different services. The economies and populations of places are interconnected.

FIGURE 4 Bangladesh is one of the most flood-prone countries in the world. This is due to a number of factors. Firstly, it is largely the floodplain for three major rivers (the Ganges, Brahmaputra and Meghna), which all carry large volumes of water and silt. Secondly, being a floodplain, the topography therefore is very flat, which allows for large-scale flooding. In addition, the country is located at the head of the Bay of Bengal, which is susceptible to typhoons and storm surges. It is expected that sea level rises associated with global warming will increase the flooding threat even further in the future.

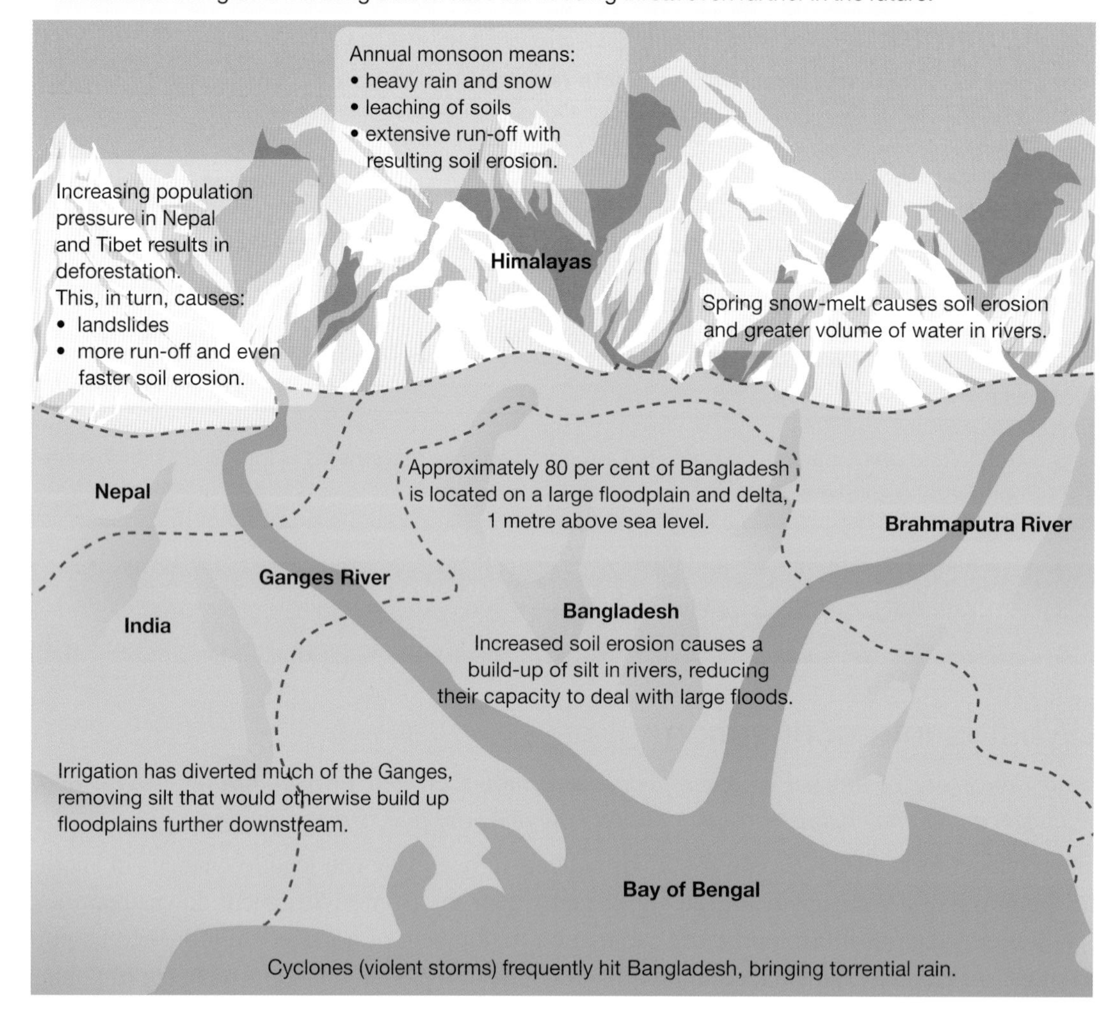
Annual monsoon means:
• heavy rain and snow
• leaching of soils
• extensive run-off with resulting soil erosion.
Increasing population pressure in Nepal and Tibet results in deforestation.
This, in turn, causes:
• landslides
• more run-off and even faster soil erosion.
Himalayas
Spring snow-melt causes soil erosion and greater volume of water in rivers.
Nepal
Approximately 80 per cent of Bangladesh is located on a large floodplain and delta, 1 metre above sea level.
Brahmaputra River
Ganges River
India
Bangladesh
Increased soil erosion causes a build-up of silt in rivers, reducing their capacity to deal with large floods.
Irrigation has diverted much of the Ganges, removing silt that would otherwise build up floodplains further downstream.
Bay of Bengal
Cyclones (violent storms) frequently hit Bangladesh, bringing torrential rain.

1.2.4 Activities

To answer questions online and to receive **immediate feedback** and **sample responses** for every question, go to your learnON title at www.jacplus.com.au. *Note*: Question numbers may vary slightly.

Refer to figure 4.

1. What is the ***interconnection*** between the physical characteristics of Bangladesh and its flood risk?
2. What is the ***interconnection*** between human activities and Bangladesh's flood risk?
3. Use information from the figure to construct a flow diagram to show the ***interconnection*** between human activities and natural processes (increased risk of flooding) in Bangladesh.
4. How might an increase in the number and severity of floods affect:
   (a) people's wellbeing
   (b) economic development in the country?
5. Considering the ***interconnections*** that you have identified, suggest some possible steps that could be taken to reduce the impact of flooding.

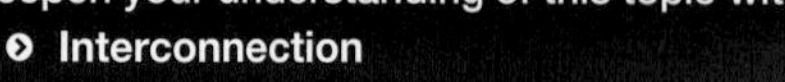
Deepen your understanding of this topic with related case studies and questions.
Interconnection

1.2.5 What is change?

The concept of change is about using time to better understand a place, an environment, a spatial pattern or a geographical problem.

Topics that are studied in Geography are in a constant process of change over time. The scale of time may be a short time period; for example, the issue of traffic congestion in peak hour or the erosion of a beach during a storm. Other changes can take place over a longer time period, such as changes in the population structure of a country, or revegetation of degraded lands.

FIGURE 5 Population pyramids for Kenya, showing the predicted changes from 2012 to 2050. The graphs represent the number of males and females in five-year age groups.

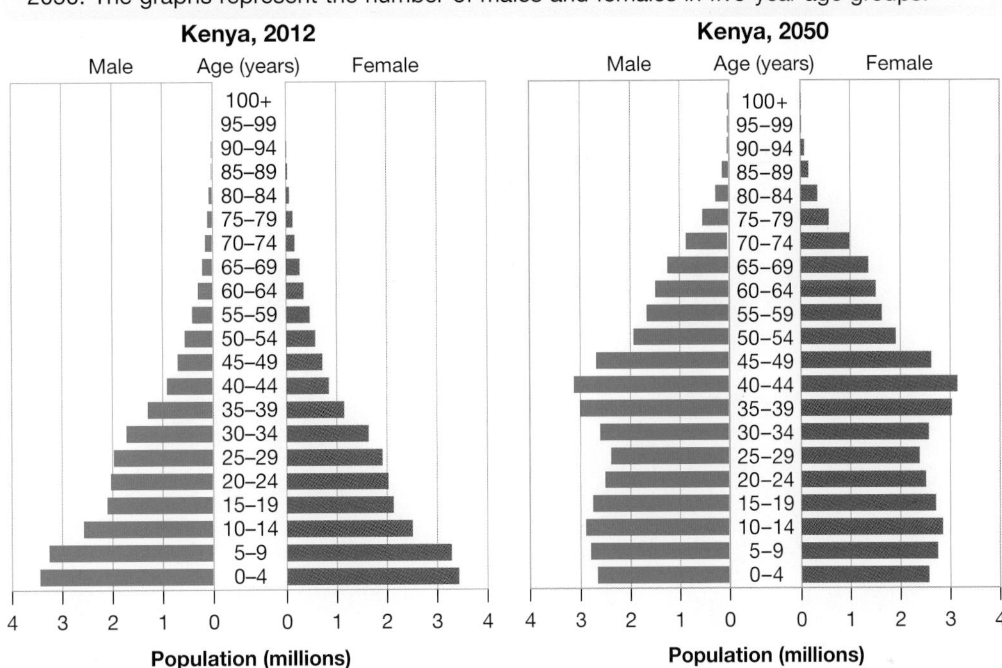

1.2.5 Activities

To answer questions online and to receive **immediate feedback** and **sample responses** for every question, go to your learnON title at www.jacplus.com.au. *Note*: Question numbers may vary slightly.

Refer to figure 5.

1. How many males and females were under 15 years of age in 2012?
2. Approximately how many males and females are predicted to be under 15 years of age in 2050?
3. Suggest some implications for a country that has a large proportion of its population as children.
4. Compare the ***change*** in the number of people between 40 and 60 years of age for both time periods.
5. What ***changes*** in Kenya's population structure might cause a possible decline in the wellbeing of its people? Give reasons for your answer.

Deepen your understanding of this topic with related case studies and questions.
Change

1.2.6 What is environment?

People live and depend on the environment, so it has an important influence on our lives.

There is a strong interrelationship between humans and natural and urban environments. People depend on the environment for the source, sink, spiritual and service functions it provides.

Humans significantly alter environments, causing both positive and negative effects. The building of dams to reduce the risk of flooding, the regular supply of fresh water and the development of large-scale urban environments to improve human wellbeing are examples. On the other hand, mismanagement has created many environmental threats such as soil erosion and global warming, which have the potential to have a negative impact on the quality of life for many people.

FIGURE 6 Lake Urmia is the largest lake in the Middle East and one of the largest landlocked saltwater lakes in the world. Since 2005, the lake has lost over 65 per cent of its surface area due to over-extraction of water for domestic and agricultural needs. The lake and its surrounding wetlands are internationally important as a feeding and breeding ground for migratory birds.

1.2.6 Activities

To answer questions online and to receive **immediate feedback** and **sample responses** for every question, go to your learnON title at www.jacplus.com.au. *Note*: Question numbers may vary slightly.

Refer to figure 6.

1. What physical features make up this ***environment***?
2. What features of the natural ***environment*** are consistent across the two images?
3. Describe the changes to this ***environment*** over the time period of 1998 to 2011.
4. Describe the distribution of salt flats around the lake in 1998 compared with their distribution in 2011.
5. How might the loss of water and increase in salt flats affect:
   (a) people
   (b) the ***environment*** in the surrounding region?
6. Suggest a possible future scenario for Lake Urmia:
   (a) if water continues to be extracted and withdrawn
   (b) if water withdrawals for irrigation are reduced, and water conservation methods are introduced in neighbouring ***places***.

Deepen your understanding of this topic with related case studies and questions.
Environment

1.2.7 What is scale?

When we examine geographical questions at different spatial levels we are using the concept of scale to find more complete answers.

Scale is a useful tool for examining issues from different perspectives; from the personal to the local, regional, national and global. It is also used to look for explanations or compare outcomes. For example, explaining the changing structure of the population in your local area may require an understanding of migration patterns at a national or even global scale.

FIGURE 7 A map of India showing the distribution of literacy levels (percentage) for 2011

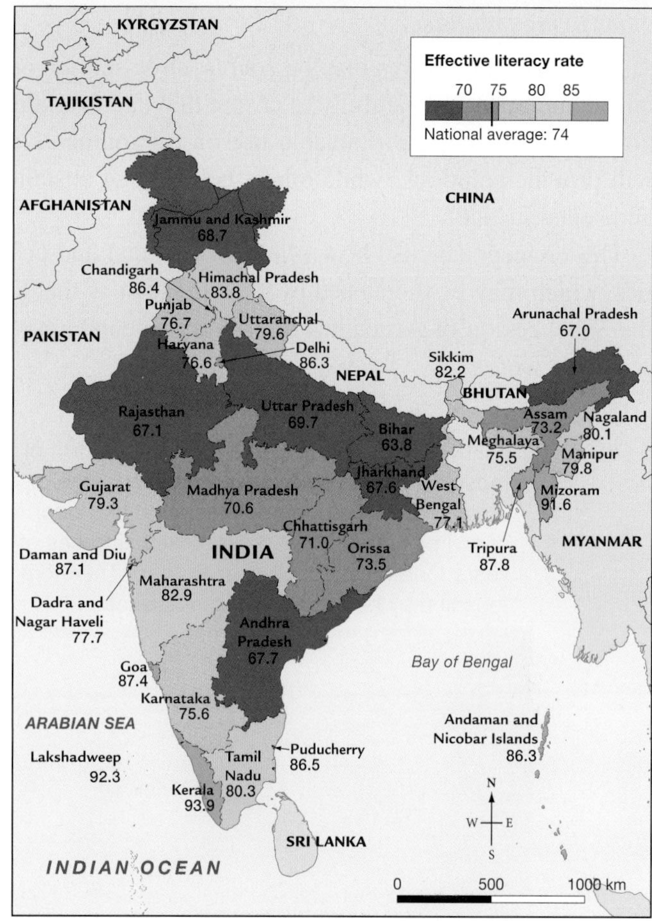

Source: Government of India, Ministry of Home Affairs, Office of Registrar General

1.2.7 Activities

To answer questions online and to receive **immediate feedback** and **sample responses** for every question, go to your learnON title at www.jacplus.com.au. *Note*: Question numbers may vary slightly.

Refer to figure 7.

1. At the national ***scale***, what is the average literacy rate for India?
2. At the regional ***scale***, which three states have the lowest literacy levels?
3. What factors might contribute to a state's low literacy level?
4. How might literacy levels affect the wellbeing of people?
5. Data, such as for literacy levels, is collected by governments during a census. How would knowing this sort of information assist a government in planning for future populations?

Deepen your understanding of this topic with related case studies and questions.

Scale

1.2.8 What is sustainability?

Sustainability is about maintaining the capacity of the environment to support our lives and the lives of other living creatures.

Sustainability ensures that the source, sink, service and spiritual functions of the environment are maintained and managed carefully to ensure they are available for future generations. There can be variations in how people perceive sustainable use of environments and resources. Some people think that technology will provide solutions, while others believe that sustainable management involves environmental benefits and social justice.

This concept can also be applied to the social and economic sustainability of places and their communities, which may be threatened by changes such as the degradation of the environment. Land degradation in the Sahel region of Africa has often forced people, especially young men, off their land and into cities in search of work.

FIGURE 8 Dust storms are an extreme form of land degradation. Dry, unprotected topsoil is easily picked up and carried large distances by wind before being deposited in other places. Drought, deforestation and poor farming techniques are usually the cause of soil being exposed to the erosional forces of wind and water. It may take thousands of years for a new topsoil layer to form. Therefore, any land practices that lead to a loss of topsoil may be considered unsustainable.

1.2.8 Activities

To answer questions online and to receive **immediate feedback** and **sample responses** for every question, go to your learnON title at www.jacplus.com.au. *Note*: Question numbers may vary slightly.

Refer to figure 8.

1. Complete the following table with examples of factors contributing to soil erosion.

Natural factors contributing to soil erosion	Human factors contributing to soil erosion

2. Explain how the ***interconnection*** of human activities and natural processes can contribute to land degradation.
3. Describe the impacts of the dust storm on people living in these two different ***places***:
   (a) rural areas (source of the soil)
   (b) the urban area shown in the image.
4. What are the long-term implications of the ***unsustainable*** use of soil?
5. How can farming be made more ***sustainable*** in terms of soil conservation?

Deepen your understanding of this topic with related case studies and questions.
- **Sustainability**

1.3 Review

1.3.1 Applying the concepts

Dharavi, home to over one million people, is considered to be the largest slum in Asia. Located on former swamp land, 80 distinct neighbourhoods have developed, with an estimated population density of 45 000 per hectare. Largely built by the residents themselves, Dharavi lacks traditional urban infrastructure of sewerage, sealed roads and public amenities. On the other hand, it is calculated that the annual value of goods produced in Dharavi's informal slum industries is US$500 million. The land alone that the slum is located on is worth over $10 billion in a city that is rapidly running out of room and where many of its citizens exist on $1 per day. The city plans to replace the slums with high-rise apartment blocks, but there is concern that this will destroy the small industries and social networks that presently exist.

FIGURE 1 Dharavi, a slum located in the middle of Mumbai, India's financial capital

1.3 Activities

To answer questions online and to receive **immediate feedback** and **sample responses** for every question, go to your learnON title at www.jacplus.com.au. *Note*: Question numbers may vary slightly.

Refer to figure 1.

1. Where is Dharavi located? (***space***)
2. How has the former ***environment*** of swamps and marshes been ***changed*** to accommodate slum neighbourhoods?
3. Describe the living conditions in the ***place*** you see in the photograph.
4. In what ways would living in this type of ***environment*** possibly impact on people's wellbeing?
5. What would be the main forms of transport in the slum? (***interconnection***)
6. List the types of urban infrastructure that are evident and not evident in this image. (***place***)
7. What would be the advantages and disadvantages of running small-***scale*** industries such as recycling, leather work and clothing in Dharavi?
8. How would such a high population density in the slum assist in the ***interconnection*** of people and businesses?
9. List the possible social, economic and ***environmental changes*** that a new housing estate would bring to the residents.
10. Dharavi is largely a self-sufficient and economically viable community. What would be needed to ensure its continual and future ***sustainability***?
11. How might the citizens' rights and wellbeing be challenged if the slum is demolished to make way for new housing?
12. Knowing the value of the land and needs of a growing city, do you think this is the most ***sustainable*** use of a scarce resource? Give reasons for your answer.

UNIT 1 ENVIRONMENTAL CHANGE AND MANAGEMENT

In the twenty-first century, the world faces many environmental challenges. These challenges can range from a local scale, for example, degradation of a nearby creek, through to a global scale, for example, the threat of global warming. Understanding how people and their environments interconnect is vital for explaining environmental changes and helps in planning effective management for a sustainable future.

The future is in our hands.

TOPIC 2 Introducing environmental change and management

2.1 Overview

Numerous **videos** and **interactivities** are embedded just where you need them, at the point of learning, in your learnON title at www.jacplus.com.au. They will help you to learn the content and concepts covered in this topic.

2.1.1 Introduction

Across the world there are many environmental changes that have been caused by humans, such as pollution, land degradation and impacts on aquatic environments. People have different points of view, or world views, on many of these changes. Climate change is a major environmental change as it impacts on all aspects of the biophysical environment, such as plants and animals; our land; inland water resources; coastal, marine and urban environments. It is vital that we respond intelligently to, and effectively manage, all future environmental changes to minimise negative social and economic impacts.

Human-induced climate change has led to increased severe weather events such as drought. Rivers can dry up, with consequent loss of plant and animal life.

Starter questions

1. The ***environment*** supports all life on Earth — humans, plants and animals. As a class, brainstorm examples of ***environmental changes*** people have caused, and discuss where these are occurring.
2. Choose one ***environmental change*** from the list your class created and discuss the various viewpoints different people, groups or organisations have about it.
3. Brainstorm specific examples of ***environmental changes*** people have caused that have been positive, and that have come about by people deliberately and efficiently managing the ***change***.

INQUIRY SEQUENCE

2.2 SkillBuilder: Evaluating alternative responses

online only

WHAT IS INVOLVED IN EVALUATING ALTERNATIVE RESPONSES?

Alternative responses are a range of different ideas or opinions on an issue. Evaluating ideas involves weighing up and interpreting your research to reach a judgement or a decision based on the information.

Go online to access:

- a clear step-by-step explanation to help you master the skill
- a model of what you are aiming for
- a checklist of key aspects of the skill
- a series of questions to help you apply the skill and to check your understanding.

FIGURE 1 Alternative responses to the question 'Should tourist numbers on Fraser Island be limited?'

A special environment

The island's special features include:

- long surf beaches and rocky headlands
- about 40 crystal-clear freshwater lakes. Some of these are perched lakes (that is, they sit, or perch, on an impermeable layer of rock or hardened organic matter lying above the watertable). There are also 'window' and barrage lakes. Window lakes appear when depressions in the land surface dip below the watertable, thus exposing part of it. Barrage lakes form when shifting sand dunes block running water and cause it to pool.
- many streams and creeks
- coloured sand cliffs, some 35 kilometres in length
- salt pans, lagoons, mangrove forests and wetlands
- thick rainforests, some of which are so dense that sunlight does not penetrate the canopy
- offshore seagrass beds to support colonies of dugong
- over 25 species of mammals, including dingoes thought to be the purest strain of the species in Australia
- over 350 species of birds. One of Australia's rarest birds, the endangered ground parrot, is found on the island.
- vast sandblows (that is, tracts of sand moved by the wind) and lofty sand dunes.

Past land uses

Fraser Island once had a sand-mining industry (mining its tracts of mineral-rich black sand). This was stopped in 1976 following a federal government inquiry. There was also a timber industry, disbanded in December 1991 by the Queensland Government after a separate inquiry.

Tourists

Hundreds of thousands of tourists now visit the Fraser coast region every year, injecting some $366 million into the region in 1999. It is estimated that 32 per cent of this visitor expenditure was contributed by tourists to Fraser Island itself. The most obvious risks that tourism brings to the national park have to do with land-clearing, waste, increased traffic and disturbance of the island's flora and fauna.

Ecotourism facilities

The island's Kingfisher Bay Resort and Village has the highest level of accreditation as an ecotourism facility. An environmental impact statement was prepared before the proposed facility was approved for construction.

Dingo management

In the past, many tourists fed the dingoes that roam the island. In April 2001, however, a young boy was tragically killed by dingoes. Tourists are now provided with a 'Dingo Smart' brochure, and are heavily fined if caught feeding a dingo or trying to encourage its attention. Any dingoes known to be a problem are culled.

Managing camping facilities

There are six government-owned camps — at Central Station, Lake Boomanjin (the largest perched lake in the world), Lake McKenzie, Dundubara, Waddy Point and Wathumba — and two that are privately owned. People can also camp on a restricted number of beach areas, but not within 50 metres of a creek, stream or lake. Beach camping areas are temporarily closed sometimes to allow vegetation to regrow or to halt erosion.

Managing four-wheel drive vehicles

Four-wheel drive vehicles are needed to travel around the island. Left unmanaged, these large vehicles could have a significant impact on the island's flora and fauna and on levels of erosion, especially because touring parties tend to drive in the same areas. It is the most attractive parts of the island that are often the most vulnerable.

All vehicles travelling on the island have to display a purchased permit and, more recently, driving and parking on sand dunes have been made illegal. The Environmental Protection Agency has started a campaign to educate four-wheel drivers about the impacts their vehicles have on the island's environment.

Government funding

In 2004, the Australian government's Natural Heritage Trust granted $300 000 to reduce road-related erosion, provide environmentally friendly amenities and better direct pedestrian movement around Fraser Island. Barriers along the sides of the island's roads and better planning have reduced the degree of erosion from run-off. The idea was partly to make pedestrian travel a more attractive option, as well as draw pedestrians away from the island's vulnerable dunes. To do this, boardwalks were built along the banks of Eli Creek (see the photograph on page 26). Stretches of dunes are also regularly closed for rehabilitation. Although the potential threat from tourism-related erosion remains, the stability of the island's sand dunes is improving.

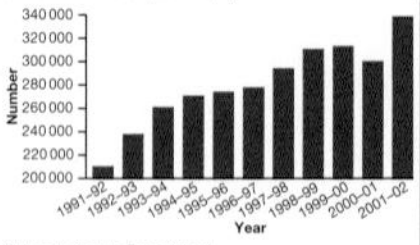

Number
340 000
320 000
300 000
280 000
260 000
240 000
220 000
200 000
1991–92 1992–93 1993–94 1994–95 1995–96 1996–97 1997–98 1998–99 1999–00 2000–01 2001–02
Year

Visitor numbers to Fraser Island

Please tourists, don't pee in the lake

So, we're sure your momma told you not to pee in the swimming pool — but did she also tell you it's bad to pee in a lake? Down in Australia, the beautiful Basin Lake on Fraser Island off Queensland isn't doing well these days, and one of the causes is high levels of urine in the water. The official word is that too many tourists are using the lake as a toilet and that's led the Queensland Parks and Wildlife Service to consider closing one of the access tracks to reduce visitor numbers. Right now 35 000 people visit the lake every year and since there's no in- or outflow from the lake; whatever goes in, stays in. Our alternative suggestion is to simply stick up a big notice advertising the current urine levels in the lake. We're fairly sure most people would skip the swim.

Source: Fraser Island Travel Guide, 17 October 2008.

learnon RESOURCES — ONLINE ONLY

Watch this eLesson: Evaluating alternative responses (eles-1744)

Try out this interactivity: Evaluating alternative responses (int-3362)

2.3 How do people interact with the environment?

2.3.1 How much space do we need?

If you gathered together all 7.4 billion humans from around the world and gave each person a space of one square metre, the island of Cyprus, which is approximately 8000 square kilometres, would provide standing room for everyone (see figure 1). Clearly this would be impractical, and providing services to ensure human wellbeing in an area with a density of almost 1 000 000 per square kilometre would be impossible.

While this idea is unrealistic in suggesting that 0.005 per cent of the total space on Earth is sufficient for humanity, it suggests we need to think about how little personal space we actually occupy, and what a large impact we have on the Earth and its biophysical systems.

FIGURE 1 Humanity crowded onto the island of Cyprus

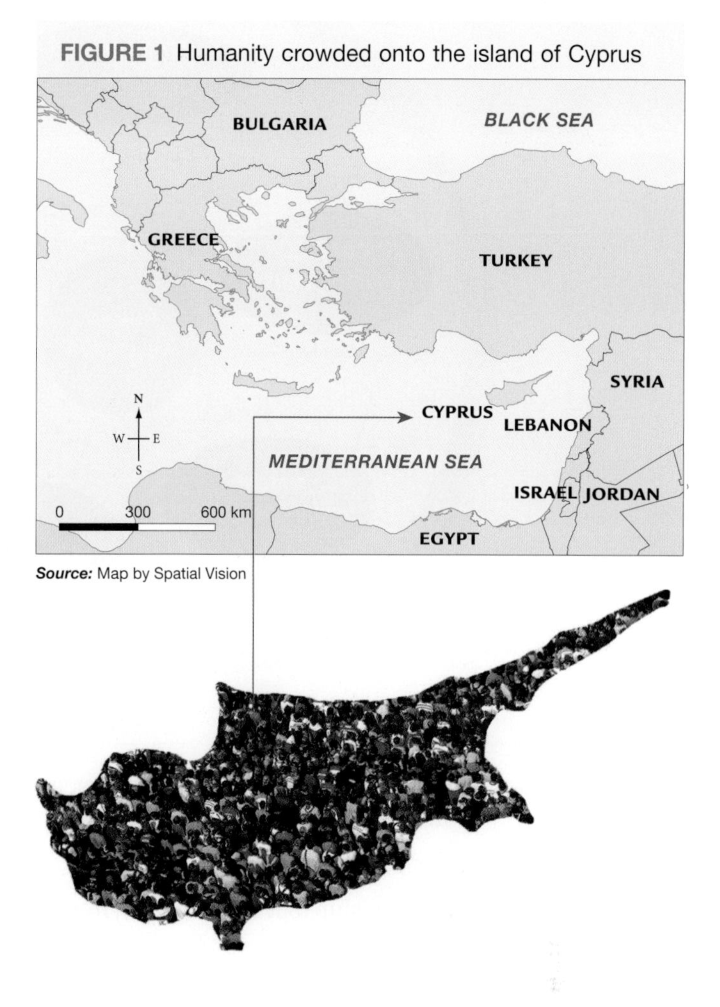

Source: Map by Spatial Vision

2.3.2 How do humans interact with the environment?

Over 200 years ago, an English scholar named Thomas Malthus proposed that England's population growth would eventually outstrip agricultural production. Malthus's earth-centred **environmental worldview** foretold of problems with supplies of food and warned that there would be more deaths due to famine and wars over resources. In 1798 he wrote, 'The power of population is so superior to the power in the earth to produce subsistence for man …' At the time Malthus wrote his thesis, England was moving into a period known as the Industrial Revolution; a time when the human-centred environmental worldview of the government and leaders of industry considered the earth's resources as limitless and that the development of the economy should take priority over the preservation of the natural world.

Today, many environments have become overloaded with the growing demands for food, land and other resources. This pressure on biomes and ecosystems has led to land degradation, with a consequent loss of habitats and biodiversity. Further consequences of this change are a reduction in human wellbeing and a struggle for social justice as land becomes unproductive due to overuse. Nevertheless, it should be remembered that change can be a natural process as well as human-induced.

Some topics that can help us explore change and the need for careful management include marine environments and coasts, the land, inland waters, and urban or built environments (see figure 2).

FIGURE 2 Interaction of environmental change with human wellbeing

LAND ENVIRONMENTS
Land degradation
COASTAL ENVIRONMENTS
INLAND WATER ENVIRONMENTS
HUMAN WELLBEING
Sea level rise
Overuse of water
Marine pollution
Loss of ecological services
MARINE ENVIRONMENTS
URBAN ENVIRONMENTS
NATURAL . . . . HUMAN-INDUCED
CHANGE . . . . CHANGE

> The Earth offers enough for everyone's need, not for everyone's greed. *Mahatma Gandhi*

What are ecological services?

A new view of the relationship between the environment and people is one of an **ecological service** or 'what nature provides for humanity'.

Ecological services can be thought of as biological and physical processes that occur in natural or semi-natural ecosystems and maintain the habitability and livelihood of people on the planet. These services are shown in figure 3.

Understanding the link (interconnection) between ecological services and human action is important as it can lead to more sustainable practices. The idea of ecological management takes an earth-centred environmental worldview, promoting **stewardship** or custodial management. This view considers caring for the land and the ecological services it provides as paramount. By applying this earth-centred viewpoint to human uses and management of the environment, future options for human wellbeing will be sustainable. The question is: how do we evaluate human impacts on the environment and what management strategies can be implemented to reverse damage and create a sustainable future? As such, we need to consider the costs and benefits, or more simply, the advantages and disadvantages of changes we make to the environment, as there will be consequences in terms of economic viability and social justice.

FIGURE 3 Ecological services

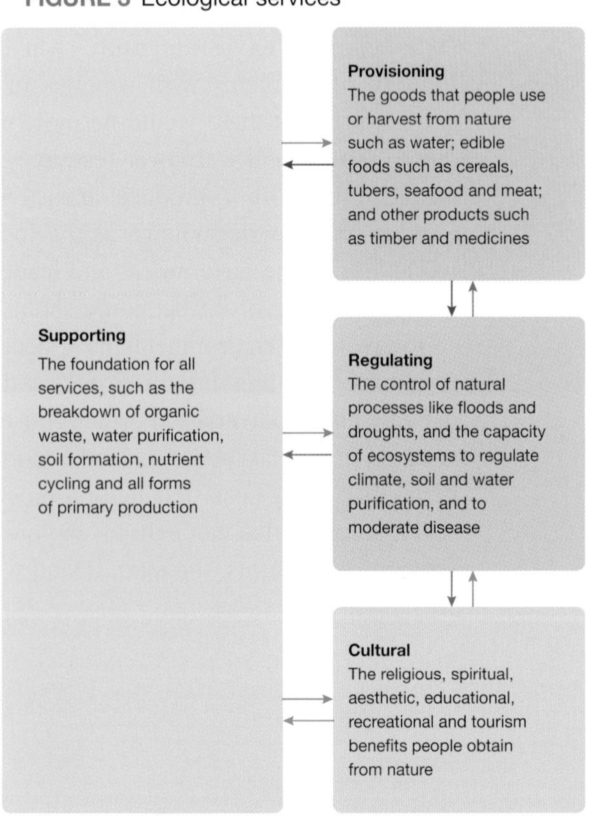

Supporting
The foundation for all services, such as the breakdown of organic waste, water purification, soil formation, nutrient cycling and all forms of primary production
Provisioning
The goods that people use or harvest from nature such as water; edible foods such as cereals, tubers, seafood and meat; and other products such as timber and medicines
Regulating
The control of natural processes like floods and droughts, and the capacity of ecosystems to regulate climate, soil and water purification, and to moderate disease
Cultural
The religious, spiritual, aesthetic, educational, recreational and tourism benefits people obtain from nature

What is the ecological footprint?

The **ecological footprint** is one means of measuring human demand for ecological services. The footprint takes into account the regenerative capacities of biomes and ecosystems, which are described as the Earth's **biocapacity**. The footprint is given as a number, in hectares of productive land and sea area, by measuring a total of six factors, as shown in figure 4. The ecological footprint is a useful indicator of environmental sustainability.

FIGURE 4 Measuring the Earth's ecological footprint

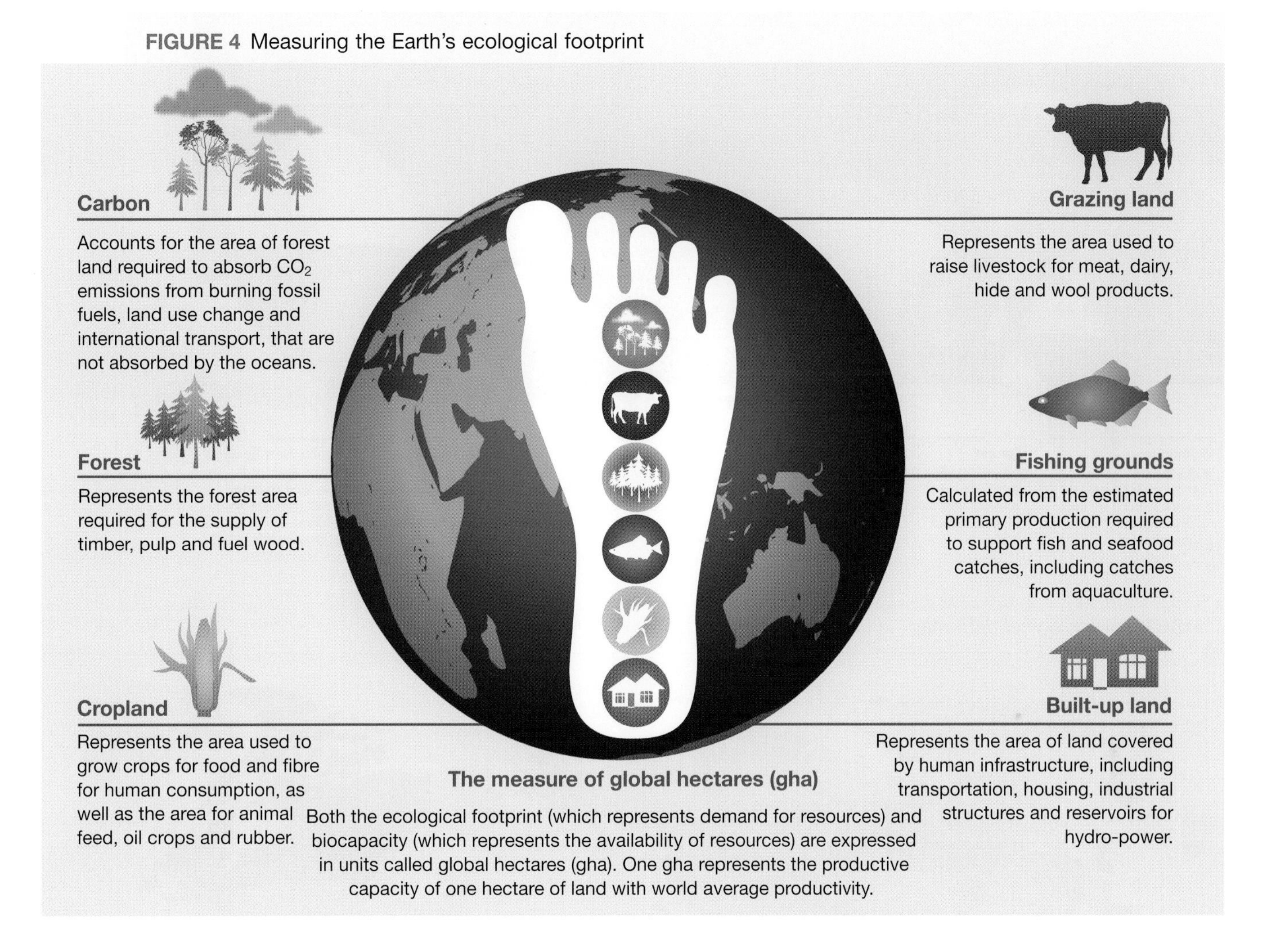

Carbon
Accounts for the area of forest land required to absorb CO_2 emissions from burning fossil fuels, land use change and international transport, that are not absorbed by the oceans.
Forest
Represents the forest area required for the supply of timber, pulp and fuel wood.
Cropland
Represents the area used to grow crops for food and fibre for human consumption, as well as the area for animal feed, oil crops and rubber.
Grazing land
Represents the area used to raise livestock for meat, dairy, hide and wool products.
Fishing grounds
Calculated from the estimated primary production required to support fish and seafood catches, including catches from aquaculture.
Built-up land
Represents the area of land covered by human infrastructure, including transportation, housing, industrial structures and reservoirs for hydro-power.
The measure of global hectares (gha)
Both the ecological footprint (which represents demand for resources) and biocapacity (which represents the availability of resources) are expressed in units called global hectares (gha). One gha represents the productive capacity of one hectare of land with world average productivity.

Figure 5 compares the ecological footprint with biocapacity. The elephants represent each region's footprint (per capita) and the balancing balls represent the size of the region's biocapacity (per capita). The dark green background represents the gross footprint of regions that exceed their biocapacity, and the light green background represents those regions that use less than their biocapacity.

In 2014 the total ecological footprint was estimated at 1.5 planet Earths, which means that humanity used ecological services at 1.5 times the biocapacity of the Earth to renew them. The 1.5 ecological footprint figure represents an average for all regions of the Earth. However, the United States and Canada, which have an ecological footprint of 7.9, are well above this average. This level of resource use is not sustainable into the future, and raises questions of economic viability, environmental benefit and social justice. Figure 6 shows a map of the Earth's ecological debt. Note that there is a strong relationship between ecological footprint and a country's wealth and/or population. For example, the United States and much of Europe and Japan are wealthy countries with large ecological footprints and small biocapacities. China and India are highly populated countries with large ecological footprints and small biocapacities. Australia and New Zealand have minimal ecological footprints because they have relatively small populations and high biocapacities.

FIGURE 5 Biocapacity and ecological footprint

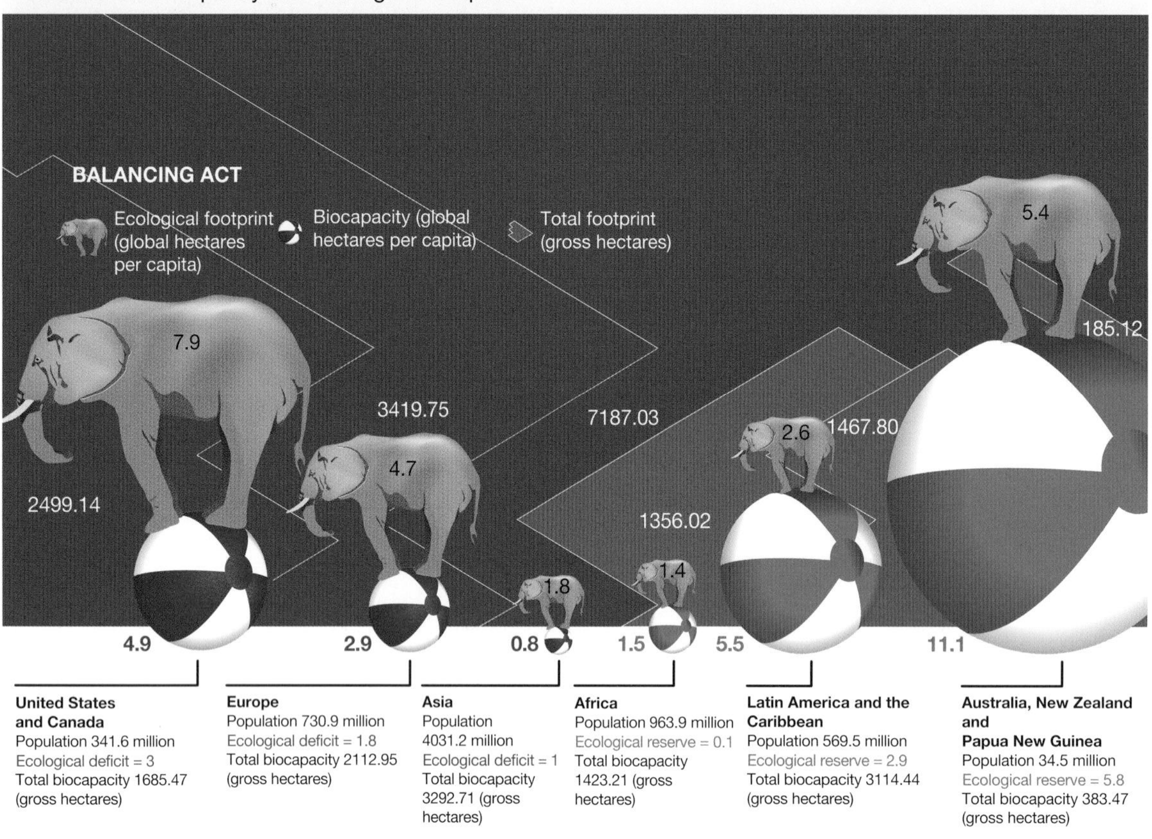

United States and Canada
Population 341.6 million
Ecological deficit = 3
Total biocapacity 1685.47 (gross hectares)

Europe
Population 730.9 million
Ecological deficit = 1.8
Total biocapacity 2112.95 (gross hectares)

Asia
Population 4031.2 million
Ecological deficit = 1
Total biocapacity 3292.71 (gross hectares)

Africa
Population 963.9 million
Ecological reserve = 0.1
Total biocapacity 1423.21 (gross hectares)

Latin America and the Caribbean
Population 569.5 million
Ecological reserve = 2.9
Total biocapacity 3114.44 (gross hectares)

Australia, New Zealand and Papua New Guinea
Population 34.5 million
Ecological reserve = 5.8
Total biocapacity 383.47 (gross hectares)

FIGURE 6 Ecological debt map

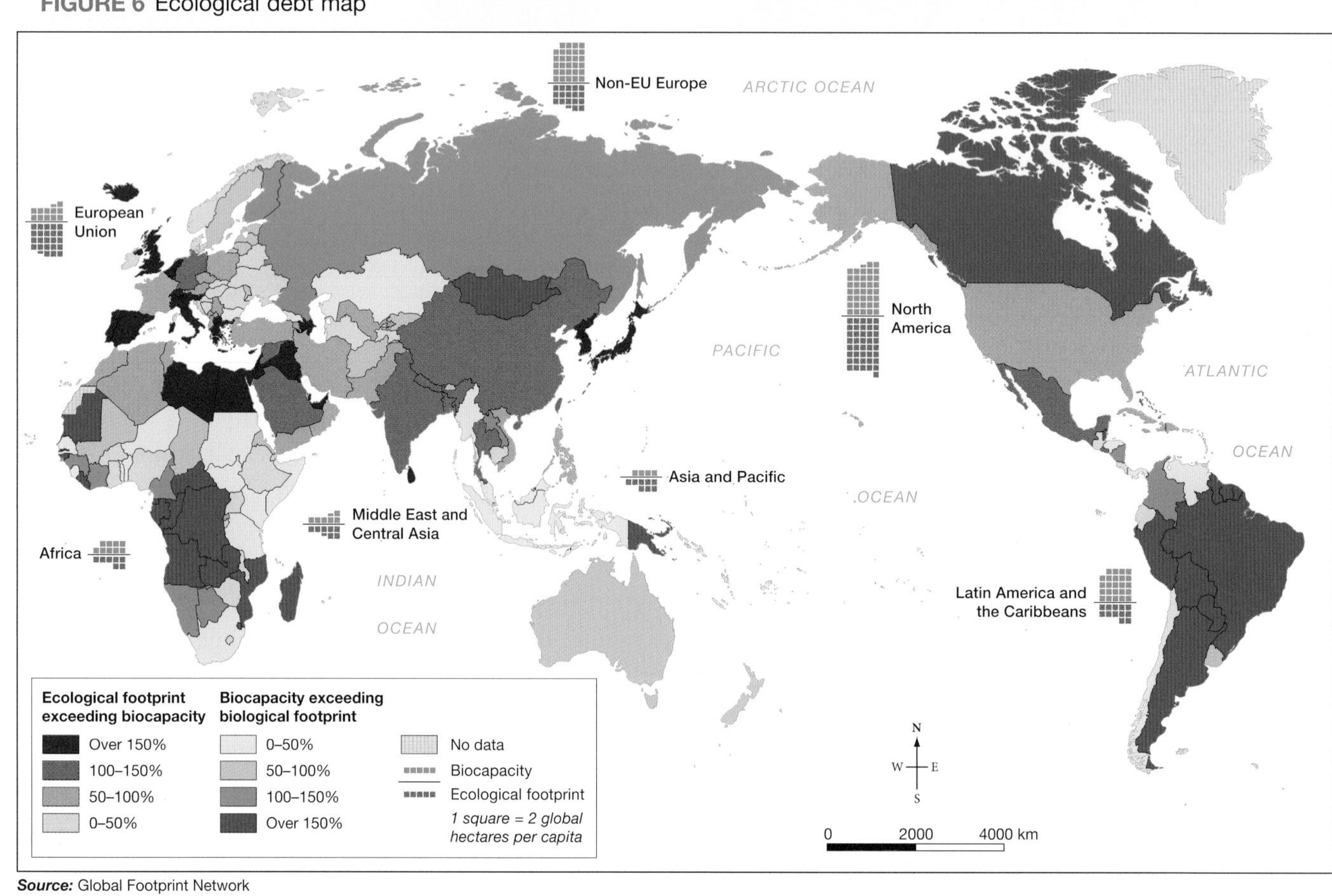

Source: Global Footprint Network

What is a sustainable world?

A range of indices have been developed in recent years to examine the link between ecological services, human wellbeing and sustainability. These include the Human Development Index (HDI), the Sustainable Society Index (SSI) and the Happy Planet Index (HPI), and each gives a slightly different perspective on human activity and/or sustainability.

The Sustainable Society Index says that sustainable human action must:

(a) meet the needs of the present generation yet not compromise the ability of future generations to meet their own needs

(b) ensure that people have the opportunity to develop themselves in a free, well-balanced society that is in harmony with nature.

It is worthwhile studying these indices as they put forward many sound ideas about human wellbeing and the sustainability of the ecological services of the natural world.

The Sustainable Society Index gives values to 21 factors across a range of social, political, economic and environmental considerations. Australia rates, for example, highly in clean air and sufficient food and lowly in renewable energy and consumption. For further details see figure 7.

FIGURE 7 Australia's situation based on the Sustainable Society Index

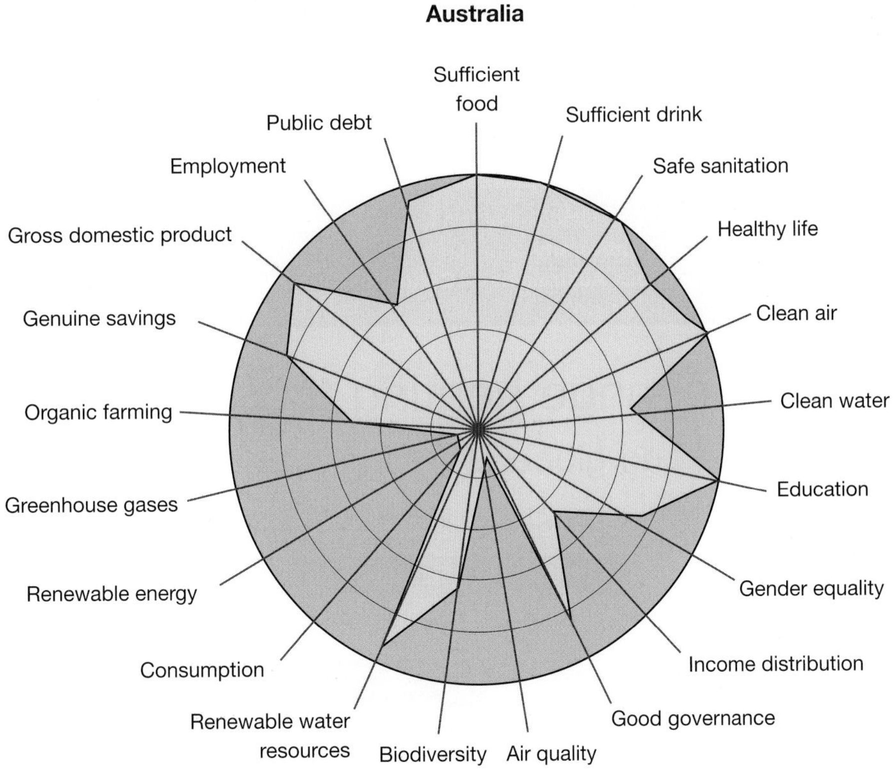

2.3 Activities

To answer questions online and to receive **immediate feedback** and **sample responses** for every question, go to your learnON title at www.jacplus.com.au. *Note*: Question numbers may vary slightly.

Explain

1. Figure 7 represents Australia's situation based on the Sustainable Society Index. Values increase from the centre of the circle outwards. Use the **Sustainable Society Index** weblink in the Resources tab to find out more about the Index. Then answer the following questions.

(a) Divide the various factors shown around the circle into the categories human wellbeing, environmental wellbeing and economic wellbeing.
(b) Explain how these factors are ***interconnected***.
(c) List the factors that Australia needs to ***change*** to be a more ***sustainable*** nation. Consider ***environmental***, social and economic criteria from the index to inform your recommendations.
(d) Suggest reasons why Australia rates poorly in some of these factors.
(e) In what factors with respect to human wellbeing does Australia rate highly? Why would this be so?

2. Refer to figure 5.
   (a) What reasons can you suggest for the very high ***environmental*** or ecological footprint for the United States and Canada?
   (b) How might the three regions with the dark green very high gross footprint improve their biocapacity?
   (c) Why is Australia in such a good position in terms of ecological footprint compared to biocapacity?

Think

3. Use the **Forest depletion** weblink in the Resources tab to explore information on this topic.
   (a) What aspects of ***sustainability*** and the concept of stewardship can you draw from this information?
   (b) Make a list of nations that have an ***unsustainable*** level of forest depletion.

Try out this interactivity: Nature's bounty (int-3287)

Explore more with these weblinks: Sustainable Society Index, Forest depletion

Deepen your understanding of this topic with related case studies and questions.

2.4 Is climate change heating the Earth?

2.4.1 Climate change and global warming

The world's climate has been changing for millions of years, but more recently there has been an increase in the concentration of greenhouse gases in the atmosphere, leading to **global warming**. It is believed that human activity, particularly the burning of fossil fuels such as coal and oil, have led to what is known as the **enhanced greenhouse effect**, which is heating the Earth and its atmosphere. The wider consequences of global warming will lead to environmental change across a wide range of biophysical systems (see figure 1).

FIGURE 1 Consequences of changes in the global climate

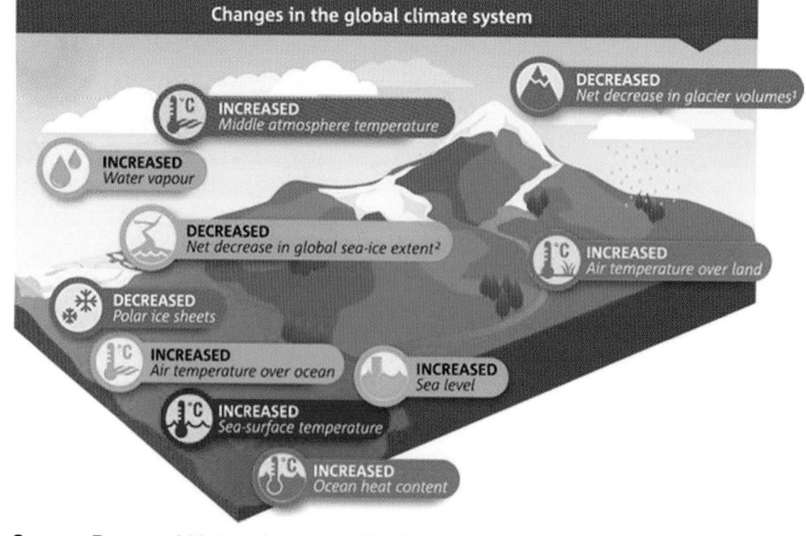

Source: Bureau of Meteorology and CSIRO

Climate, which can be defined as the yearly changes in the Earth's atmosphere, is highly variable over the Earth's surface. As such, climates in the tropics contrast markedly with climates near the poles. Climate also varies over extensive periods of time, and scientists have described these changes, which date back millions of years, long before the emergence of the human species, as warm periods and ice ages. Currently the Earth is in a warm period, having moved out of ice age conditions as recently as 6000 years ago. Today it is realised that human activity is increasing the rate of global warming leading to **climate change**, particularly in the past few hundred years, and this can have serious consequences for the planet (see figure 2).

FIGURE 2 Average global temperature, 1880–2014, with projection to 2100

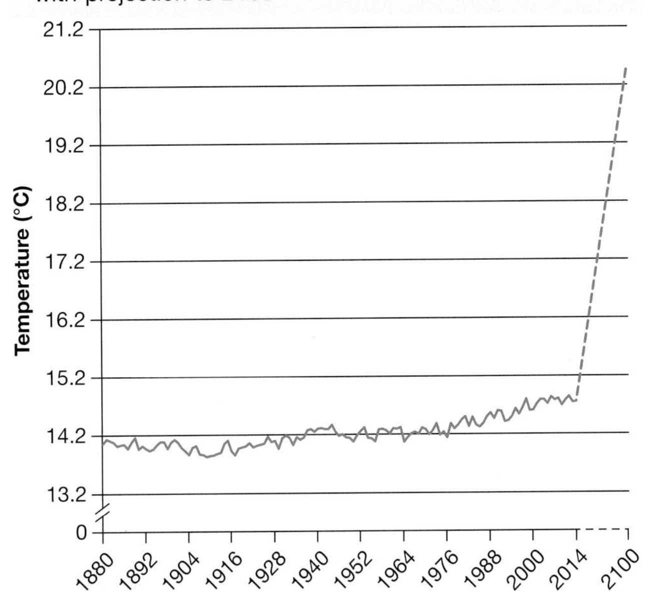

Source: University Corporation for Atmospheric Research

The greenhouse effect

The greenhouse effect is the mechanism whereby solar energy is trapped by water vapour and gases in the atmosphere, thereby heating the atmosphere and helping to retain this heat, as in a glasshouse. The three most important gases responsible for the greenhouse effect are carbon dioxide, nitrous oxide and methane. Without this greenhouse effect the atmosphere would be much cooler, and ice age conditions would prevail over the planet, making life as we know it impossible (see figure 3).

FIGURE 3 How the greenhouse effect works

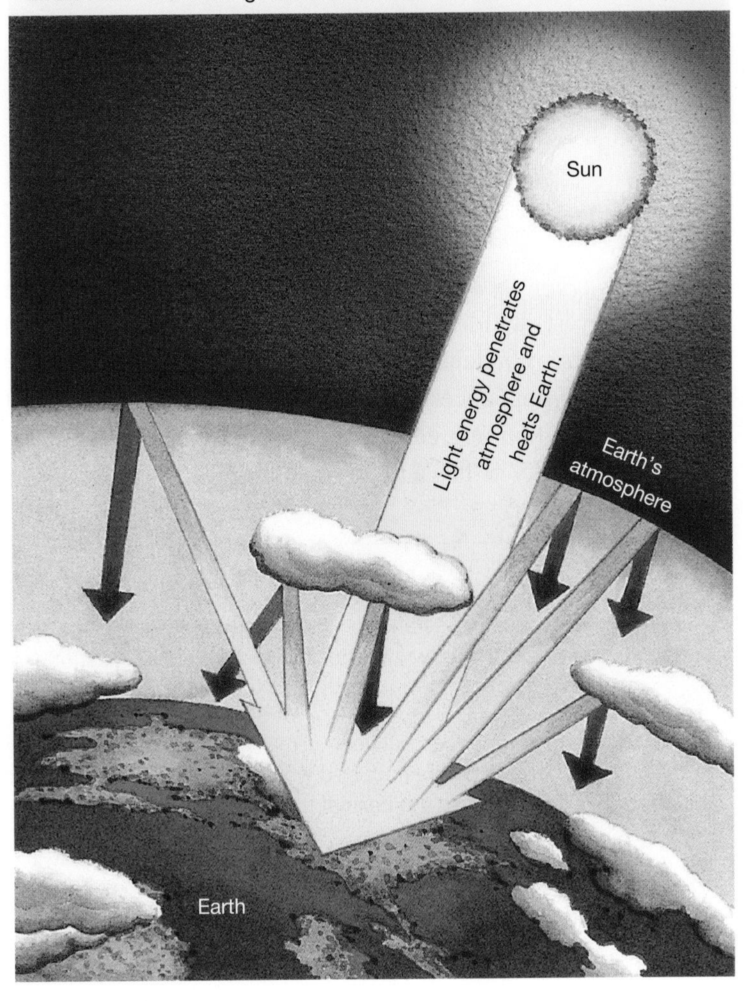

Human activity and the enhanced greenhouse effect

Changes in the balance of the greenhouse gases are a natural event, leading to the different climatic conditions on the planet as experienced over geological time. The issue today is how much impact human activity is having on the natural cycle of events, and how this activity is leading to climate change and global warming.

The term 'the enhanced greenhouse effect' has been developed to show that heating of the atmosphere is moving at a rate that is above what could be expected by natural processes of change (see figure 4). Recent research by government and non-government organisations has indicated that all parts of the world are vulnerable to the impacts of the enhanced greenhouse effect and associated climate change. Six key risks that have been identified in Australia alone include higher temperatures, sea level rise, heavier rainfall, greater wildfire risk, less snow cover, reduced run-off over southern and eastern Australia, and more intense tropical cyclones and storm surges along the coast.

FIGURE 4 The enhanced greenhouse effect

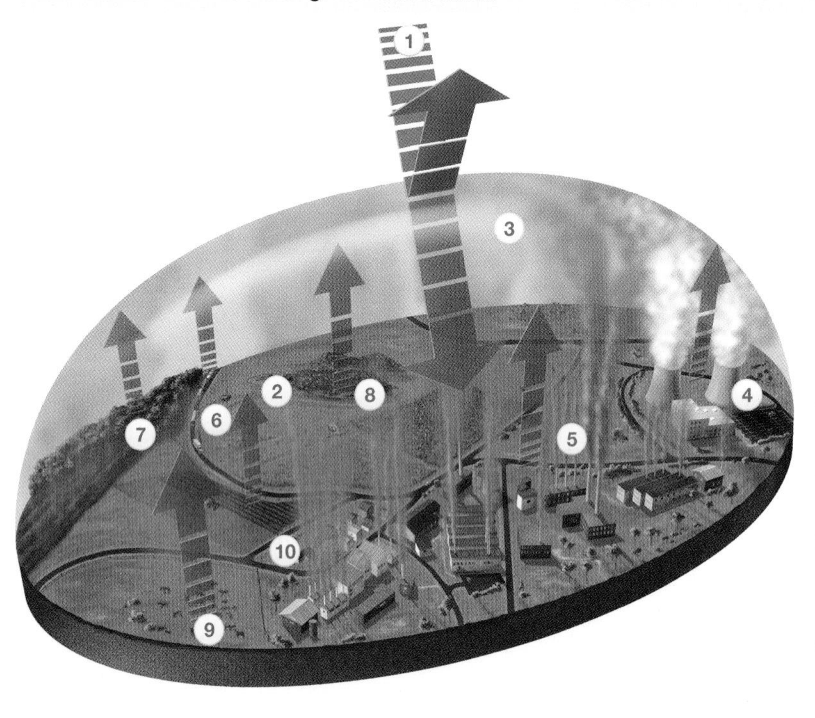

1. Heat from the sun
2. Heat trapped by greenhouse gases
3. Heat radiating back into space
4. Greenhouse gases produced by power stations burning fossil fuels
5. Greenhouse gases produced by industry burning fossil fuels
6. Greenhouse gases produced by transport burning fossil fuels
7. Greenhouse gases released by logging forests and clearing land
8. Methane escaping from waste dumps
9. Methane from ruminant (cud-chewing) livestock, e.g. cattle, sheep
10. Nitrous oxide released from fertilisers and by burning fossil fuels

2.4.2 What can we do?

A switch to renewable energy sources such as solar, wind, water (hydro) and geothermal (heat from inside the Earth's crust) will lead to sustainable energy use in the future, reduce carbon emissions into the atmosphere and thereby reduce the enhanced greenhouse effect. At the household level, using energy-efficient light bulbs and appliances and purchasing solar panels to produce hot water and electricity can lead to a significant reduction in greenhouse gas emissions. You could even think of purchasing a new motor vehicle that uses electricity or has a higher fuel efficiency rating.

2.4 Activities

To answer questions online and to receive **immediate feedback** and **sample responses** for every question, go to your learnON title at www.jacplus.com.au. *Note*: Question numbers may vary slightly.

Remember

1. What are the differences between climate change and global warming?
2. What is the greenhouse effect and what are the three atmospheric gases responsible for this effect?
3. What would happen to the Earth if there was no greenhouse effect?

Explain

4. What changes have occurred to the Earth's climate over geological time?
5. Why would sea levels be much lower in an ice age period?

Discover

6. What role do trees play in the carbon cycle and in controlling the level of greenhouse gases?
7. What impacts will global warming, and in particular higher water temperatures, have on a marine ecosystem such as the Great Barrier Reef?

Predict

8. Refer to figure 2, which shows average global temperatures.
   (a) What is the time period shown in the graph?
   (b) In which year did the highest average temperature and lowest average temperature occur?
   (c) What is the projected temperature in 2100?
   (d) What is the general trend shown by the graph?

Think

9. In groups, prepare a report that explains how the enhanced greenhouse effect operates, based on the information in figure 4. Prepare a presentation for the class that includes your suggestions about what we can do to reduce the impacts of the enhanced greenhouse effect.

Deepen your understanding of this topic with related case studies and questions.
- **Causes of climate change**
- **Larsen Ice Shelf break-up**
- **Impacts on polar bears**
- **Climate change and Australia**
- **Global warming and Antarctica**

2.5 Can we slow climate change?

2.5.1 Global action

Climate change is a global phenomenon. The greenhouse gases produced in one country spread through the atmosphere and affect other countries. Action by only a few countries to reduce greenhouse gases will, therefore, have little impact — it requires international cooperation, especially by the largest polluters.

Since the 1990s, countries have met at United Nations Intergovernmental Panel of Climate Change (IPCC) conferences and agreed to take steps to reduce emissions of greenhouse gases. An early conference developed the **Kyoto Protocol**, an agreement that sets targets to limit greenhouse gas emissions, and 128 countries have agreed to this Protocol. Further conferences in 2009 in Copenhagen, Denmark, 2010 in Cancun, Mexico and in 2015 in Paris led to an important new direction, with all countries agreeing to contain global warming within 2 °C. This means that emissions of CO_2, which were at 395 parts per million (ppm) in 2013, must be kept below 550 ppm to reach this target. If no actions (mitigation measures) are taken, temperatures could increase by 5 °C, as shown in figure 1. To date, 192 of the world's 196 countries have signed the Kyoto Protocol,

FIGURE 1 Global average temperature outcomes for three emissions cases, 1990–2100

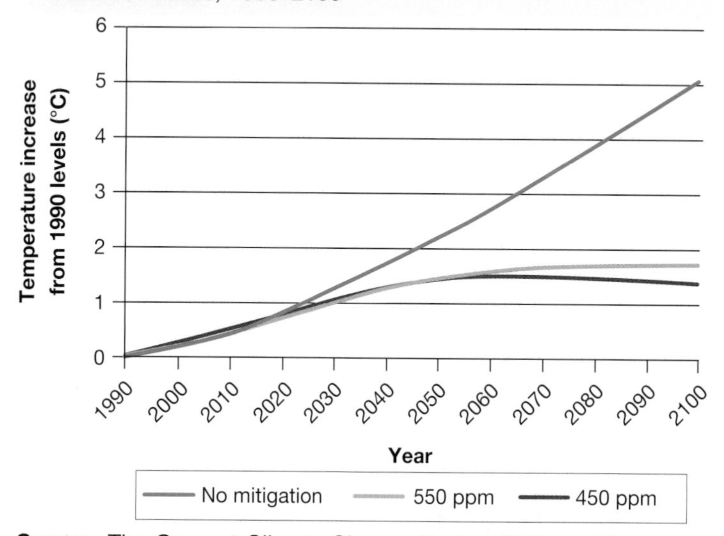

Source: The Garnaut Climate Change Review 2008, p. 88

however, close to half have modified their commitment to reach targets for greenhouse emission reductions set for 2020. The United States has signed the Protocol but has not ratified emission targets and Canada has withdrawn from the Protocol.

To meet the greenhouse gas emissions targets defined by these agreements, countries must make changes that reduce their level of emissions. They can also meet the targets in two other ways:

1. A country can carry out projects in other countries that reduce greenhouse gas emissions and offset these reductions against their own target.
2. Companies can buy and sell the right to emit carbon gases. For example, a major polluter, such as a coal power station, is allowed to emit a certain amount of greenhouse gases. If it is energy efficient, and emits less than its limit, it gains **carbon credits**. It has the right to sell these credits to another company that is having difficulty reducing its emissions. Companies can also gain credits by investing in projects that reduce greenhouse gases (such as renewable energy), improve energy efficiency, or that act as carbon sinks (such as tree planting and underground storage of CO_2).

2.5.2 Australia's action

The Garnaut Report 2011 and the findings of the IPCC state that it is in Australia's national interest to do its fair share in a global effort to mitigate climate change (see table 1). The findings of this report were confirmed at the IPCC meeting in Paris in 2015. The introduction in 2012 by the Australian Government of an **emissions trading scheme** with a fixed price on carbon for three years and then a floating price led to the introduction of a carbon tax set at $23 per tonne of carbon dioxide emissions. Big businesses and industries that use large amounts of fossil fuels have complained that the tax will affect profits and force the price of goods and commodities higher. In 2015, the new Australian Government reversed the decision to support the previously established emissions trading scheme, based on social and economic criteria stating that the carbon tax had increased costs of power to households and businesses.

TABLE 1 Potential impacts for each of the three emissions cases by 2100

Emissions case	450 ppm	550 ppm	No action
Likely range of temperature increase from 1990 level	0.8–2.1°C	1.1–2.7°C	3–6.6°C
Percentage of species at risk of extinction	3–13%	4–25%	33–98%
Area of reefs above critical limits for coral bleaching	34%	65%	99%
Likelihood of starting large-scale melt of the Greenland ice sheet	10%	26%	100%
Threshold for starting accelerated disintegration of the West Antarctic ice sheet	No	No	Yes

Source: The Garnaut Climate Change Review 2008, p. 102

2.5.3 Taking personal action

Australian households produce about one-fifth of Australia's greenhouse gases through their use of transport, household energy and the decay of household waste in landfill. This amounts to about 15 tonnes of CO_2 per household per year. (A tonne of CO_2 would fill one family home.) The Australian Conservation Foundation has suggested a 10-point plan (see figure 2) that every Australian household can follow to reduce its level of greenhouse gas pollution.

FIGURE 2 The Australian Conservation Foundation Plan

1 Switch to green power
Choose renewable energy from your electricity retailer and support investment in sustainable, more environmentally friendly energies. Make sure it is accredited GreenPower [electricity produced using renewable resources] — see www.greenpower.gov.au for a list of who qualifies.
2 Get rid of one car in your household
A car produces seven tonnes of greenhouse pollution each year (based on travelling 15 000 kilometres per year).

This does not include the energy and water used to build the car — 83 000 litres of water and eight tonnes of greenhouse pollution. So share a car with your family.

3 Take fewer air flights
A return domestic flight in Australia creates about 1.5 tonnes of greenhouse emissions (based on Melbourne to Sydney return). A return international flight creates about 9 tonnes (based on Melbourne to New York return). Holiday closer to home.

4 Use less power to heat your water
A conventional electric household water heater produces about 3.2 tonnes of greenhouse pollution in a year. Using less hot water will reduce your pollution. Using the cold cycle on your washing machine will save 3 kg of greenhouse pollution. Switching off your water heater when you're away will also reduce your energy use.

5 Eat less meat
Meat, particularly beef, has a very high environmental impact, using a lot of water and land to produce it, and creating significant greenhouse pollution. If you reduce your red meat intake by two 150-gram serves a week, you'll save 20 000 litres of water and 600 kg of greenhouse pollution a year.

6 Heat and cool your home less
Insulate your walls and ceilings. This can cut heating and cooling costs by 10 per cent. Each degree change can save 10 per cent of your energy use. A 10 per cent reduction is 310 kg of greenhouse pollution saved.

7 Replace your old showerhead with a water-efficient alternative
This will save about 44 000 litres of water a year and up to 1.5 tonnes of greenhouse pollution from hot water heating (on average).

8 Turn off standby power
Turning appliances off at the wall could reduce your home's greenhouse emissions by up to 700 kg a year.

9 Cycle, walk or take public transport rather than drive your car
Cycling 10 kilometres to work (or school) and back twice a week instead of driving saves about 500 kg of greenhouse pollution each year and saves you about $770. Besides, it's great for your health and fitness!

10 Make your fridge more efficient
Ensure the coils of your fridge are clean and well ventilated — that will save around 150 kg of greenhouse pollution a year. Make sure the door seals properly — this saves another 50 kg. Keep fridges and freezers in a cool, well-ventilated spot to save up to another 100 kg a year. If you have a second fridge, turn it off when not in use.

2.5 Activities

To answer questions online and to receive **immediate feedback** and **sample responses** for every question, go to your learnON title at www.jacplus.com.au. *Note*: Question numbers may vary slightly.

Remember

1. Where has the United Nations convened conferences on climate ***change*** over the past 20 years?
2. What is the Kyoto Protocol?

Explain

3. Explain why the two basic strategies developed by the Kyoto Protocol can ***sustainably*** reduce the amount of greenhouse gases in the atmosphere.
4. Explain why organisations such as the Conservation Council of Australia would have different views from business companies that produce electricity on the topic of 'climate ***change*** and global warming'.

Discover

5. Investigate the idea of an eco-friendly house using the internet. Start with the **Eco system homes** weblink in the Resources tab. In a team of four, design an eco-friendly house of the future. Present your design to the class.
6. Find what the Garnaut Report 2011 and the State of the Climate 2014 report, produced by the Australian Government and the CSIRO, have to say about the impacts of climate ***change*** on Australia's ***environment***.
7. Use the **Ecological footprint** weblink in the Resources tab to find out more about an ecological footprint. Which of your activities could be changed to reduce your footprint?
8. Find out how the carbon tax works using the **Carbon tax** weblink in the Resources tab.
9. Find out more about Australia's situation with respect to climate change by looking at the State of the Climate 2014 report, which can easily be located on the internet.

Predict

10. Refer to figure 1. How much will temperatures increase by 2070 with no mitigation? Which action will reduce temperature ***change*** the most by 2100?
11. Refer to table 1 and the 450 ppm case. What is the percentage of species at risk of extinction? What might happen to the Great Barrier Reef under the 550 ppm case?

Think

12. What part do international forums play in helping to solve climate ***change***?
13. Create a poster to communicate the main points of the Australian Conservation Foundation's 10-point strategy to reduce greenhouse gases.

RESOURCES — ONLINE ONLY

Explore more with these weblinks: Eco system homes, Ecological footprint calculator, Carbon tax

2.6 How can we reduce the impacts of climate change?

2.6.1 Fossil fuels

It has been recognised by climate authorities that global warming is possibly the most important issue impacting on life on Earth at this time and into the future. The burning of **fossil fuels**, which generate greenhouse gases, is causing the atmosphere to heat up, and it is believed that a sustainable future, in terms of energy use, can be achieved only by reducing the consumption of energy and/or switching to renewable energy forms. While use of fossil fuels is a significant

FIGURE 1 World electricity generation by fuel, 2005–30

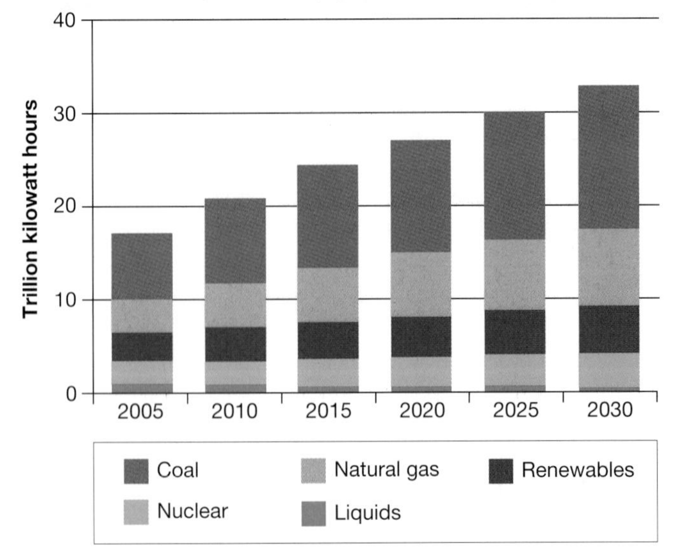

Source: Energy Information Administration (EIA)

FIGURE 2 Australia's primary energy consumption

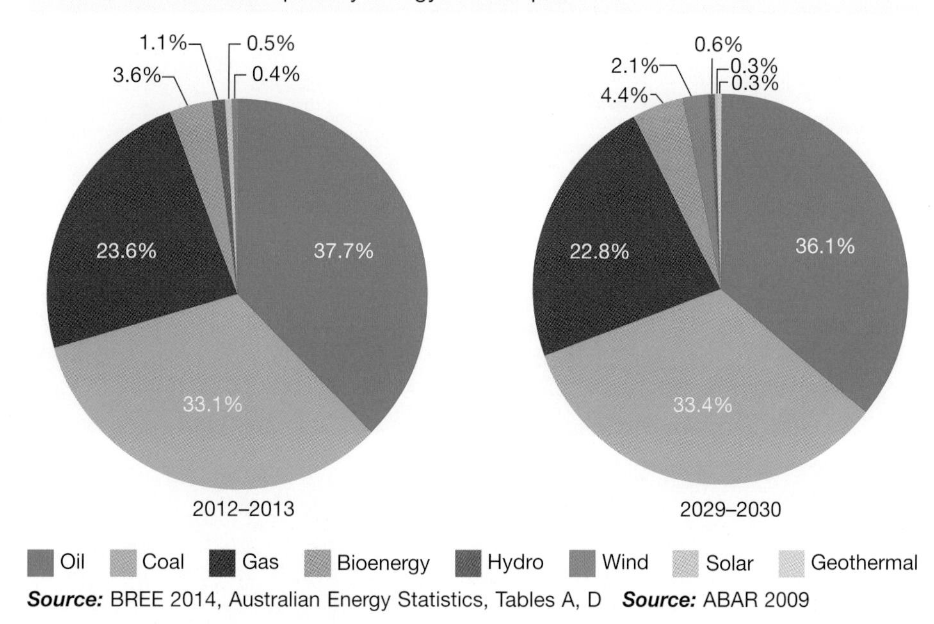

Source: BREE 2014, Australian Energy Statistics, Tables A, D ***Source:*** ABAR 2009

factor in global warming, it should also be realised that there are a number of other human activities that lead to greenhouse gas emissions.

Fossil fuels have been widely used for energy production by human societies since the **Industrial Revolution**. Burning of wood in fires was the earliest use of fuels, and today coal, oil and gas are the fossil fuels of choice. Much of the energy used in society today for transport, domestic use and all forms of industry is from electricity generated by power stations that are fired by fossil fuels (see figures 1 and 2).

2.6.2 Renewable energy

The alternative and environmentally friendly option to fossil fuels is renewable energy. This includes hydropower, solar, wind, wave and tidal, **geothermal**, and bioenergy sources to generate electricity. These sources do not produce greenhouse gases and are replenished in relatively short periods of time (see figure 3). These actions represent a significant change from a human-centred to an earth-centred worldview. This change will lead to a more sustainable use of energy with a significantly lower impact of greenhouse gas emissions on the environment.

Many countries throughout the world are now using or developing sustainable energy industries. The USA, for example, has established the Clean Energy Plan and currently produces 0.54% of its energy needs from solar power, with renewable energy sources comprising 13% of its total electricity generation. In Europe, Germany has made great progress in harnessing renewable energy sources, which today provide 78% of its power needs. Solar energy alone provides 50% of this amount.

In Australia, with expansive desert regions, there is huge potential to generate solar power. In recent years, the installation of solar panels for domestic households has increased, and this has been supported by a Federal Government subsidy scheme, however, currently solar energy accounts for only 0.1% of Australia's total energy requirements. In other renewable energy fields, wind farms have become more widespread in southern Australia, and there are companies investigating the potential for geothermal energy production.

FIGURE 3 Some sources of renewable energy: (a) solar, (b) wind, (c) hydro-electric and (d) geothermal

2.6.3 What can be done in the future?

In 2015, the IPCC confirmed the 2007 recommendations to reduce greenhouse gas emissions. The recommendations cover a wide range of human activities, with suggestions for management to mitigate global warming (see table 1).

For each of the mitigation actions shown in table 1 there are economic, social and environmental consequences. For example, considering the 'developing safer and cleaner nuclear energy' action, there may be positive economic consequences, such as the creation of energy security and job opportunities, but also negative consequences, such as the cost of waste disposal. Similarly, the social and environmental consequences may be positive, such as reduced air pollution, and negative, such as nuclear accidents.

TABLE 1 Reducing greenhouse gas emissions

	Ways to reduce greenhouse gas emissions
Energy supply	• Switching from coal to gas • Developing safer and cleaner nuclear energy • Increasing use of renewables such as hydro-power, solar, wind, wave and tidal, geothermal and bioenergy • Carbon Capture and Storage (CCS) at fossil fuel electricity generating facilities
Transport	• More fuel-efficient vehicles such as electric, hybrid, clean diesel and biofuels • Changing from road to rail and bus transport systems • Promoting cycling and walking to work

	Ways to reduce greenhouse gas emissions
Buildings	• Installing more efficient lighting and day-lighting systems and electrical appliances for heating and cooling, cooking, and washing • Increased use of photovoltaic (PV) solar panels • Improved refrigeration fluids including the recovery and recycling of fluorinated gases
Industry	• More efficient electrical equipment • Heat and power recovery • Material recycling and substitution • Control of gas emissions
Agriculture	• Improved crop yields and grazing land management • Increased storage of carbon in the soil and reduction of methane gas emissions from livestock manure • Restoration of cultivated soils and degraded lands • Improved nitrogen fertiliser application techniques to reduce nitrous oxide emissions • New bioenergy crops to replace fossil fuels
Forestry/forests	• Planting new forests • Better harvested wood management • Use of forestry products for bioenergy to replace fossil fuel use • Better remote sensing technologies for analysis of vegetation and mapping land-use change
Waste	• Landfill methane recovery • Waste incineration with energy recovery • Composting of organic waste • Controlled waste water treatment • Recycling and waste minimisation

Source: UN IPCC Report 2007

2.6 Activities

To answer questions online and to receive **immediate feedback** and **sample responses** for every question, go to your learnON title at www.jacplus.com.au. *Note*: Question numbers may vary slightly.

Remember

1. What is meant by the term *fossil fuel*?
2. List some major renewable energy sources.

Explain

3. What would be the negative impacts if all fossil fuels were banned tomorrow?
4. What would be the best renewable energy source for the future? Give reasons for your selection.

Discover

5. How is ethanol produced as a renewable energy source for power production, and is it ***sustainable*** as a renewable energy source?
6. Use the internet to find out about geothermal energy and its potential as a future energy source.

Predict

7. Refer to figure 2. What percentage of Australia's energy currently comes from renewable sources, and by how much is this projected to ***change*** by 2019–20?
8. Considering the range of nuclear power plant accidents in the past and their impacts on the ***environment***, how might nuclear energy be managed as a safe energy source into the future?

Think

9. Why isn't the use of fossil fuels ***sustainable***?
10. Select examples of recommendations from the UN IPCC Report in table 1 that you think you could apply in your everyday living to reduce global warming by greenhouse gas emissions.

11. Use the internet to access and peruse the IPCC's Climate Change 2014 Mitigation of Climate Change Report. Consider the environmental, social and economic impacts of climate change mitigation for one of the following: transport, buildings, energy systems or industry.
12. What would be the environmental, social and economic consequences of the different management strategies adopted for renewable energy use in Australia, the USA and Germany, outlined in section 2.6.2.

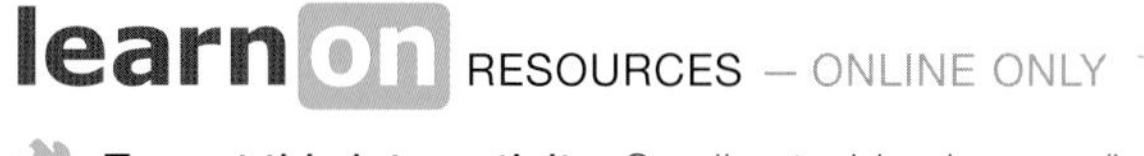

Try out this interactivity: Small acts, big changes (int-3288)

2.7 Is Australia's climate changing?

2.7.1 Impacts of climate change in Australia

Research by government and non-government organisations, such as the Bureau of Meteorology (BOM), the Commonwealth Scientific and Industrial Research Organisation (CSIRO) and the IPCC in 2014–15, has indicated that Australia is particularly vulnerable to climate change. The consequent changes that will affect all Australian biophysical systems have been identified as eight key risks, which are outlined in figure 1.

FIGURE 1 Key climate and severe weather risks for Australia

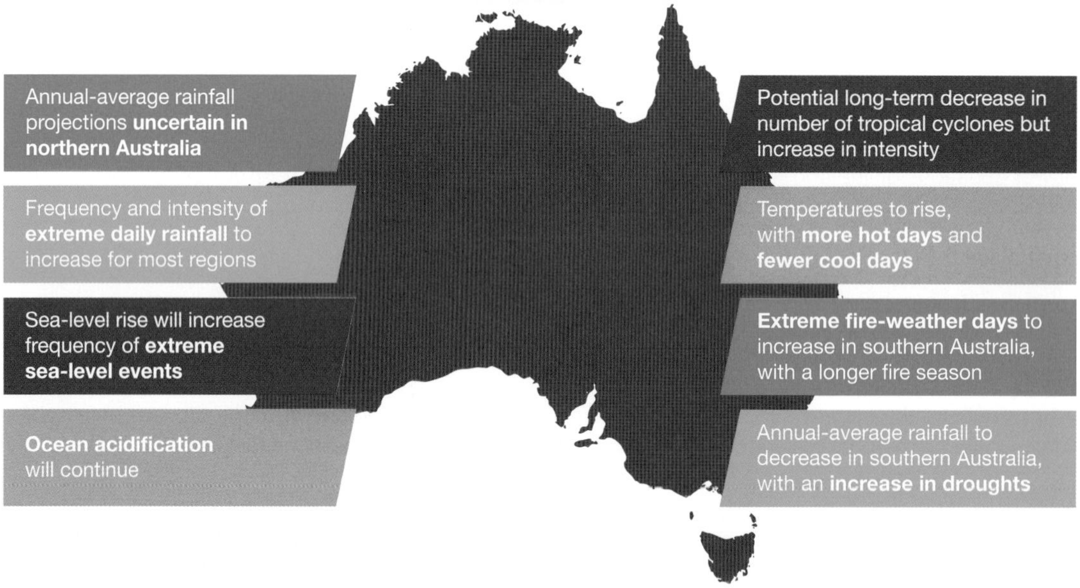

Annual-average rainfall projections **uncertain in northern Australia**

Frequency and intensity of **extreme daily rainfall** to increase for most regions

Sea-level rise will increase frequency of **extreme sea-level events**

Ocean acidification will continue

Potential long-term decrease in number of tropical cyclones but increase in intensity

Temperatures to rise, with **more hot days** and **fewer cool days**

Extreme fire-weather days to increase in southern Australia, with a longer fire season

Annual-average rainfall to decrease in southern Australia, with an **increase in droughts**

Source: Bureau of Meteorology and CSIRO

The challenge for the future is how to manage these risks to minimise negative consequences for the Australian environment, economy and social systems.

2.7.2 How has Australia's climate changed and what will it be like in the future?

Australia's climate is quite variable from one year to the next, and floods and droughts have always occurred. The concern raised by global warming and climate change is the degree of climate variability and the likelihood of more extreme weather events. For instance, will we experience worse floods and droughts and more bushfires and severe cyclones, tornadoes and the like? Scientific evidence supports the view that there have been more extreme weather events in recent years and that the climate of Australia has undergone significant regional change (see figure 2).

FIGURE 2 *State of the Climate 2014* report

The *State of the Climate 2014* report, produced by the BOM and CSIRO made the following summary points.

- Australia's climate has warmed by 0.9 °C since 1910, and the frequency of extreme weather has changed, with more extreme heat and fewer cool extremes.
- Rainfall averaged across Australia has slightly increased since 1900, with the largest increases in the northwest since 1970.
- Rainfall has declined since 1970 in the southwest, dominated by reduced winter rainfall. Autumn and early winter rainfall has mostly been below average in the southeast since 1990.
- Extreme fire weather has increased, and the fire season has lengthened, across large parts of Australia since the 1970s.
- Australian temperatures are projected to continue to increase, with more extremely hot days and fewer extremely cool days.
- Average rainfall in southern Australia is projected to decrease, and heavy rainfall is projected to increase over most parts of Australia.
- Sea-level rise and ocean acidification are projected to continue.

2.7.3 Recent severe weather events

Each year, climatic records are being broken and severe weather is seen to be on the increase. The general consensus is that these events are due to global warming and climate change impacting on regional weather patterns. From November 2012 to March 2013 alone, more than 120 records were broken. Figure 3 outlines a number of these.

FIGURE 3 Climatic records broken in Australia

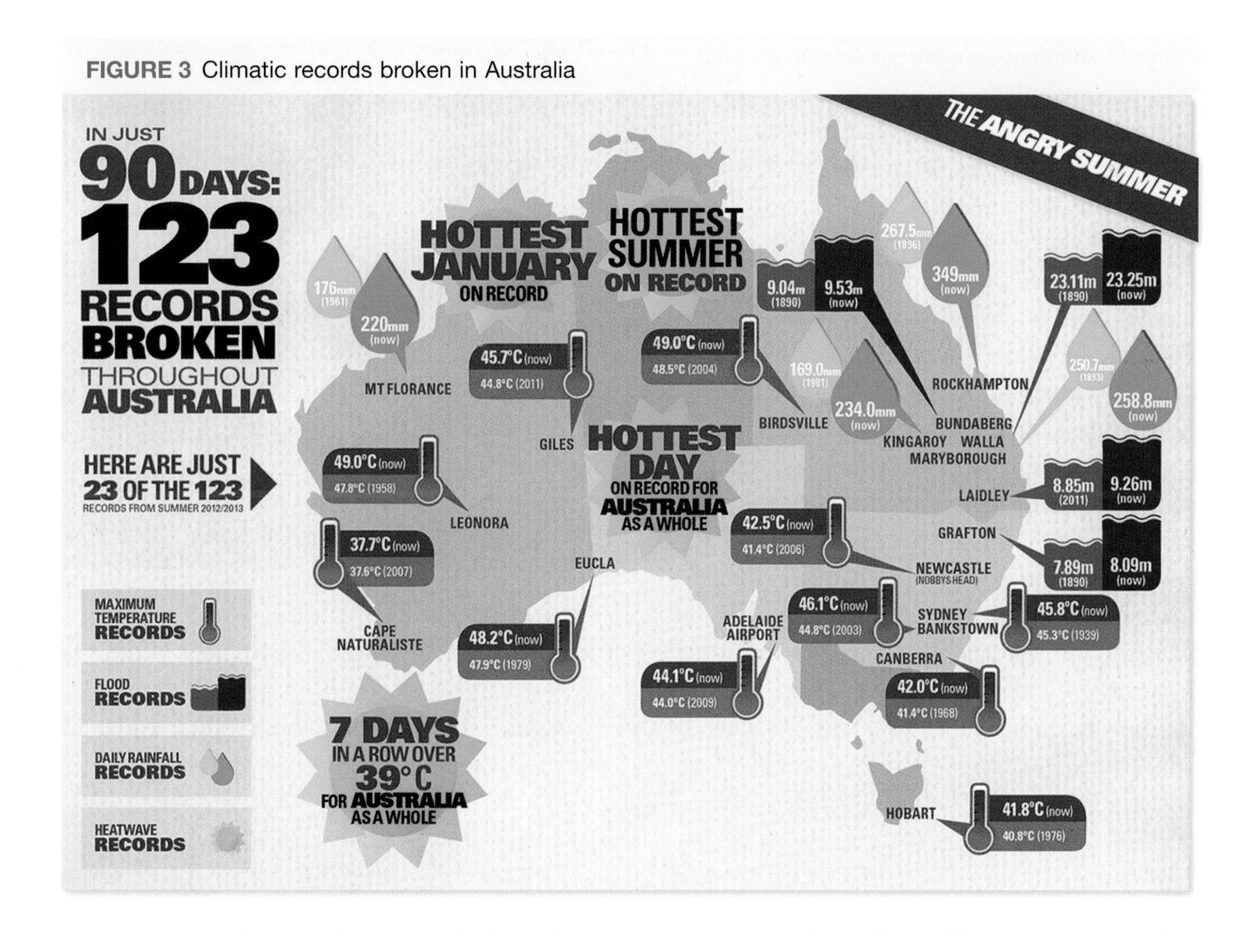

In 2015–16, a number of serious floods occurred after torrential downpours. Places that experienced these extreme events included the east coast of Australia, including Sydney; Bendigo, Victoria; the Kimberley region's Fitzroy Valley and the Dampier Peninsula in Western Australia; and Oodnadatta and Marree, in northern South Australia. Figure 4 shows the floodwaters experienced in Sydney in 2016.

In 2015–16 drought prevailed in many regions throughout Australia and this, combined with heatwaves, created conditions conducive to bushfires in places such as Wye River and Separation Creek, Victoria; the Hamley Bridge area and Mt Compass, South Australia; and Rockingham, Western Australia (see figure 5).

FIGURE 4 Floods in Sydney, June, 2016

FIGURE 5 Remains of a house destroyed by a bushfire in the outer suburbs of Adelaide in January, 2015

2.7.4 How will climate change affect the environment, economy and social systems?

Some of the impacts of climate change that will require management by governments and communities include:

- impacts on fragile and diverse biomes and ecosystems, for example the Great Barrier Reef, where warming of 1 °C is expected to have significant impacts on biodiversity, with losses of species and associated coral communities and the potential for up to 97% of the reef to be subject to coral bleaching
- changed temperatures and rainfall regimes affecting the potential of agriculture and forestry to maintain crop yields such as wheat, and timber yields from forests
- reduced river flows in the Murray–Darling Basin with significant impacts on agriculture, industry and urban household use
- more extreme weather events such as heatwave conditions, with an increase in the number of days when the forest fire index rating is very high or extreme
- more severe tropical cyclones, with associated property damage due to strong winds and flooding
- spread of tropical diseases such as dengue fever and malaria to southern regions.

FIGURE 6 The fragile Great Barrier Reef ecosystem may be significantly affected by climate change.

2.7.5 How can the impacts of severe weather be managed?

Scientific experts agree that environments will change due to global warming and climate change and there will be a range of economic and social consequences, to which society will need to adapt. Where particular industries such as agriculture and forestry may be impacted, there could be a need for governments and other agencies to encourage and facilitate the development of employment opportunities in alternative industries, such as renewable energy.

In dealing with the potential impact of severe weather events, a number of approaches may be taken. The redesign of urban infrastructure to improve storm water drainage is a management strategy to reduce the threat of flooding. If redesign is not able to solve the problem, there may be a need for some people to consider relocating away from the flood-prone coastal and riverine locations in which they currently live.

Successful management strategies in relation to events such as cyclones and bushfires include the development of improved tropical cyclone warning systems, with monitoring conducted and warnings issued by the Bureau of Meteorology, and bushfire warnings, issued by relevant state fire authorities. National and state-based agencies such as Emergency Management Victoria, Emergency New South Wales and the Department of Community Safety in Queensland provide a range of information and resources aimed at minimising the impacts on communities of severe weather events, and assisting with management strategies such as emergency evacuation planning. Improved building design to withstand these severe weather events is another successful form of management strategy.

Government Disaster Relief programs that offer financial and other assistance to individuals and communities to recover after events such as flood, fire and drought are further examples of impact management.

Perhaps most importantly, the root causes of severe weather events as a consequence of global warming and climate change need to be addressed. The Australian Conservation Foundation's 10-point plan (see figure 2 in section 2.5.3) suggests a range of personal energy use management strategies that aim to minimise individuals' contribution to greenhouse gas emissions, such as switching to solar energy and other renewables. If adopted by businesses and the general community, these strategies will go a long way towards reducing the environmental impacts of climate change and global warming, thereby mitigating the social and economic impacts.

FIGURE 7 The FireReady app provides warnings and other information about bushfires in Victoria

2.7 Activities

To answer questions online and to receive **immediate feedback** and **sample responses** for every question, go to your learnON title at www.jacplus.com.au. *Note*: Question numbers may vary slightly.

Remember

1. How are Australia's temperatures expected to change due to climate change?
2. Name three extreme weather events that are expected to increase in frequency due to climate change.

Explain

3. Study figure 3, which outlines climatic records broken in Australia. Describe the general pattern of temperature and rainfall extreme weather events for the 2012–13 period outlined.
4. What types of temperature and rainfall changes were experienced where you live?

Discover

5. Develop an evacuation plan to save life and property for a house or town in a bushfire-prone area.
6. Use the internet to find out about Pacific Island nations that are threatened by rising sea levels due to climate change.

▶

Predict

7. How might climate change affect tourism in the Snowy Mountains region of Australia?
8. How will rising sea levels affect Australia's state capital cities that are located on the coast?

Think

9. How might people who live in tropical cyclone-prone areas cope with increased severe weather events?
10. List three positives, or benefits, of climate change to a particular region of Australia.

2.8 SkillBuilder: Drawing a futures wheel

online only

WHAT IS A FUTURES WHEEL?

A futures wheel is a series of bubbles or concentric rings with words written inside each to show the increasing impact of change. It helps show the consequences of change.

Go online to access:

- a clear step-by-step explanation to help you master the skill
- a model of what you are aiming for
- a checklist of key aspects of the skill
- a series of questions to help you apply the skill and to check your understanding.

FIGURE 1 Possible responses by the ski and alpine resort industry to climate change

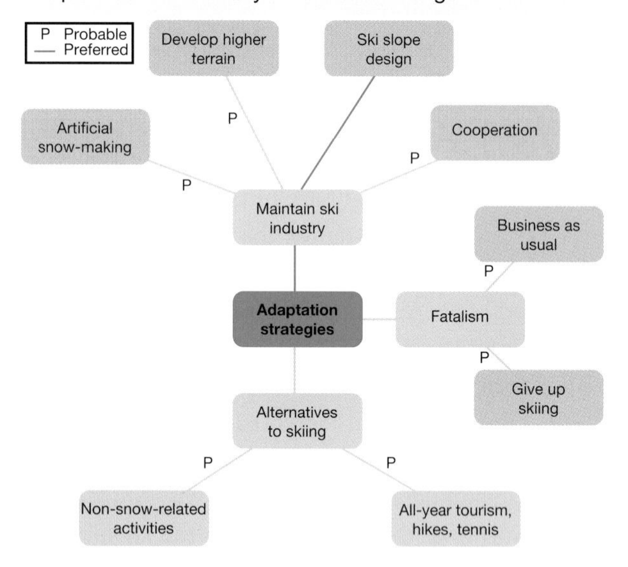

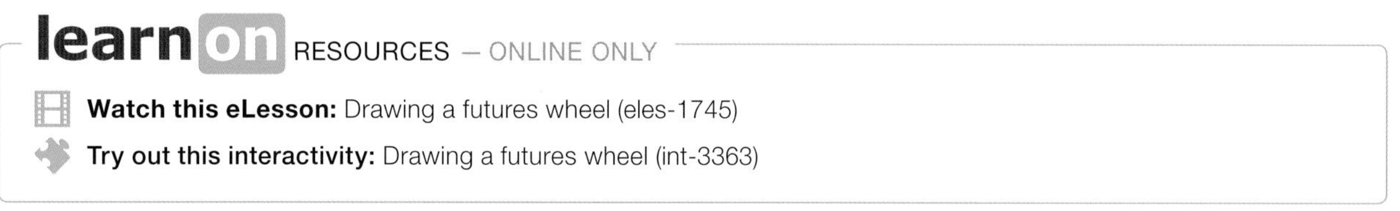

learnON RESOURCES — ONLINE ONLY

Watch this eLesson: Drawing a futures wheel (eles-1745)

Try out this interactivity: Drawing a futures wheel (int-3363)

2.9 Review

2.9.1 Review

The Review section contains a range of different questions and activities to help you revise and recall what you have learned, especially prior to a topic test.

2.9.2 Reflect

The Reflect section provides you with an opportunity to apply and extend your learning.
Access this subtopic at **www.jacplus.com.au**

TOPIC 3
Land environments under threat

3.1 Overview

Numerous **videos** and **interactivities** are embedded just where you need them, at the point of learning, in your learnON title at www.jacplus.com.au. They will help you to learn the content and concepts covered in this topic.

3.1.1 Introduction

Land is one of our most valuable resources. Left alone it exists in a state of balance, and if managed wisely can continue to do so. However, the land is under increasing pressure as a direct result of population growth — agriculture, mining and the expansion of settlements — all of which have the potential to interfere with natural processes.

A severely degraded landscape in southern Queensland. Although we have a better understanding of factors that contribute to land degradation, the challenge is to manage land sustainably for the future and reverse the trends.

Starter questions

1. What do you think is meant by the term *natural balance*?
2. Copy the following table.

How we use the land	Sustainable land use	Cause of land degradation

In column 1, list the ways in which we use the land. Then use columns 2 and 3 to record whether you think each use is a ***sustainable*** land use (compatible with your definition of natural balance), or a cause of land degradation.

3. Briefly explain how human activity can have a negative impact on natural processes.

INQUIRY SEQUENCE

3.2 Why does the land degrade?

3.2.1 How is land degraded?

Land degradation is a serious problem all over the world. Increasingly, valuable land is becoming less productive because of a decline in its quality.

Everyone on Earth relies on the land. Apart from providing us with a place to live, the land also provides most of our food and products such as oil and timber. With the world's population expected to reach 9.7 billion by 2050 and 11.2 billion by the end of the century, the land and its resources will be placed under even more pressure. Global food production, for example, is already being undermined by land degradation and shortages of both farmland and water resources, making feeding the world's rising population even more daunting.

Today there are seven times more people living on the Earth than at the mid-point of the twentieth century. Our primary energy use is five times higher and our use of fertilisers has increased eightfold. In addition, the amount of nitrogen pumped into our oceans has quadrupled.

The United Nations estimates that 25 per cent of the world's farmland is highly degraded. Although we have a better understanding of factors that contribute to land degradation, the challenge is to manage the land sustainably for the future and reverse the trends.

3.2.2 What are the main issues?

Land can be degraded in many ways, but most of the causes can be traced back to the influences of human activity on the natural environment. Figure 1 outlines these activities and their impacts.

FIGURE 1 Why land degrades

A When land is cleared or overgrazed, it becomes vulnerable to erosion by wind and water. The nutrient-rich soil is either washed or blown away, reducing the quality and quantity of crop yields. Dust storms result and sediment transported to rivers smothers marine species.

B Introduced species such as rabbits eat grass, shrubs and young trees (saplings) down to the soil, thus exposing it to erosion. Their burrows increase erosion as they destabilise the soil. Rabbits also compete with native animals for food and burrows.

C Tourism encourages the clearing of sand dunes for high-density housing and mountain slopes for ski runs, leaving the surface exposed to erosion.

D Overgrazing leads to nutrient-rich soil being washed or blown away. Animals with hard hoofs such as sheep and cattle trample vegetation and compact the soil. This leads to increased run-off after heavy rain.

E Climate change will affect land degradation in the future. Higher sea levels will flood low-lying coastal areas. Expanding cities, removal of vegetation and use of concrete reduces the ability of the land to absorb moisture. This not only increases erosion, but can reduce the amount of rainfall in an area.

F Urban communities produce large quantities of waste which is deposited in landfills. Much of the rubbish remains toxic or, in the case of plastic bags, takes hundreds of years to break down. Liquid and solid waste seeps into groundwater and runs off into rivers and eventually into the sea, killing marine species.

G Introduced plant species such as blackberries and Paterson's Curse (Salvation Jane) choke the landscape and compete with native vegetation. Their dense ground cover prevents light from reaching the soil.

H Salinity occurs naturally in areas where there is low rainfall and high evaporation and also where the land was below sea level millions of years ago. Salinity is also caused by excess irrigation and clearing natural vegetation. In some cases the watertable rises, bringing salt to the surface.

FIGURE 2 A former freshwater lake affected by dryland salinity. The high salt levels have killed the native eucalypts; the smaller plants are more salt tolerant.

3.2 Activities

To answer questions online and to receive **immediate feedback** and **sample responses** for every question, go to your learnON title at www.jacplus.com.au. *Note*: Question numbers may vary slightly.

Remember

1. List the different ways in which the land can become degraded.
2. Outline the impact of land degradation on water resources.

Explain

3. Explain why land degradation is a current geographical issue.

Discover

4. Investigate a particular type of land degradation and produce an annotated visual display to show the parts of Australia affected by it. Cover major contributing factors and possible management strategies. Add an inset diagram that examines a particular ***place***, ***scale*** and rate of ***change*** associated with this type of degradation. Include your own recommendations for ***sustainable*** use of the ***environment*** to combat the issue you have investigated.
5. (a) Investigate your school ***environment*** or visit a popular parkland near your school. Take photographs of your observations. Prepare a map of the area and annotate it to show areas where land degradation is evident and other areas where strategies have been employed to protect the ***environment***. Add your photos to provide a visual representation and give meaning to your annotations.
   (b) Draw a second map or plan of your area. Devise your own ***sustainable*** management strategy. Include photos of your original observations.

Predict

6. (a) In small groups, prepare a fold-out educational pamphlet outlining the damage caused by the waste produced by urban communities each year. Make sure you clearly outline the ***interconnection*** between human activity and ***environmental*** harm. Devise a strategy to reduce this waste and estimate the difference that this would make to the amount of waste generated.

(b) Evaluate your own and others' contributions to this group task, critiquing roles including leadership, and provide useful feedback to your peers.
(c) Evaluate task achievement and make recommendations for improvements in relation to team goals.

Think

7. Land degradation is often the result of lots of little actions and events, the effects of which interact and build up over time. Identify some things that you do, consciously or unconsciously, that might be helping to degrade the land where you live. Give reasons for your answer.

Try out this interactivity: Destroying the land (int-3289)

3.3 What is land degradation?

3.3.1 Explaining land degradation

Land degradation is the process that reduces the land's capacity to produce crops, support natural vegetation and provide fodder for livestock. Of the 5 million square kilometres used for agricultural and pastoral activities in Australia, more than half has been affected by, or is in danger of, degradation (see figure 1).

3.3.2 What are the effects of land degradation?

Even small changes can have dramatic effects on the land. The shortcut students take from the oval to the classroom can soon reduce a grassy area to dust. Drought can quickly reduce the productivity of an area used for farming. A farmer who neglects the land after one growing season may still be able to raise a good crop the following season, but if the land is neglected year after year it will eventually become unproductive.

The effects of land degradation are far-reaching. The productivity of farming land diminishes and yields drop because the soil becomes exhausted through overuse or deforestation (see figure 2). Expenditure increases as the land requires more treatment with fertiliser, and **topsoil** and nutrients in the soil need to be replaced. Valuable topsoil is often washed away into rivers and out to sea. Nutrients cause foul-smelling blue–green **algal blooms** that choke waterways. These blooms decrease water quality, poison fish and pose a direct threat to other aquatic life. A dog that licks itself after swimming in affected water can die. The factors affecting land degradation can be seen in figures 3 and 4.

FIGURE 1 Land degradation causes physical, chemical and biological changes. The natural environment deteriorates and the landscape undergoes a dramatic change. Common causes of land degradation include soil erosion, increased salinity, pollution and desertification.

FIGURE 2 Land clearing and deforestation leave the land vulnerable to erosion. When rain falls on a hillside that is well vegetated, it is absorbed by plant roots and held in the soil. However, if the vegetation is removed, there is nothing to stabilise the soil and hold it together — rills and gullies form (see Rill erosion and Gully erosion in subtopic 3.5) where the unprotected soil is washed away, and landslides may occur.

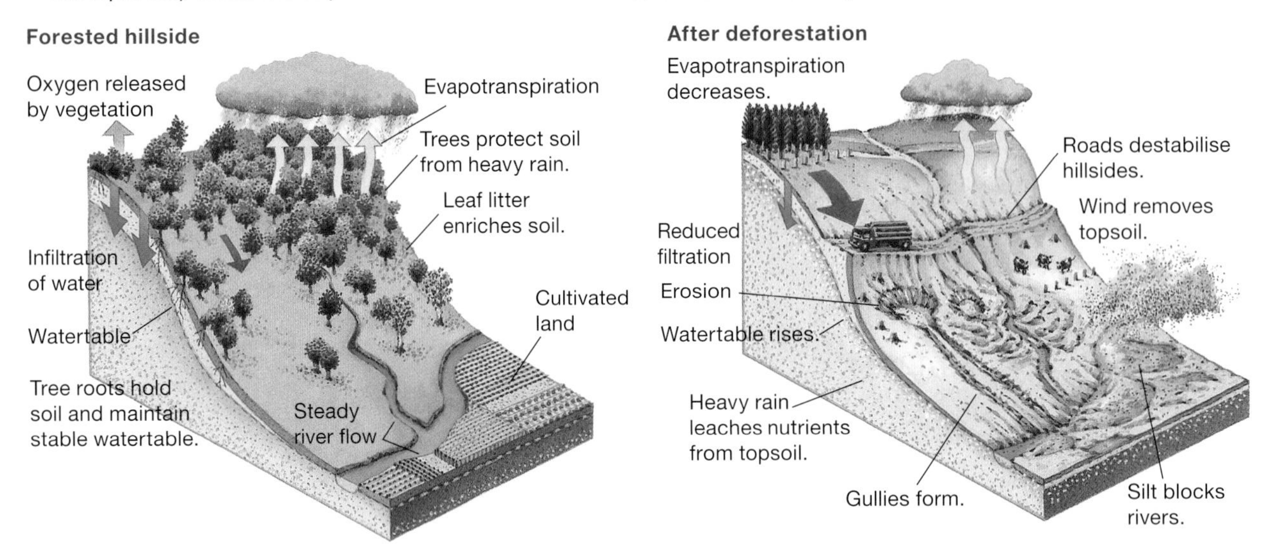

FIGURE 3 Causes of land degradation in the Asia–Pacific region. Australia is ranked fifth in clearing of native vegetation.

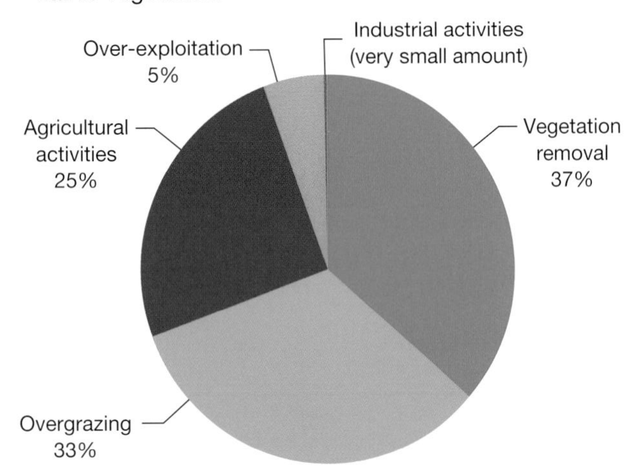

FIGURE 4 Main causes of degradation globally

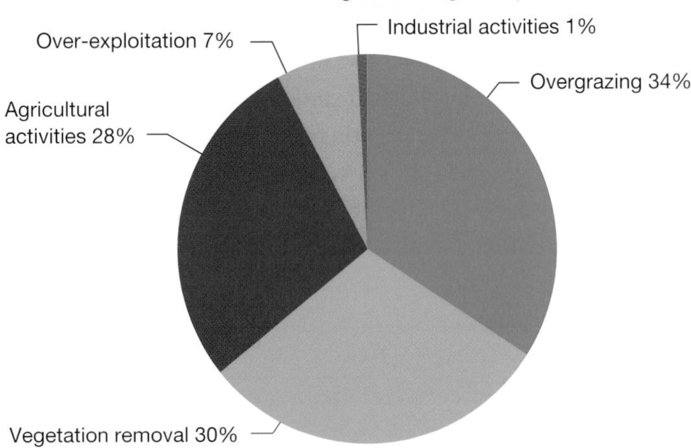

FACTORS THAT CONTRIBUTE TO LAND DEGRADATION

- Poor management leads to the loss of nutrients vital for plant growth.
- Removal of vegetation makes the land vulnerable to erosion by wind and water.
- When urban development encroaches on agricultural land, vegetation is removed and the waste generated is disposed of in landfill.
- Poor agricultural practices, especially related to irrigation and the use of chemical fertilisers, can lead to the soil becoming saline or acidic.

Globally, around 52 per cent of the world's agricultural land is considered degraded. However, in Australia this figure is estimated to be close to two-thirds. Twenty-five years ago only around 15 per cent was considered degraded. Both figures 3 and 4 show that agricultural activities and overgrazing combined account for more than 50 per cent of this problem.

Land degradation is a global problem. If the current trends continue, our ability to feed a growing world population will be threatened.

3.3.3 How has agriculture degraded the Australian landscape?

Climate, topography, water supply and soil quality are the major physical factors that determine how land can be used. When white settlers first colonised Australia they brought with them seeds and animals from Europe. They intended to farm here as they had always done at home; they undertook large-scale clearing of trees and shrubs and planted crops and pasture. However, the Australian landscape is much different from what they had left behind. Australia's soils are naturally low in nutrients and have a poor structure. Much of the vegetation is shallow-rooted and easily disturbed when the land is ploughed and made ready for cultivation. Even in areas where the soil is fertile, over-irrigation can raise the **watertable** and bring salt to the surface, decreasing soil fertility. Australia also has variable rainfall, and drought can last for years. This leaves the earth dry, parched, barren and unproductive. Floods can wash away a farmer's livelihood and leave the land flooded.

3.3.4 Where is the land degrading?

Forty years ago, on a per capita basis, there were 0.5 hectares of **arable** (productive) land available to grow food for every man, woman and child. Today this figure has more than halved to 0.2 hectares. This is due to factors such as population growth, urban sprawl, land degradation and climate change. Figures 5 and 6 show the severity of land degradation in Australia and globally.

In 2011 the world's population reached 7.34 billion, and it is expected to reach 9.7 billion by 2050. While global population is increasing, the land upon which food is grown to feed this population is degrading. Almost one-quarter of the global land is affected in some way by land degradation. Managed sustainably, this land could produce 20 million tonnes of grain each year.

FIGURE 5 Severity of soil degradation in Australia

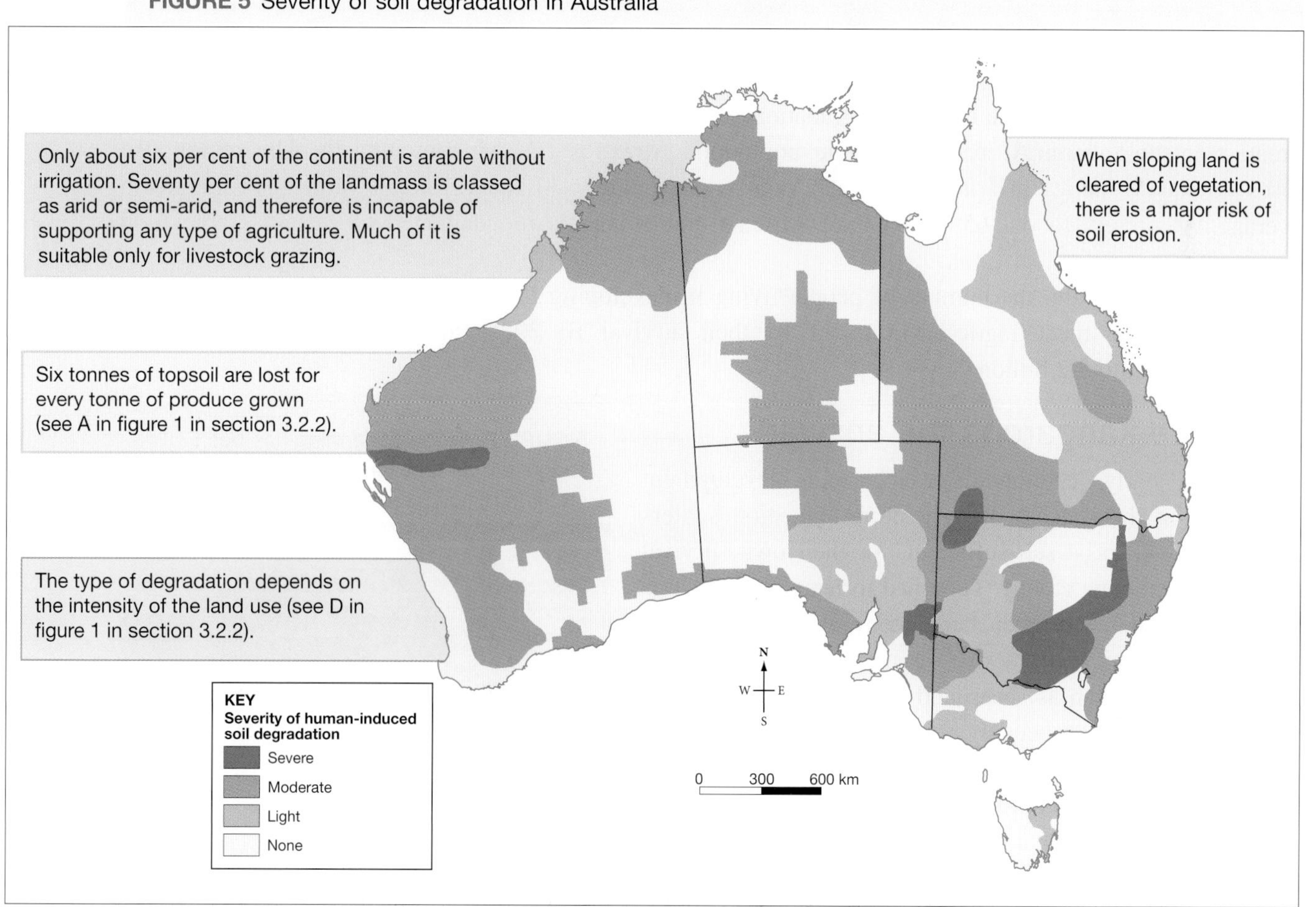

Only about six per cent of the continent is arable without irrigation. Seventy per cent of the landmass is classed as arid or semi-arid, and therefore is incapable of supporting any type of agriculture. Much of it is suitable only for livestock grazing.
When sloping land is cleared of vegetation, there is a major risk of soil erosion.
Six tonnes of topsoil are lost for every tonne of produce grown (see A in figure 1 in section 3.2.2).
The type of degradation depends on the intensity of the land use (see D in figure 1 in section 3.2.2).
KEY
Severity of human-induced soil degradation
Severe
Moderate
Light
None
N
W
E
S
0
300
600 km

Source: MAPgraphics Pty Ltd Brisbane

FIGURE 6 Soil degradation is a global problem affecting every permanently inhabited continent.

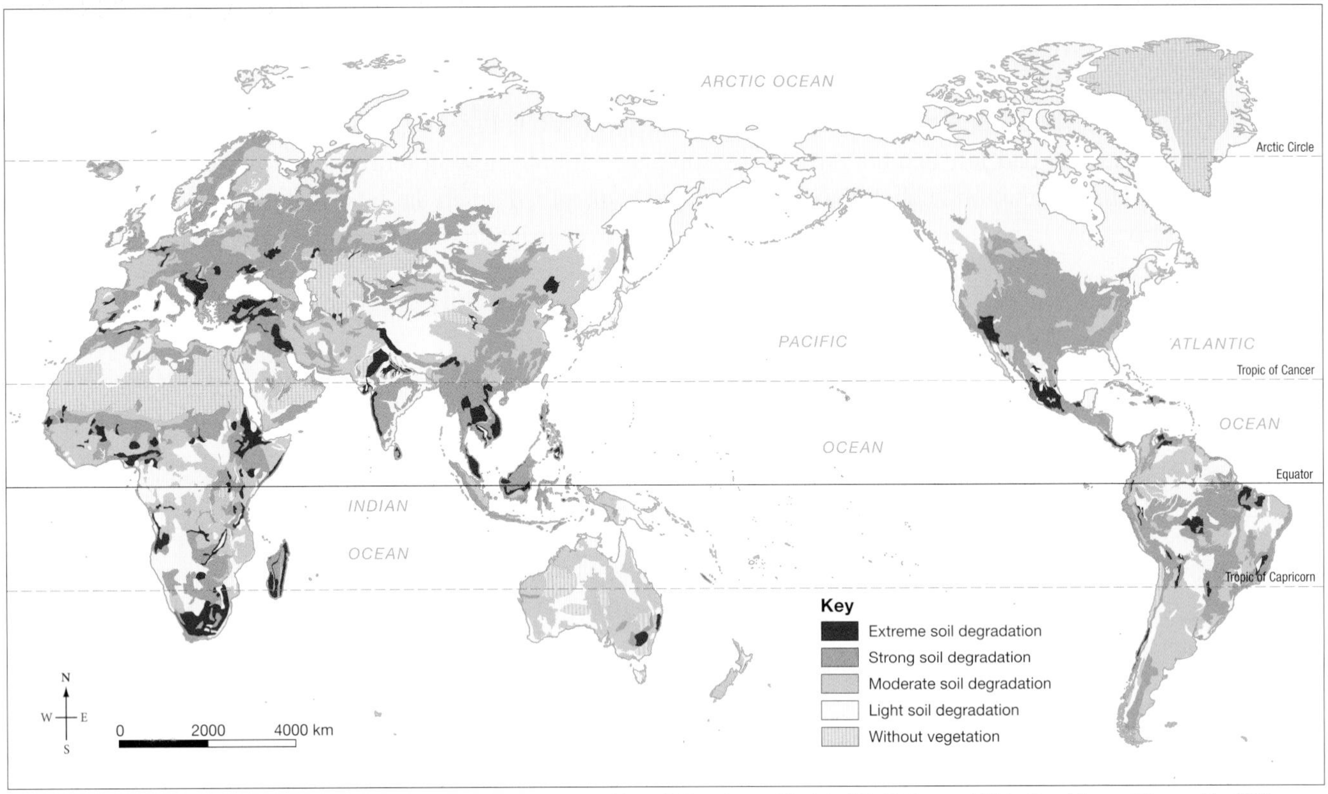

Source: © Commonwealth of Australia Geoscience Australia 2013. © Commonwealth of Australia Department of Sustainability, Environment, Water, Population and Communities 2013. Map by Spatial Vision

According to the United Nations, around 42 per cent of the world's poorest people live on the most degraded lands. Areas where the land degradation is most rapid are also those where population growth is greatest. In sub-Saharan Africa, for example, population growth is 2.74 per cent annually. This region also loses 76 square metres of arable land each year, approximately 30 square metres more than the world average. At present, about 65 per cent of Africa's arable land is too damaged to sustain viable food production.

In the last 60 years, the number of people living in developing lands has doubled. These people are dependent on the 'most fragile environment' for their survival. By 2025, this number is expected to rise from 1.3 billion to 3.2 billion.

3.3.5 Are kangaroos the answer?

Australia's early economic growth and development depended on the success of agriculture. The first settlers knew they had to be self-sufficient, for their own survival and that of the new colony. They had to learn quickly how to farm soil that was often hard, stony and exposed to a variety of climatic extremes. Overgrazing by heavy, hard-hoofed animals such as sheep and cattle increased the rate of land degradation, especially in arid and semi-arid regions. Kangaroo farming has been presented as an alternative sustainable solution to this problem (see figure 7).

FIGURE 7 Kangaroo farming has been suggested as an alternative to sheep and cattle farming.

Those in favour of kangaroo farming claim it would be more environmentally friendly as they are not hard-hooved (see D in figure 1 in section 3.2.2), and that there are added

health benefits as kangaroo meat contains less fat and fewer calories than both lamb and beef. Those against the idea argue that kangaroo farming is not commercially viable in the long term (see figure 8).

FIGURE 8 Comparing commercial viability of kangaroo farming with sheep farming

- Young dependent on mother for 14 months
- Cannot be sold live
- One-off use (meat and skin)
- 18 months before meat can be harvested
- A 60 kg kangaroo yields 6 kg of prime meat; the rest is suitable only as pet food
- Can meet only 0.5 per cent of current needs

- Young dependent on mother for a few months
- Can be sold live
- Multiple uses (wool, meat and skin)
- Breed from 12 months; multiple births possible
- Meat can be harvested from 3–6 months
- Yields 20 kg of prime meat
- Easier to herd and care for

3.3 Activities

To answer questions online and to receive **immediate feedback** and **sample responses** for every question, go to your learnON title at www.jacplus.com.au. *Note*: Question numbers may vary slightly.

Remember

1. Describe in your own words what land degradation is.
2. Why were European farming methods unsuitable for the Australian ***environment***?
3. Do you think land degradation is happening on a small or large ***scale***? Explain.

Explain

4. Study figures 1 and 2.
   (a) In your own words, describe the damage that has occurred to the ***environment***.
   (b) Suggest how these ***changes*** have come about.
   (c) How would you go about trying to restore this ***place*** and manage its resources in a ***sustainable*** manner?

Discover

5. Create an overlay theme map. Prepare a base map that shows the extent of land degradation around the world. Prepare an overlay map showing land use. Annotate your overlay with any similarities and differences between the two maps.
6. (a) Investigate alternatives to traditional livestock farming of sheep and cattle, such as kangaroos or emus. Use the information presented in this section as a starting point. Present a reasoned argument for or against this type of farming as a ***sustainable*** alternative.
   (b) Evaluate emotional responses and the management of emotions in terms of this type of farming.

Think

7. Describe an area or ***place*** that is near where you live, that you have visited recently or that you have heard about in the media, and that you think is degraded. Give reasons for your choice and suggest how and why you think this degradation came about.
8. Working in pairs, create a presentation showing the different ways people use and manage the land in another country.
   (a) Design a suitable symbol for land degradation and use this to highlight any uses you think might result in land degradation.
   (b) Add annotations to explain how highlighted activities might degrade ***environments***, and the ***scale*** of this ***change***.
   (c) Suggest a possible ***sustainable*** solution for each type of degradation identified.

3.4 SkillBuilder: Interpreting a complex block diagram

WHAT IS A COMPLEX BLOCK DIAGRAM?

A complex block diagram is a diagram that is made to appear three-dimensional. It shows information about a number of aspects of a topic or location, such as what is happening at the surface of the land or water, what is happening above the land or water, and what is happening beneath the soil or water at a number of different locations across an area.

Go online to access:

- a clear step-by-step explanation to help you master the skill
- a model of what you are aiming for
- a checklist of key aspects of the skill
- a series of questions to help you apply the skill and to check your understanding.

FIGURE 1 Saltbush Farm, land audit, 2012. Saltbush Farm is in the catchment of the Naangi River, a tributary of the Murray.

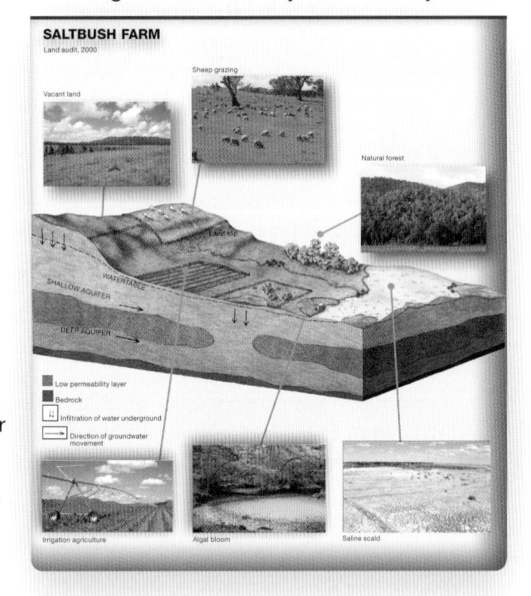

Source: CSIRO Land and Water/John Coppi (vacant land), CSIRO Land and Water/Greg Heath (sheep grazing), CSIRO Land and Water (natural forest), Alberto Loyo/Shutterstock (saline scald), CSIRO Land and Water/Willem van Aken (irrigation agriculture), Getty Images/Science Photo Library/Michael Marten (algal bloom)

RESOURCES — ONLINE ONLY

Watch this eLesson: Interpreting a complex block diagram (eles-1746)

Try out this interactivity: Interpreting a complex block diagram (int-3364)

3.5 Where has the soil gone?

3.5.1 What is soil?

Soil formation is a complex process brought about by the combination of time, climate, landscape and the availability of organic material. In some areas it takes hundreds of years to develop, while in others soil can form in a few decades. While erosion is a natural process, human activity due to farming, land clearing and the construction of roads and buildings can accelerate the process.

Soil is a mixture of broken-down rock particles, living organisms and **humus**. Over time, as surface rock breaks down through the process of **weathering** and mixes with organic material, a thin layer of soil develops and plants are able to take root (see figure 1). These plants then attract animals and insects and when these die their dead bodies decay, making the soil rich and thick.

How is soil being lost?

Sheet erosion

Sheet erosion (see figure 2) occurs when water flowing as a flat sheet flows smoothly over a surface, removing a large, thin layer of topsoil. Sheet erosion might happen down a bare slope. It occurs when the amount of water is greater than the soil's ability to absorb it.

Strategies to combat this form of erosion include planting slopes with vegetation and adding **mulch** to the exposed soil so that it can absorb greater volumes of water. Alternatively, the landcape could be terraced, whereby the landscape will resemble a series of steps rather than a steep slope.

FIGURE 1 Wild flowers taking root in cracks in the rocks

FIGURE 2 What evidence of sheet erosion can you observe?

Rill erosion

Rill erosion (see figure 3) often accompanies sheet erosion, occurring where rapidly flowing sheets of water start to concentrate in small channels (or rills). These channels, less than 30 centimetres deep, are often seen in open agricultural areas. With successive downpours, rills can become deeper and wider, as fast-flowing water scours out and carries away more soil.

Strategies to combat rill erosion include tilling the soil (turning it over before planting crops) to slow the development of the rills. Building contours in the soil and planting a covering of grass can help slow the flow of water and hold the soil in place.

FIGURE 3 In which direction do you think the water is flowing? How can you tell?

Gully erosion

Gully erosion (see figure 4) often starts as rill erosion. Over time, one or more rills may deepen and widen as successive flows of water carve deeper into the soil. Gully erosion may also start when a small opening in the surface such as a rabbit burrow or a pothole is opened up over time. Soil is often washed into rivers, dams and reservoirs, muddying the water and killing marine species. Large gullies need bridges or ramps to allow vehicles and livestock to cross.

Strategies to combat gully erosion largely involve stopping large water flows reaching the area at risk, such as planting vegetation or crops to soak up the water. Other strategies include building diversion banks to channel the water away from the area, and constructing dams.

FIGURE 4 What impact will the falling water have? If you owned this land, what would you do to prevent more damage?

Tunnel erosion

Sometimes water will flow under the soil's surface (for example, under dead tree roots or through rabbit burrows), carving out an underground passage or tunnel (see figure 5). The roof of the tunnel may be thin and collapse under the weight of livestock or agricultural machinery. When these tunnels collapse they create a pothole or gully.

Strategies to combat tunnel erosion include planting vegetation both to absorb excess water and to break up its flow. Sometimes major earthworks are needed to repack the soil in badly affected areas.

Wind erosion

When the surface of the land is bare of vegetation, the wind can pick up fine soil particles and blow them away (see figure 6). It is more common during periods of drought or if the land has been overgrazed. The soil can be transported large distances and deposited in urban areas.

Strategies to combat wind erosion include planting bare areas with vegetation, mulching, planting wind breaks and avoiding overgrazing.

FIGURE 5 What do you notice about the ground around these tunnels? What do you think will happen if water flows through these tunnels? Would this ground be suitable for livestock grazing? Why?

Source: © John Ivo Rasic

FIGURE 6 Did you know that soil from China has been deposited in the United States?

3.5 Activities

To answer questions online and to receive **immediate feedback** and **sample responses** for every question, go to your learnON title at www.jacplus.com.au. *Note:* Question numbers may vary slightly.

Explain

1. With the aid of a flow diagram, show the ***interconnection*** between sheet, rill and gully erosion. Use the captions and the questions that appear with each image to help you.

Discover

2. Working with a partner, use the internet to investigate an international ***environment*** such as the Dust Bowl in the United States or the Yellow River in China that has been degraded due to soil erosion.
   (a) Annotate a sketch of this ***environment*** to explain what has happened to the area. Include an inset sketch map that shows the location of this ***place***, and describe the ***scale*** and rate of ***change***.
   (b) Swap your eroded ***environment*** sketch with another pair who will devise a series of management strategies to rehabilitate the ***environment*** and allow it to be used in a ***sustainable*** manner. Add these to your annotated sketch.

Think

3. Why do you think soil erosion in all its forms is such a significant cause of land degradation?
4. (a) Look at figure 7, which depicts the types and ***scale*** of soil erosion in Victoria. In which parts of the state is erosion highest resulting from (i) wind and (ii) water?
   (b) Compare this map with a relief map of Victoria in your atlas. What conclusions can you draw about the ***interconnection*** between topography and erosion caused by water?
   (c) Use your atlas to find a map showing vegetation in Victoria. Explain why wind erosion is more common in north-west Victoria than south-east Victoria.

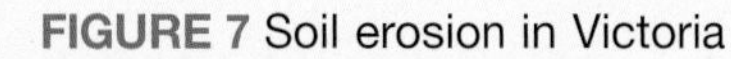

FIGURE 7 Soil erosion in Victoria

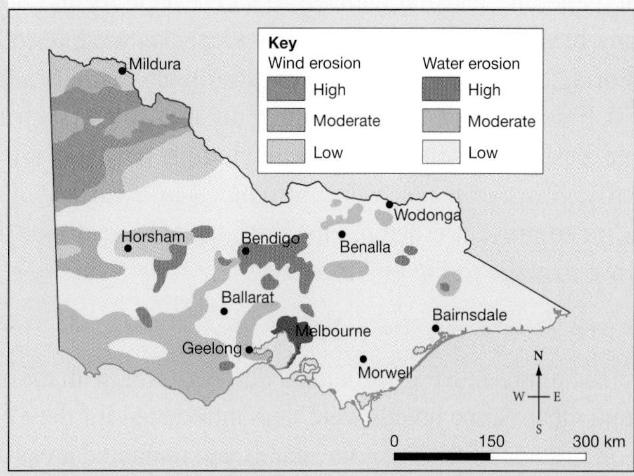

Source: MAPgraphics Pty Ltd Brisbane

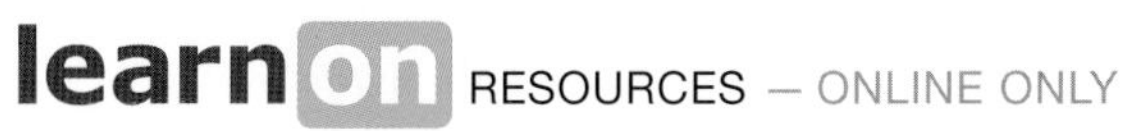

RESOURCES — ONLINE ONLY

Try out this interactivity: Down in the dirt (int-3290)

Deepen your understanding of this topic with related case studies and questions.

Desertification in Mauritania

3.6 Why are drylands drying up?

Access this subtopic at **www.jacplus.com.au**

3.7 How do we manage land degradation?

Access this subtopic at **www.jacplus.com.au**

3.8 Who are the invaders?

3.8.1 What is an invasive species?

Invasive species (sometimes referred to as **exotic species**) are a major cause of land degradation. Often introduced for a specific reason, they can soon take over the environment, threatening indigenous plant and animal species and taking over what was once valuable farming land.

An invasive species is any plant or animal species that colonises areas outside its normal range and becomes a pest. Such species take over the environment at the expense of those that occur naturally in the region, generally causing damage to native habitats and degrading the landscape.

Many of the most damaging invasive species were introduced either for sport (rabbits and foxes), as pets (cats) or as livestock (goats) and pack animals (camels and horses). Some such as the cane toad and mosquito fish were introduced to control other species (such as the cane beetle and mosquitoes), and instead became pests themselves. Others, including rats and mice, arrived accidentally as stowaways on ships. Similarly, invasive plants were introduced in a variety of ways; for example, as crops, pasture or garden plants, or to prevent erosion. However, some spread into the bush where they continued to thrive, causing immense damage to the environment.

Why are goats an issue?

Goats (see figure 1) were introduced into Australia with the arrival of the First Fleet in 1788 as a source of both milk and meat. Some breeds were later introduced for their hair. During the nineteenth century sailors released goats on to some of the offshore islands and mainland areas as an emergency food source. Over time, however, domestic goats escaped, were abandoned or were deliberately released and became feral, posing a threat to inland pastoral areas and native forests. More recently, they have been utilised as a method of weed control in plantation forests and in limited numbers on large pastoral runs. They are also still kept as livestock.

FIGURE 1 Goats were a food source for early settlers and sailors alike. Their size, hardiness and ability to eat a range of plants made them an ideal source of both meat and milk.

It is now estimated that at least 2.6 million feral goats occupy approximately 28 per cent of Australia (see figure 2), in concentrations of up to 40 animals per square kilometre. They are found in all states and territories and on offshore islands, but are most common in semi-arid regions. The absence of predators and the establishment of a water supply for sheep grazing has created ideal conditions in which goats can thrive. Their numbers have been adversely affected by drought and eradication programs; however, high fertility levels have meant they are difficult to control.

FIGURE 2 Spread of feral goats in Australia

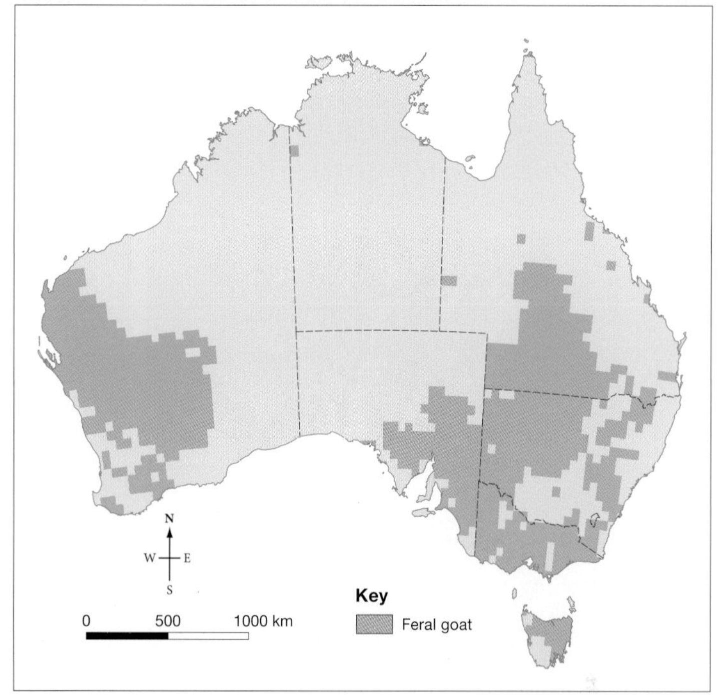

Source: © Commonwealth of Australia Geoscience Australia 2013. © Commonwealth of Australia Department of Sustainability, Environment, Water, Population and Communities 2013. Map by Spatial Vision

Feral goats cause widespread damage to native vegetation. They damage the soil and overgraze native grasses, herbs, trees and shrubs, causing erosion and preventing plant regeneration. They introduce weeds through seeds contained in their dung, and pollute water courses. They dramatically increase the rate of erosion on steep hillsides, where widespread gullying can quickly develop. By focusing on a favoured food source and preventing its regrowth, goats can totally remove some species of vegetation from an area, allowing more invasive plant species to take over.

During times of drought they also compete with native wildlife and domestic livestock for food, water and shelter, creating an additional imbalance in the food chain.

The impact of feral goats is worse in regions where rabbits are also out of control; together they can reduce to bedrock what was once a well-vegetated environment, leaving it open to erosion by both wind and water.

Are introduced plants a problem?

Invasive plant species are often referred to as **weeds**. While many were introduced as garden plants, they soon spread to other areas and now pose a significant threat to both the natural environment and agricultural industries.

Paterson's Curse and Viper's Bugloss

Both Paterson's Curse and Viper's Bugloss (see figure 3) are similar in appearance and are often found in similar regions, if not together. Both were introduced in the 1850s as garden plants because of their attractive flowers. However, as their seeds germinate earlier than native plants, they are able to establish extensive root systems and spreading leaves which crowd out other plant species. Their seeds are further spread in the fur of livestock or by water in areas where erosion is already present and run-off levels are high. They thrive in areas where rainfall is high in winter and have adapted to cope with dry summers. The nutritional value of Paterson's Curse is low, and when eaten in significant quantities can be toxic to livestock, especially horses and small animals. Their stomachs cannot fully process the plant, and this leads to liver damage, loss of condition and eventually, in extreme cases, death.

Contact can also cause skin irritations and other allergic reactions in both humans and livestock. Once Paterson's Curse colonises an area, soil fertility is reduced. It has been estimated that 33 million hectares of land is infested across the nation, and that the cost to Australia's grazing industry is in excess of $250 million annually.

FIGURE 3 (a) Paterson's Curse has two long stamens protruding from the flower plus two shorter ones, and its flowers are more purple. (b) Viper's Bugloss has four long stamens protruding from the flower. Its flowers are more blue, and prickles are visible on the stem.

3.8 Activities

To answer questions online and to receive **immediate feedback** and **sample responses** for every question, go to your learnON title at www.jacplus.com.au. *Note*: Question numbers may vary slightly.

Remember

1. Using information in this section and your own general knowledge, copy and complete the table below. List as many species (both plant and animal) as you can that were introduced into Australia, and why they might have been introduced. Use your atlas as another source of information.

Introduced species	Reasons for introduction

Discover

2. Use the **Weed species** weblink in the Resources tab to prepare an educational leaflet that will assist people in recognising one of these plant species.
3. Visit a local river or creek near your school and make a field sketch of the area. Survey the area around the creek and annotate your sketch to show the location of areas where there are invasive plant species. Add additional annotations to suggest a ***sustainable*** solution to this problem.
4. Refer to the table you completed for question 1. Find an image of one exotic plant and one exotic animal species in your table. Annotate your images with reasons for their introduction, and their impact on the ***environment***. Compare your findings with those of other members of the class.

learnon RESOURCES — ONLINE ONLY

Explore more with this weblink: Weed species

Deepen your understanding of this topic with related case studies and questions:
Introduced species in Australia

3.9 How can we control invasive species?

3.9.1 Can one problem be part of the solution for another?

While we might want to remove invasive species from Australia and other parts of the world, in many cases this is not possible. Total eradication may be feasible in island communities where the risk of re-infestation is limited; however, on a large scale such as mainland Australia control appears to be the best option.

Introduced species pose a serious threat to the productivity of land and diversity of natural environments. In Western Australia, trials have discovered that goats, which themselves pose a threat to both native vegetation and pasture land, can be used to control a wide variety of **invasive plant species**, such as saffron thistle (see figure 1).

While it is known that some weeds are spread by viable seeds passing through the digestive systems of animals, this has not been the case with goats. Less than 1 per cent of the saffron thistle seeds were found in the dung of goats, and these would not germinate. Similar results were found in test sites for the control of blackberries (see figure 2). Within 12 months of goats being allowed to feed on both weed types, there was a notable reduction in their spread. Goats can also be used to control hundreds of different invasive plant species such as English Ivy, Paterson's Curse and Viper's Bugloss (see section 3.8.1), which are toxic to grazing livestock.

Goats have the added advantage of being an environmentally friendly method of weed control. They eliminate the need for using herbicides and fertilisers. Soil quality is improved naturally by goat droppings. Fossil fuel burning machinery is not needed to remove the weeds, and goats can be used in environments where other control methods are not viable; for example, steep slopes.

FIGURE 1 Within 3 years, saffron thistle had been almost completely eradicated by goats. Careful monitoring is needed as seeds can lie dormant for up to 10 years.

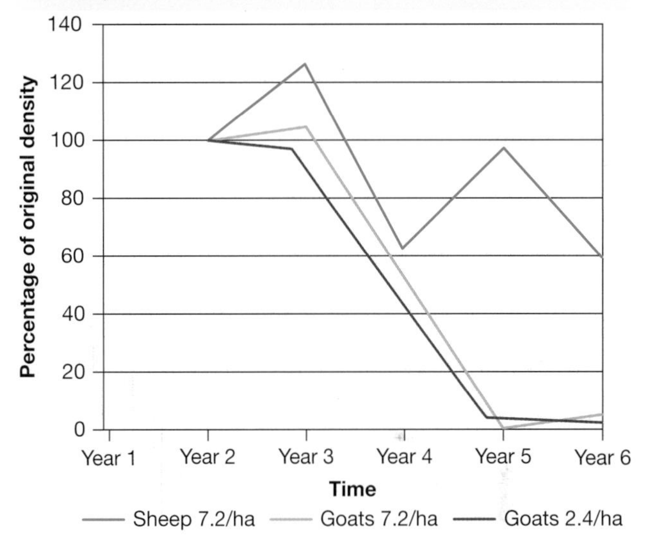

FIGURE 2 Goats have been used successfully in the Tolt River Dam region in Seattle, Washington. There, a herd of 200 goats are used to control the spread of blackberries on ground that is too steep and uneven for mowing by machinery.

Can we control foxes and rabbits?

Both foxes and rabbits were introduced into Australia by the early settlers. With no natural predators, each species spread rapidly.

Left unchecked, foxes pose a significant threat to agriculture and native fauna. Fox predation accounts for one-third of new lamb deaths, and native animals such as the bandicoot are easy prey. Foxes carry a wide range of diseases and parasites such as hepatitis, distemper, mange and rabies.

Feral rabbits not only degrade the environment but also compete with native wildlife for food, and damage vegetation — eating the roots as well as the foliage. They **ringbark** trees and eat seeds and seedlings, preventing the regeneration of plants. Once eradicated from an area, the environment can regenerate (see figure 4).

FIGURE 3 Distribution of red foxes and feral European rabbits in Australia

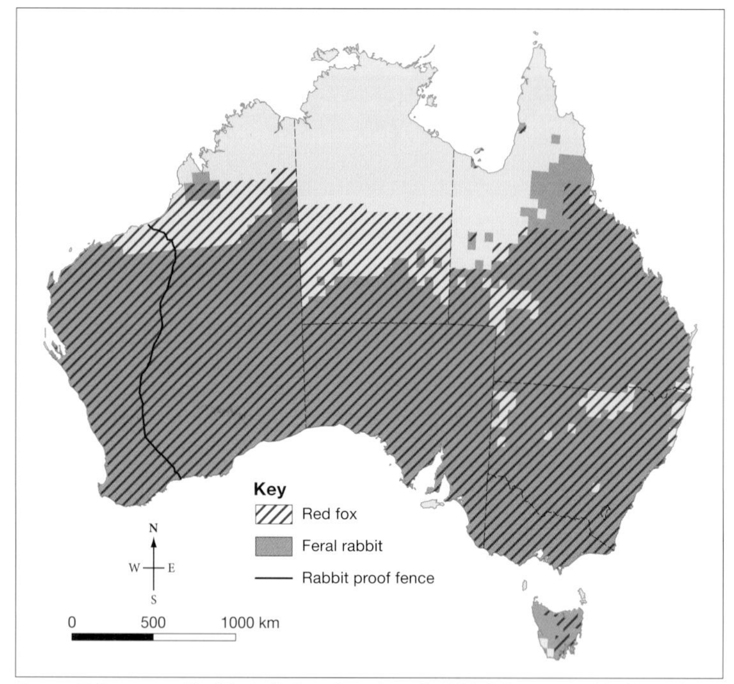

Source: © Commonwealth of Australia Geoscience Australia 2013. © Commonwealth of Australia Department of Sustainability, Environment, Water, Population and Communities 2013. Map by Spatial Vision

Both foxes and rabbits have proven difficult to control and pose the same risks today as they did in the past. Foxes are the only natural predator of rabbits. Currently rabbits are controlled through the use of biological (introduction of viruses such as myxomatosis and calicivirus), chemical (baits and poisons) and mechanical (destroying warrens, shooting and laying traps) methods.

Hunting, baiting and shooting reduce adult fox populations in the short term; however, their populations soon recover. Scientists are now trialling biological controls and working on the development of some form of virus or birth control that will interfere with the reproductive system of the fox, making them infertile and incapable of breeding.

FIGURE 4 (a) Rabbits reduced Phillip Island (near Norfolk Island, off the east coast of Australia) to a wasteland. (b) After the rabbits were eradicated, the recovery of the island was spectacular.

Middle Island Maremma Project

Middle Island, a small rocky island about 2 hectares in size off the Victorian coast near Warrnambool, is home to a Little Penguin colony. At low tide, less than 12 centimetres of water separates the island from the mainland, providing easy access for predators such as foxes.

In 1999 Middle Island had a thriving colony of Little Penguins, comprised of about 600 birds. By 2005, foxes had reduced the population to fewer than 10, with only two breeding pairs remaining. In 2006 an ambitious experiment was launched using Maremma dogs to guard and protect the remaining penguins. This breed of dog has long been used to guard livestock, including chickens, with reports that once the dogs were on duty, fox kills stopped.

After some initial teething problems the program has proved highly successful. Within a short space of time there was evidence of the penguins breeding. By 2013 the Little Penguin population had rebounded to 180.

FIGURE 5 Using Maremma dogs to guard the penguins on Middle Island from fox attacks proved to be a highly successful strategy. The island's penguin population, previously close to extinction, continues to grow under the protection of the dogs.

3.9 Activities

To answer questions online and to receive **immediate feedback** and **sample responses** for every question, go to your learnON title at www.jacplus.com.au. *Note*: Question numbers may vary slightly.

Remember

1. Why are goats an effective method of controlling invasive plant species?
2. Why is it said that rabbits caused the demise of native plant and animal species?
3. Describe the distribution over ***space*** of foxes and rabbits in Australia.

Explain

4. Explain why goats would be considered an ***environmentally*** friendly method of controlling invasive plant species.
5. Why do you think that foxes are not found in Australia's tropical region?
6. Examine figure 4. Describe the appearance of the ***environment*** in each image. Do these images represent the same ***place***? Suggest reasons for the ***changes*** that have occurred in this ***environment***.
7. Explain why it is easier to eradicate invasive species from island communities than from mainland Australia.

Think

8. The invasive animal species described in this section have proven more difficult to control than the plant species. Suggest a reason for this.
9. (a) Copy the table below. In the first column (after the name of the control method), write your own definition for each of the control methods.

Method	Advantage	Disadvantage
Biological		
Chemical		
Mechanical		

   (b) Compare the advantages and disadvantages of the three main methods of rabbit control.
   (c) Which method do you think is the most effective? Give reasons for your answer.
10. Working in teams, devise your own ***sustainable*** and ***environmentally*** friendly strategy for controlling an invasive species.

3.10 Can native species create environmental change?

3.10.1 Protecting biodiversity

To protect native species, **national parks** have been established. The world's first of these, Yellowstone National Park in the United States, was established in 1874. The second was established in Sydney, Australia five years later. These parks are intended to provide safe habitats for native plant and animal species and thus help safeguard **biodiversity**. However, biodiversity is not just under threat from introduced species; left unchecked, native species can also cause widespread damage to the landscape.

3.10.2 Should we cull iconic Australian natives?

Many native Australian species have been threatened by the spread of urban settlements. As human populations expand, the natural range of animals such as koalas, kangaroos and wallabies are diminished. Despite the intricate pattern of National Parks that exist, native species can be found not only on the fringe of urban areas, but also taking refuge in our backyards where they face increasing risk from vehicles, domestic pets and poisons.

FIGURE 1 A koala drinking from a dog's water bowl in a suburban backyard

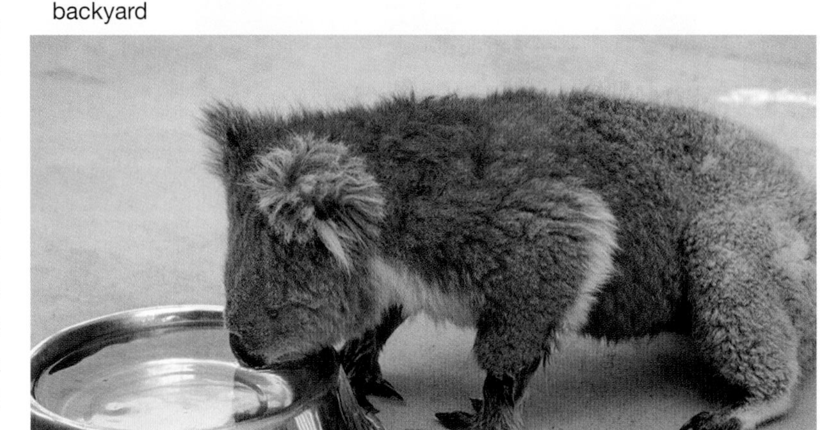

In some protected regions, natural increase can put entire colonies at risk. Koalas are preferential feeders and have only one source of food — eucalyptus leaves. In times of drought and following extended favourable breeding conditions there is simply not enough food to sustain the entire population. In the Otway Ranges and on French Island, for example, the situation has at times become so dire that koalas have faced starvation.

In 2013 and 2014, 700 koalas were killed in what the Victorian Government described as humane euthanasia to prevent the animals from starving to death. Opponents of the move have described it as a secret **cull**.

In late 2015, researchers called for koalas to be culled in parts of Victoria, New South Wales and Queensland, in areas where the local populations are infected with the disease **Chlamydia**. In koalas, Chlamydia can lead to a range of issues such as conjunctivitis, urinary tract infections, reproductive tract infections and pneumonia. Culling, the researchers suggest, will prevent further spread of the disease to healthy animals and allow the population to rebound over the next 10 years. Others have argued that the disease, in its early stages, can be treated with antibiotics.

3.10.3 What about the African elephant?

Over the last decade elephant numbers have declined by about 64% (see figure 2). At this rate, it is possible that they could become extinct in the wild within the next decade. Some estimates put the number of animals killed at up to 100 per day, as poachers seek to make their fortune selling meat, ivory and body parts to the lucrative Asian market. From 2011 to 2014 the price of ivory alone tripled in China.

FIGURE 2 African elephant range and estimated population

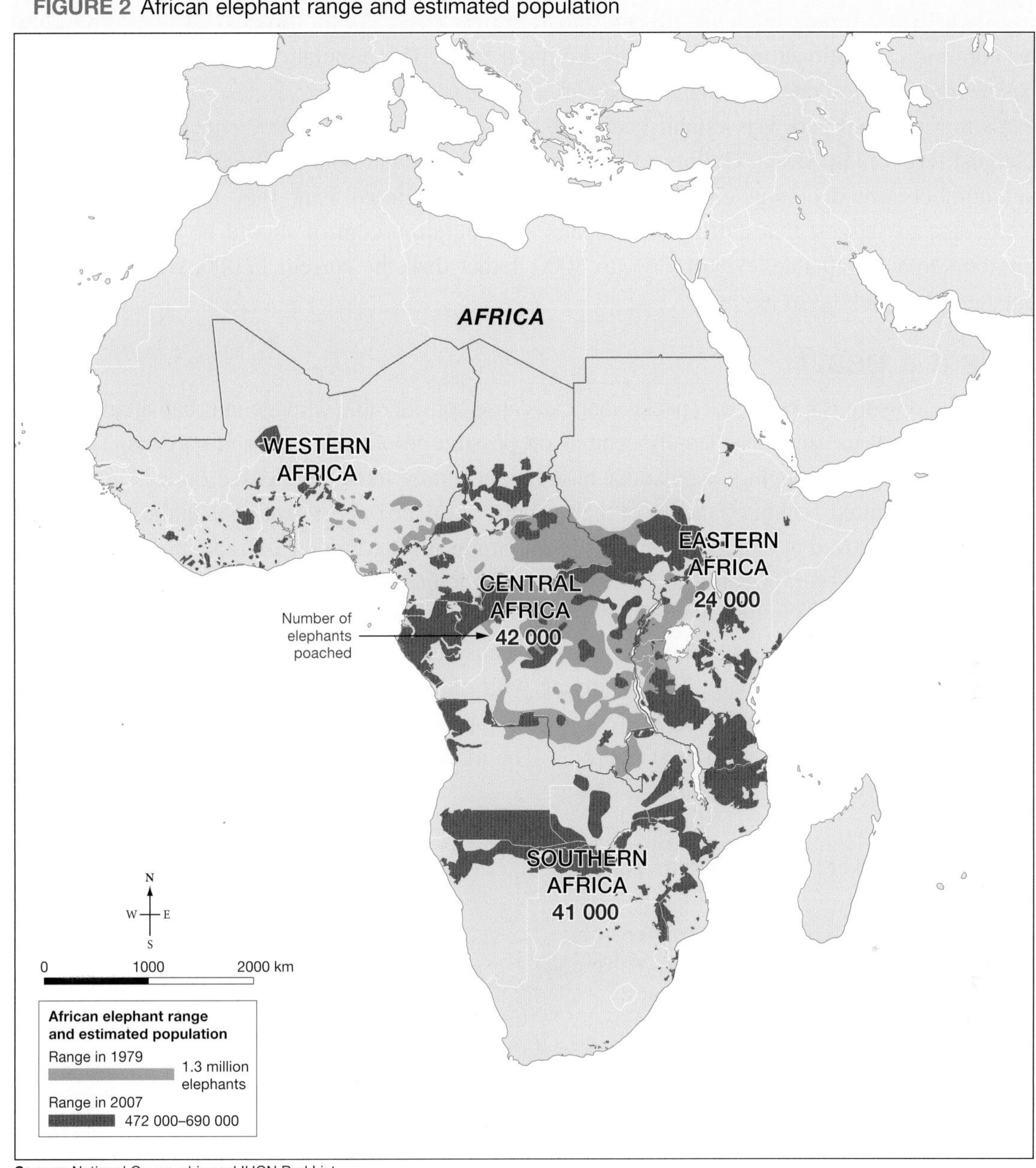

Source: National Geographic and IUCN Red List

Elephants are also under threat from expanding human populations. As the number of people increases so too does the need for land to grow the crops and raise the herds needed to sustain them, and people encroach further and further into the elephants' rangelands. Struggling farmers can also earn more from a single elephant kill than from a year of toiling on the land. When elephants enter these newly created farmlands and damage crops, the temptation to kill them is intensified.

Elephants are a key ecological species, sometimes referred to as the caretakers of the environment. They create and maintain their ecosystem and in the process create the habitat for a wide range of plant and animal species with which they coexist. The loss of elephants poses a significant threat to local ecologies.

However, the practice of confining large animals such as elephants to national parks free from predation and with an abundant water supply can result in a population explosion. An adult elephant consumes up to

136 kilograms of food in a single day. The search for food can see them cover vast distances, in the process stripping bark from, ripping branches off and pushing over trees. As the process of confining animals to reserves continues, traditional migration paths are interrupted. The population flourishes and the landscape becomes degraded as the rate of change speeds up.

Africa's first national park was established in 1926. Covering almost 20 000 square kilometres it is one of the largest reserves in Africa. Culling of elephants within the national park was banned in 1994. While elephant numbers are declining elsewhere, within Kruger National Park they are increasing. This has prompted some to suggest that culling as a means of population control should be reinstated in order to limit numbers to a sustainable level of 7000 to 8000, rather than the current 17 000, which places both the elephants and their habitat in jeopardy.

3.10.4 Is it a pest?

In Australia, the possum is a protected species that can cause considerable damage in urban areas. Because they are protected, control measures are largely centred on possum-proofing homes and gardens. There are strict regulations relating to the trapping of possums, and they generally must be released on the same property on which they were captured, within 50 metres of their capture site. The only exception is in the state of Tasmania, where they can be removed in order to protect crops and trapped for commercial trade in meat and pelts.

In New Zealand, however, possums are considered 'public enemy number one'. Originally introduced in 1837 to establish a fur trade, with no natural predators they spread rapidly and today some 30 million possums occupy 90% of the landmass. The damage they cause to native forests is unmistakable, laying large expanses of new forest growth bare. They compete with native birds for habitat and food and have been observed raiding nests. Additionally, they are known to spread bovine tuberculosis, thus posing a significant threat to dairy and deer farmers (see figure 3). The main methods of possum control in New Zealand include trapping, baiting (poisoning) and shooting.

FIGURE 3 The spread of possums in New Zealand

Possum invasion
The introduced species has a population of 30 million in New Zealand, outnumbering humans almost seven fold

1930 — Possum infested

1950

1974

2000

The invasion

- Introduced from Australia in 1830s, to establish a fur industry
- By the 1930s many batches of possums let free
- Damage to environment recognised by the 1950s
- 1951 government puts bounty on animals
- By 1963 had spread to 84% of country

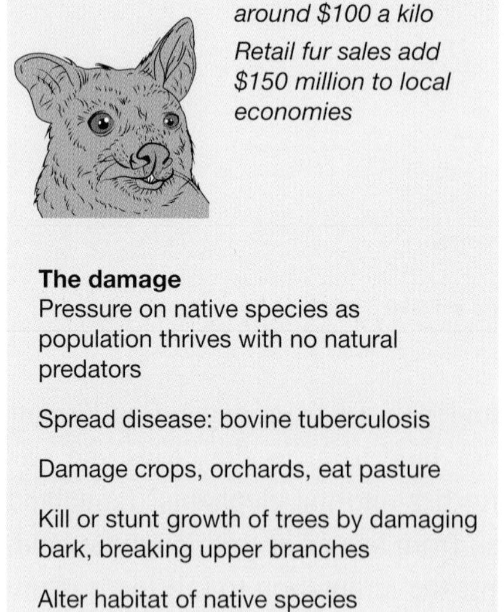

The possum fur fetches around $100 a kilo
Retail fur sales add $150 million to local economies
The damage
Pressure on native species as population thrives with no natural predators
Spread disease: bovine tuberculosis
Damage crops, orchards, eat pasture
Kill or stunt growth of trees by damaging bark, breaking upper branches
Alter habitat of native species
Prey on native birds, insects

FIGURE 4 Feral cats kill more than twice as many species in Australia than on any other island in the world.

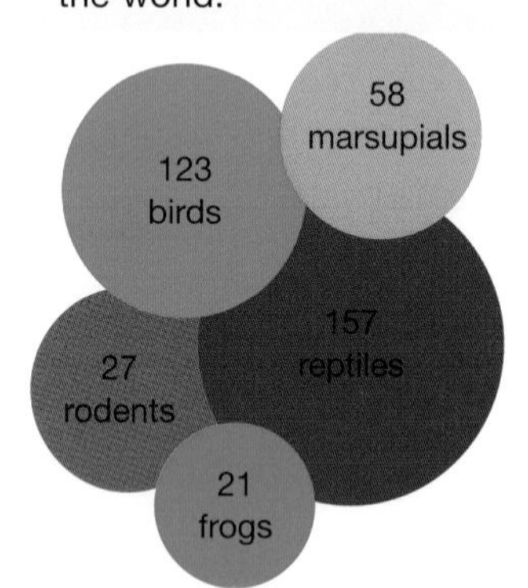

Source: Landcare Research/ Hawkes Bay regional council

3.10 Activities

To answer questions online and to receive **immediate feedback** and **sample responses** for every question, go to your learnON title at www.jacplus.com.au. *Note*: Question numbers may vary slightly.

Understand

1. Describe the spread of possums over time in New Zealand.
2. Why do you think possums are considered a pest in New Zealand but are a protected species in Australia?

Explain

3. Explain how native species can cause the land to be degraded.
4. How might an introduced species, such as cats, cause damage to the ***environment?***

Think

5. Conduct a class debate on the issue of culling as a method of controlling different species, both native and introduced.
6. (a) Feral cats have proven to be as difficult to control as other introduced species (see figure 4). Working in groups, investigate what is being done in Australia and abroad to control an invasive species. Consider social, economic and ***environmental*** factors in your investigation.
   (b) Evaluate the consequences of management responses on the ***environment*** and ***places***, comparing examples from Australia and at least one other country.
   (c) Within your group, reach consensus on an appropriate method for controlling a species. Present the reasons for your findings to the class.
   (d) Evaluate your own and others' contribution to the group tasks, critiquing roles including leadership, and provide feedback to your peers. Evaluate your task achievement and make recommendations for improvement in relation to team goals.
   (e) Analyse how divergent values and beliefs contribute to different perspectives on social issues.

3.11 Would you like salt with that?

3.11.1 Where does the salt come from?

Salinity is not a new problem. In fact, it was an environmental issue in the earliest civilisations some 6000 years ago. Historical records indicate that the Sumerians, who farmed the land between the Tigris and Euphrates rivers in the area known as Mesopotamia, ruined their land as a result of their poorly managed irrigation practices.

Salt has become a major contributor to land degradation in Australia. Rising up from below the land surface, it is destroying native vegetation and threatening the livelihood of many Australians. As plants die as a result of salinity, other problems emerge: the soil no longer has a protective cover of vegetation, which means it is more easily blown away or eroded.

Some 140 million years ago, parts of the Australian continent were covered by shallow seas and salt-water lakes. The salt stores from these waters have lain dormant below the surface of the land, much of them in the **groundwater**. In addition, salt continues to be deposited on the land's surface by rain and winds blowing in from the oceans, and by the weathering of mineral-carrying rocks.

Australia's native vegetation had built up some tolerance to the salt levels in the soil. The deep-rooted vegetation also soaked up water in the soil before it could seep down into the groundwater. This meant that the watertable stayed at a fairly constant level, and that the concentrated salt stores stayed where they were. This natural balance changed with the arrival of European settlers. The farming and land-clearing practices they introduced were, and still are, according to many experts, unsuited to Australia's generally harsh, dry climate, as well as to its geological history.

Salt has now become a serious problem. There are two ways in which the soil can become too salty: these are called dryland salinity and irrigation salinity.

3.11.2 What is dryland salinity?

Dryland salinity occurs in areas that are not irrigated. When settlers cleared the land, they replaced deep-rooted native vegetation with crop and pasture plants. These plants generally have shorter roots and cannot soak up as much rainfall as native vegetation. Excess moisture seeped down into the groundwater, raising the watertable and bringing concentrated saline water into direct contact with plant roots (see figure 2). Vegetation, even salt-tolerant plants, started dying as the salt concentrations rose. Once the vegetation dies off, the soil is left bare and is prone to erosion. Often layers of salt, known as **salt scald**, are visible on the surface of the land. The areas in Australia affected by dryland salinity are shown in figure 1.

FIGURE 1 Salinity distribution

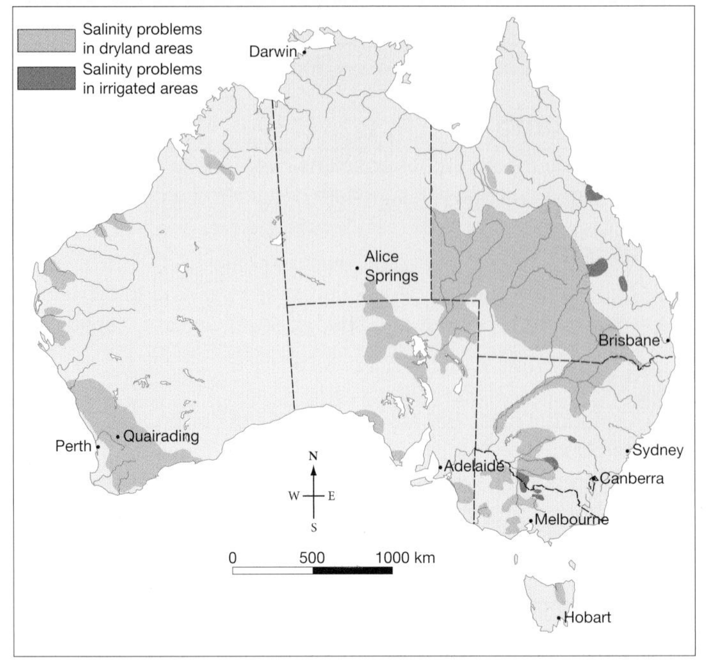

Source: Spatial Vision

FIGURE 2 The effects of a rising watertable

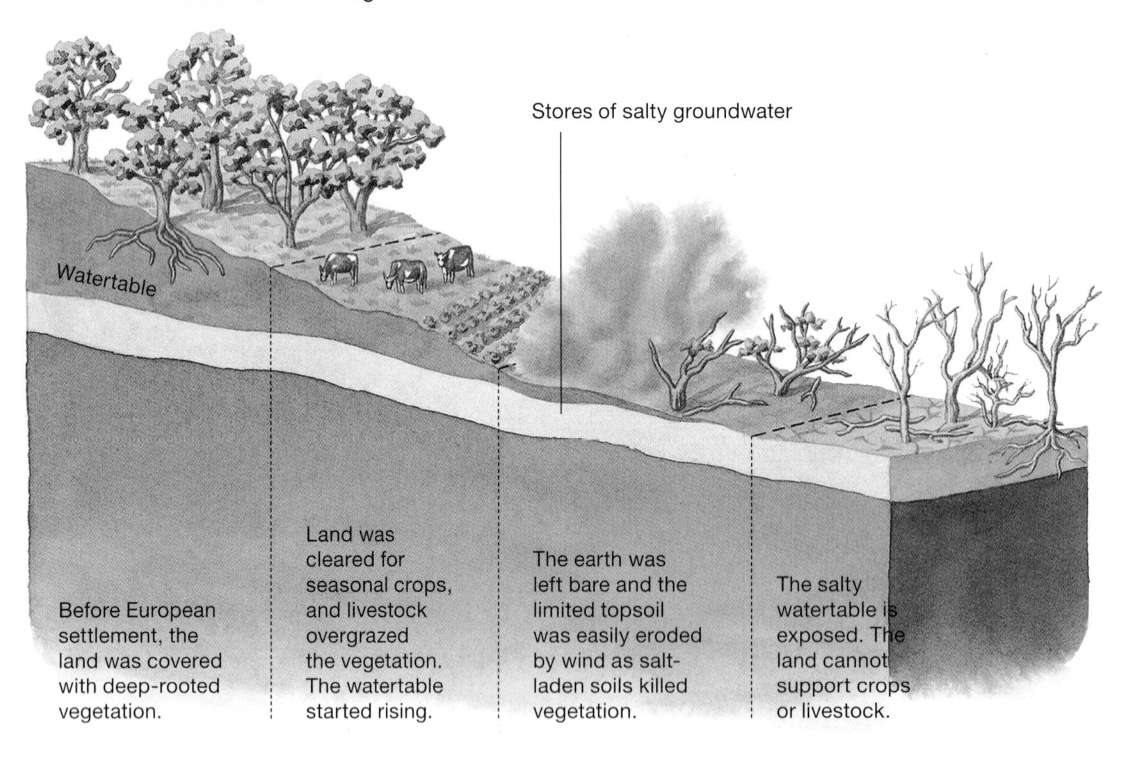

Stores of salty groundwater
Watertable
Before European settlement, the land was covered with deep-rooted vegetation.
Land was cleared for seasonal crops, and livestock overgrazed the vegetation. The watertable started rising.
The earth was left bare and the limited topsoil was easily eroded by wind as salt-laden soils killed vegetation.
The salty watertable is exposed. The land cannot support crops or livestock.

3.11.3 What is irrigation salinity?

Irrigation salinity occurs in irrigated regions (see figure 1) and is a direct result of over-watering. When more water is applied to crops or pasture plants than they can soak up, the excess water seeps down through the soil into the groundwater, causing the salty watertable to rise to the surface. Some of this salt is washed into rivers, either as run-off or groundwater seepage, and transported to other places.

HOW MUCH LAND IS AFFECTED BY DRYLAND AND IRRIGATION SALINITY AROUND THE WORLD?

- Africa: 2 per cent of Africa's landmass
- China: 21 per cent of arid lands or around 30 million hectares of land
- Western Europe: 10 per cent of the land area
- United States: 17 states
- South America: most countries
- Australia: 2.5 million hectares
- Worldwide: It is estimated that 10 million hectares of arable land succumbs to the effects of irrigation-related salinity each year. It is estimated that, without intervening action, by 2050 the affected area might triple.

How do we solve the problem?

Many programs are in place to identify and monitor problem areas. Action being taken includes:

- changing irrigation practices to reduce over-watering
- planting deep-rooted native trees and shrubs in open areas
- developing new crops that are more salt tolerant, such as new strains of wheat
- replacing introduced pasture grasses with native vegetation such as saltbush (see figure 3)
- using satellite technology to map areas at risk to enable early intervention.

FIGURE 3 Native plants such as saltbush help solve the problem on Australian grazing lands.

3.11 Activities

To answer questions online and to receive **immediate feedback** and **sample responses** for every question, go to your learnON title at www.jacplus.com.au. *Note*: Question numbers may vary slightly.

Explain

1. Explain the ***interconnection*** between soil salinity and land degradation.
2. Why would planting deep-rooted trees help solve the problem?
3. What actions could an irrigation farmer take to reduce the risk of salinity?

Discover

4. Investigate the history of agriculture in an ancient civilisation, such as Mesopotamia.
   (a) Include a sketch map of the area. Annotate this map to show how the region was affected and why.
   (b) What lessons might modern farmers learn from ancient practices?
5. In groups, investigate a method of combating salinity and ***sustainable*** practices that will improve the productivity of agricultural land. Before you begin, decide as a class which groups will cover dryland salinity and which will focus on irrigation salinity. Present your findings as a news report.

Predict

6. Find out the total land area of Australia and the world. If areas affected by irrigation salinity are expected to triple by 2050, estimate the proportion of land that will be affected on a national and global ***scale***. Use your findings as the basis for writing a letter to the Editor, urging governments to take action and halt this trend.

Think

7. Salinity was not an ***environmental*** issue in Australia when Indigenous people were its sole inhabitants. With the aid of diagrams, explain how land-use practices have ***changed*** over time. Make sure you include references to Indigenous practices that promoted ***sustainable*** use of the ***environment***. Include links to how these ***changes*** would have resulted in salinity and degraded the ***environment***.

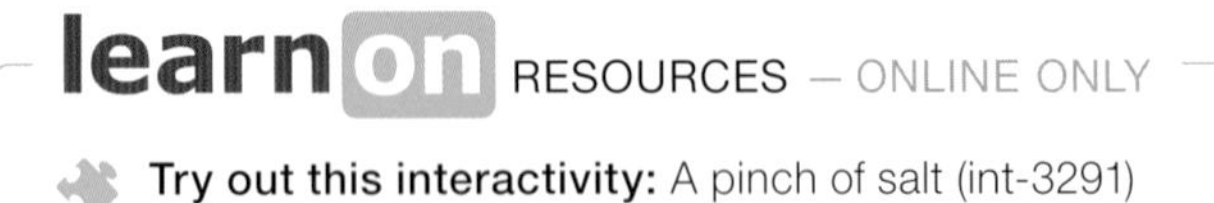

RESOURCES — ONLINE ONLY

Try out this interactivity: A pinch of salt (int-3291)

3.12 How do we deal with salinity?

Access this subtopic at **www.jacplus.com.au**

3.13 SkillBuilder: Writing a fieldwork report as an annotated visual display (AVD)

WHAT IS A FIELDWORK REPORT?

A fieldwork report helps you process all the information that you have gathered during fieldwork. You sort your data, create tables and graphs, and select images, and then interpret the data as text or annotated images and synthesise all the data in a logical presentation.

Go online to access:

- a clear step-by-step explanation to help you master the skill
- a model of what you are aiming for
- a checklist of key aspects of the skill
- a series of questions to help you apply the skill and to check your understanding.

FIGURE 1 An annotated visual display (AVD) completed from secondary sources

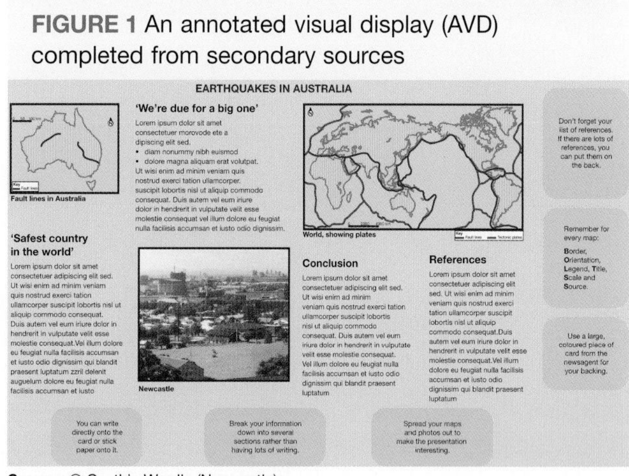

Source: © Cynthia Wardle (Newcastle)

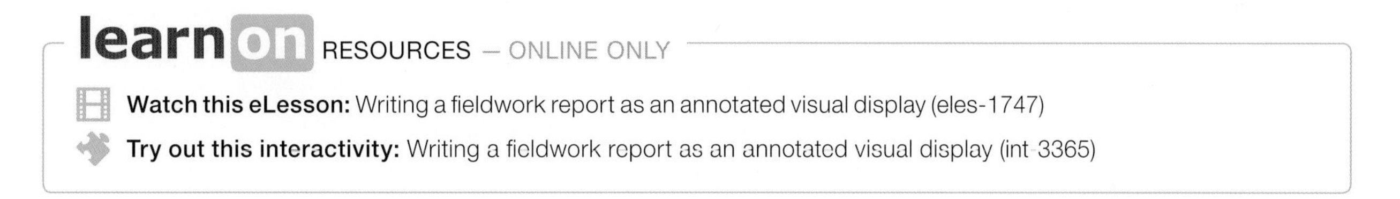

learnON RESOURCES — ONLINE ONLY

Watch this eLesson: Writing a fieldwork report as an annotated visual display (eles-1747)

Try out this interactivity: Writing a fieldwork report as an annotated visual display (int-3365)

3.14 How do Indigenous communities manage the land?

3.14.1 Traditional land management

Before the arrival of European settlers, Indigenous communities had created their own system of land management. They maintained grasslands through the use of fire, which encouraged plant regrowth and attracted a variety of animals. Their life was governed by the seasons, with each change dictating a change in the use of the land and its management.

Environmental change is not new. Indigenous communities around the world, including the Australian Aborigines, have had to manage their environments carefully. In figure 1 you can see how fire was used to manage the landscape; however, it was only one of a variety of strategies that was used to ensure the land was used in a sustainable manner.

FIGURE 1 Indigenous Australians' traditional land management practices and connection to the land

Indigenous people have adapted to environmental change over the last 50 000 years.
Aboriginal land practices involve working with the land and its elements rather than seeking to make dramatic changes.
Indigenous Australians took only what they needed and little was wasted.
Habitat loss, soil erosion and weed infestation were unknown until the time of European settlement.
Evidence also shows that fuel reduction (back burning) was used to prevent bushfires. This practice prevented large bushfires that could burn for months and permanently damage the landscape.
They take collective ownership of the land.
In some Indigenous communities some native species such as the kangaroo and platypus are considered sacred.
Their technology was simple. They used spears and fire sticks, designed to minimise environmental impact.
Fire is used to control plant growth and maintain a grassland environment. Many native seeds need fire in order to germinate.
Although nomadic, there was a pattern to their movement across the land, designed to coincide with the seasons.
Every aspect of their life is governed by the land and the seasons. The land provided all their needs — they had no need to grow crops or raise livestock.
Their spiritual and cultural connections to the land, the health of the land and its water are central to their own wellbeing.

3.14.2 What happens when?

Figure 2 is an example of an Aboriginal seasons calendar. It is for the Yolngu people who live in north-east Arnhem Land. The calendar relates the months of the year to aspects of the environment, although a traditional Indigenous community had no use for the months as we know them. Look carefully and you will see that this particular calendar includes information about the weather and the plants and animals that thrive across the year. Traditional communities were made up of hunters and gatherers. They hunted and fished for particular species and gathered bulbs, fruits and other edible vegetation at different times of the year. The calendar varied from place to place, but whatever the location it enabled Indigenous people to predict seasonal events; for instance, the arrival of march flies signalled the time to collect crocodile eggs and bush honey.

FIGURE 2 Indigenous communities are defined by observable changes in the seasons.

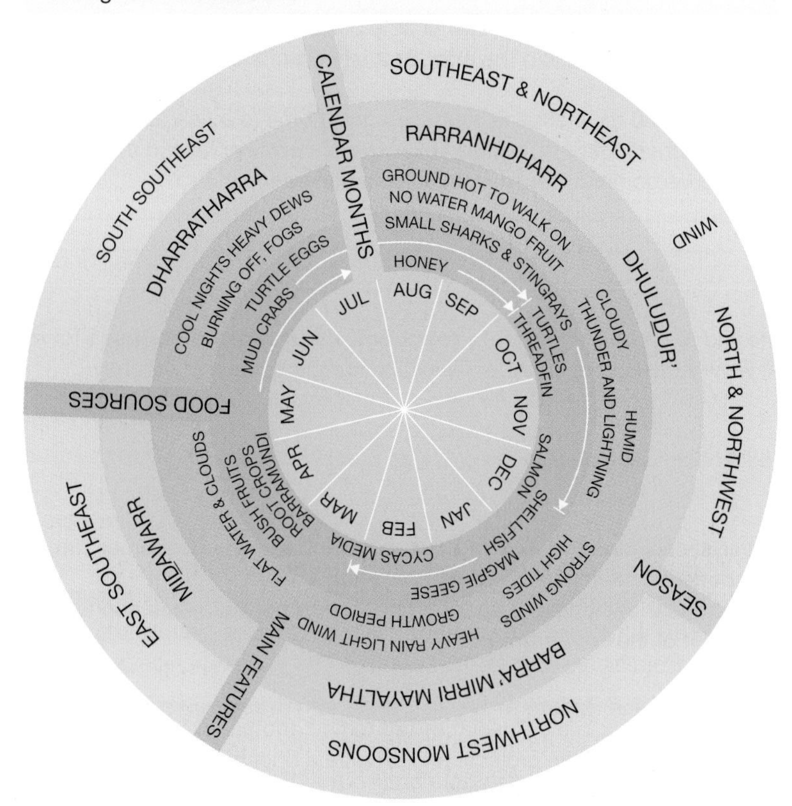

CASE STUDY

How do Indigenous communities manage wetlands in Kakadu?

FIGURE 3 Wetlands after removal of buffalo and before burning

FIGURE 4 Wetlands after burning

Kakadu is a kaleidoscope of both cultural and ecological biodiversity. The landscape varies from savannah and woodlands to escarpments and ridges as well as wetland, flood plains and tidal flats. The region includes more than 2000 plant species that have provided food, medicine and weaving materials for the Indigenous communities that have inhabited the region for some 50 000 years. Over this time they have defined six distinct seasons, all signalled by subtle changes in the weather patterns that mark the transition from one season to the next. They managed and maintained the landscape through the use of fire.

The arrival of European settlers saw a massive change in the region. Buffalo were introduced in the early to mid 1800s to serve as a food supply for new settlers. However once these new settlements were abandoned in the mid 1900s the buffalo population expanded from a modest population of less than 100 animals to more than 350 000. The impact on local habitats was extreme.

Natural habitats were devastated. The now feral buffalo took over wetland areas, disturbed native vegetation, caused significant soil erosion and changed the characteristics of the region's floodplains. Saltwater intrusion of freshwater wetlands caused the region to become further degraded, leading to a rapid decline in the flora and fauna including waterbirds that had sustained Indigenous communities.

In the late twentieth century a massive culling program was commenced to remove the feral buffalo and allow the region to regenerate. However, an invasive native plant species that had once been the main food source of the buffalo spread unchecked. It choked the wetlands and prevented waterbirds from feeding and recolonising the region.

The CSIRO undertook extensive research into the sustainable practices of the region's traditional landowners. A joint management initiative was introduced into the area. At the heart of the initiative was the traditional method of fire management.

The results have been dramatic, and the wetlands are once again home to a rich assortment of flora and fauna. The project provides an internationally recognised example of sustainable land management utilising the practices carried out by Indigenous communities across multiple generations.

3.14 Activities

To answer questions online and to receive **immediate feedback** and **sample responses** for every question, go to your learnON title at www.jacplus.com.au. *Note*: Question numbers may vary slightly.

Remember

1. Why wasn't land degradation an issue prior to the arrival of European settlers?

Explain

2. Using information from this section and the **Wetland burning** weblink in the Resources tab, explain why fire is such an important component of caring for the ***environment***.
3. How does the Aboriginal calendar demonstrate an ***interconnection*** between their connection with the land and ***sustainable*** management of the ***environment***?

Discover

4. Use the **Indigenous landcare** weblink in the Resources tab to find another example of Indigenous involvement in landcare projects. Select one of the projects and prepare a brief report on the ***scale*** of the project, what is involved and the benefits to the ***environment***.

Think

5. The Aboriginal calendar (figure 2) demonstrates an intricate understanding of the ***environment***.
   (a) How long do you think it would have taken the Aboriginal people to have developed this understanding?
   (b) How do you think this knowledge would have been passed from generation to generation?
   (c) It has been suggested that the four seasons currently used in Australia do not adequately reflect the changing nature of our seasons. Do you agree or disagree with this suggestion? Give reasons for your opinion based upon the area in which you live.
   (d) Develop your own calendar that reflects the ***interconnection*** between the seasons and changes in your life.
   (e) Do you think the Aboriginal seasons calendar should be adopted and used as an additional strategy for the ***sustainable*** management of ***environmental*** issues? Justify your point of view.

Explore more with these weblinks: Wetland burning, Indigenous landcare

3.15 Review

3.15.1 Review

The Review section contains a range of different questions and activities to help you revise and recall what you have learned, especially prior to a topic test.

3.15.2 Reflect

The Reflect section provides you with an opportunity to apply and extend your learning.

Access this subtopic at **www.jacplus.com.au**

TOPIC 4 Inland water — dammed, diverted and drained

4.1 Overview

Numerous **videos** and **interactivities** are embedded just where you need them, at the point of learning, in your learnON title at www.jacplus.com.au. They will help you to learn the content and concepts covered in this topic.

4.1.1 Introduction

Water makes life on Earth possible, and rivers are like blood running through the veins of a body. Over time we have dammed, diverted and drained water, and this has brought about significant environmental change. Careful stewardship of these resources will provide a health insurance policy for a sustainable future.

Inland waters are important sources of water for both environments and people.

Starter questions

1. How many different types of freshwater bodies can you think of within 100 kilometres of where you live?
2. Where does your fresh water come from? Name and describe the location of the freshwater bodies that supply your house. You may need to refer to an atlas.
3. If you didn't have shops, supermarkets, water taps and pipes where you live, what water sources would you get your daily fresh water from?

INQUIRY SEQUENCE

4.2 What is inland water?

4.2.1 Defining inland water

Have you ever stopped to think that the water flowing down a river or rippling across a lake is providing us with a life support system? The rivers, lakes and wetlands that make up our inland water are important for supplying water for our domestic, agricultural, industrial and recreational use. They provide important habitats for a wide range of terrestial and aquatic life.

Inland water systems cover a wide range of landforms and environments, such as lakes, rivers, floodplains and wetlands. The water systems may be **perennial** or **ephemeral**, flowing (such as rivers), or standing water (such as lakes) (see figure 1). There are interconnections between surface water and groundwater, and between inland and coastal waters. Inland water is an important link in the water cycle, as water evaporates from its surface into the atmosphere. In return, rainfall can be stored in rivers and lakes, or soak through the soil layers to become groundwater.

4.2.2 Why is inland water important?

Inland water provides both the environment and people with fresh water, food and habitats. It provides environmental services; for example, it can filter pollutants, store floodwater and even reduce the impacts of climate change. The economic value of these services cannot easily be measured. Their importance, however, can be taken for granted and not appreciated until the services are lost or degraded.

4.2.3 What are the threats to inland water?

Inland water is extremely vulnerable to change. It has been estimated that in the last century over 50 per cent of inland water (excluding lakes and rivers) has been lost in North America, Europe and Australia. Those systems remaining are often polluted and reduced in size. The loss is largely a result of human-induced environmental changes. Table 1 illustrates some of the reasons for changes to inland water systems, and their possible impacts on the environment and people. As water is such a valuable resource, much of our inland waterways have been dammed, diverted or drained to meet the needs of people.

FIGURE 1 The Parana River floodplain in northern Argentina shows a variety of different types of inland water.

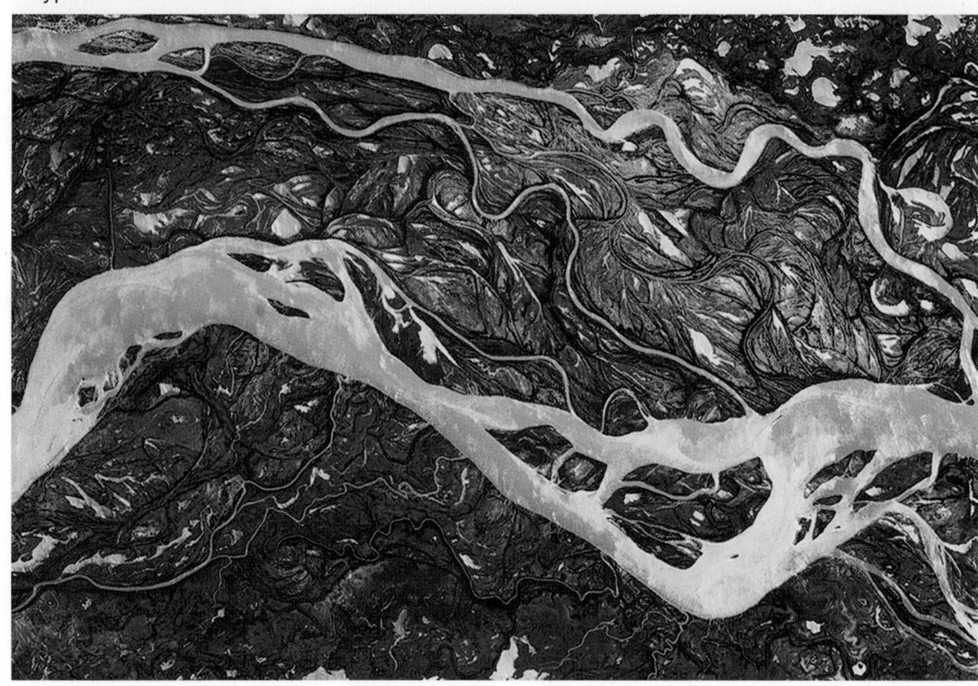

TABLE 1 Threats to inland water

Cause of change to inland water systems	Environmental functions threatened	Impacts of change
Increasing population and increasing demand for water across space	Most services (e.g. fresh water, food and biodiversity) Regulatory features such as recharging groundwater and filtering pollutants	Increased withdrawal of water for human and agricultural use Large-scale draining of wetlands to create farmland
Construction of infrastructure including dams, weirs and levee banks, diverting water to other drainage basins	Services supporting the quality and quantity of water Biodiversity, habitat, river flow and river landforms	Changes to the amount and timing of river flow. The transportation of sediment can be blocked and dams can restrict fish movements.
Changing land use (e.g. draining of wetlands, urban development on floodplains)	Holding back floodwaters and filtering pollutants Habitats and biodiversity	Alters run-off and infiltration patterns Increased risk of erosion and flood
Excessive water removal for irrigation	Reduced water quantity and quality Less water available for groundwater supply	Reduced water and food security Loss of habitat and biodiversity in water bodies
Discharge of pollutants into water or on to land	Change in water quality, habitat Pollution of groundwater	Decline in water quality for domestic and agricultural use Changes ecology of water systems

FIGURE 2 Wetlands are an example of inland water systems that are vulnerable to human-induced damage.

4.2 Activities

To answer questions online and to receive **immediate feedback** and **sample responses** for every question, go to your learnON title at www.jacplus.com.au. *Note*: Question numbers may vary slightly.

Explain

1. Match the following terms with their correct definition in the table below. You may need to use a dictionary to help. *main channel, tributary, anabranch, meander, oxbow lake (or billabong), floodplain*

Term	Definition
	a smaller stream that flows into a larger stream
	bend in the river
	area of relatively flat, fertile land on either side of a river
	main river
	a cut-off meander bend
	where a river branches off and joins back into itself

2. Make a simplified sketch of figure 1 and clearly label an example of each of the features listed in question 1 on your sketch.
3. The Parana River is 4880 kilometres long, making it the second longest river in South America. The river flows from the south-east central plateau of Brazil south to Argentina. Figure 1 is a small section of this river. Locate the river in your atlas. What evidence is there to suggest that this river frequently floods?
4. Refer to figure 1. The brown shading visible in the water and on the land represents the river's muddy sediment. This is material such as sand and silt carried and deposited by a river.
   (a) Where has this sediment come from?
   (b) How does the sediment get onto the floodplain?
   (c) If the river is dammed upstream, what ***changes*** are likely to happen to the sediment carried and to the floodplain?
5. Suggest two short-term and two long-term examples of human-induced ***changes*** that could have an impact on the wetland in figure 2.

4.3 Dam it?

4.3.1 Why dam rivers?

Are dams marvellous feats of modern engineering or environmental nightmares? Without them we would not have a dependable supply of water or electricity, nor would we feel relatively safe from floods. For many decades, dams have been seen as symbols of a country's progress and economic development. But more and more the true costs, socially, economically and environmentally, are emerging.

A reliable water supply has always been critical for human survival and settlement; however, water is not evenly distributed over the world by either time or space. Some places suffer from regular droughts, while others experience massive flooding. As a result, people have learned to store, release and transfer water to meet their water, energy and transport needs. This could be in the form of a small-scale farm dam or a large-scale multi-purpose project such as the Snowy River Scheme. Constructing dams is one of the most important contributors to environmental change in river basins. Globally, over 60 per cent of the world's major rivers are controlled by dams.

Figure 1 shows the degree of **river fragmentation**, or interruption, in the world's major drainage basins. River fragmentation is an indicator of the degree to which rivers have been modified by humans. Highly affected rivers have less than 25 per cent of their main channel remaining without dams, and/or the annual flow pattern has changed substantially. Unaffected rivers may have dams only on tributaries but not the main channel, and their discharge has changed by less than 2 per cent.

FIGURE 1 Degree of river fragmentation in the world's major drainage basins

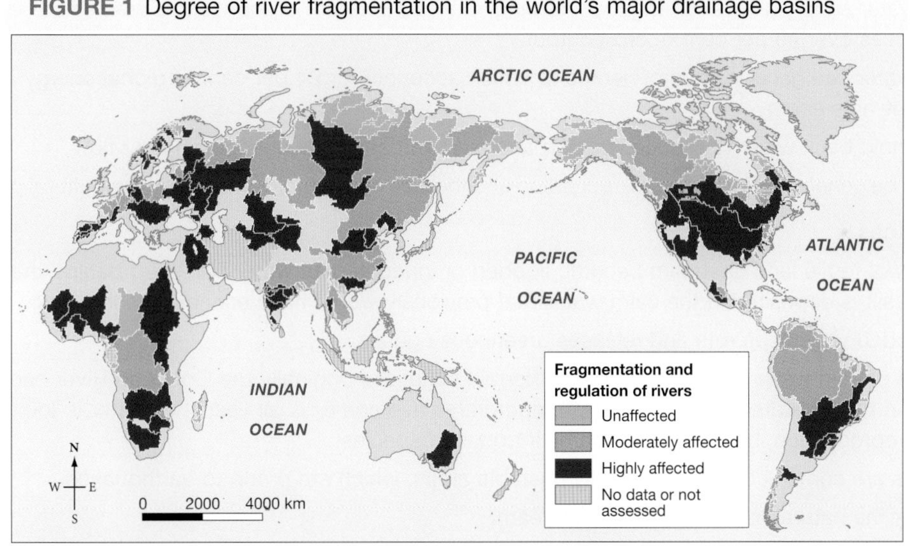

Source: Made with Natural Earth. University of New Hampshire UNH/Global Runoff Data Centre GRDC http://www.grdc.sr.unh.edu/ Map by Spatial Vision

Dams, **reservoirs** and **weirs** have been constructed to improve human wellbeing by providing reliable water for agricultural, domestic and industrial use. Dams can also provide flood protection and generate electricity.

4.3.2 What changes do dams bring?

While there are many benefits, large-scale or mega dams bring significant changes to the environment and surrounding communities (see figure 2).

FIGURE 2 The advantages and disadvantages of large-scale dams

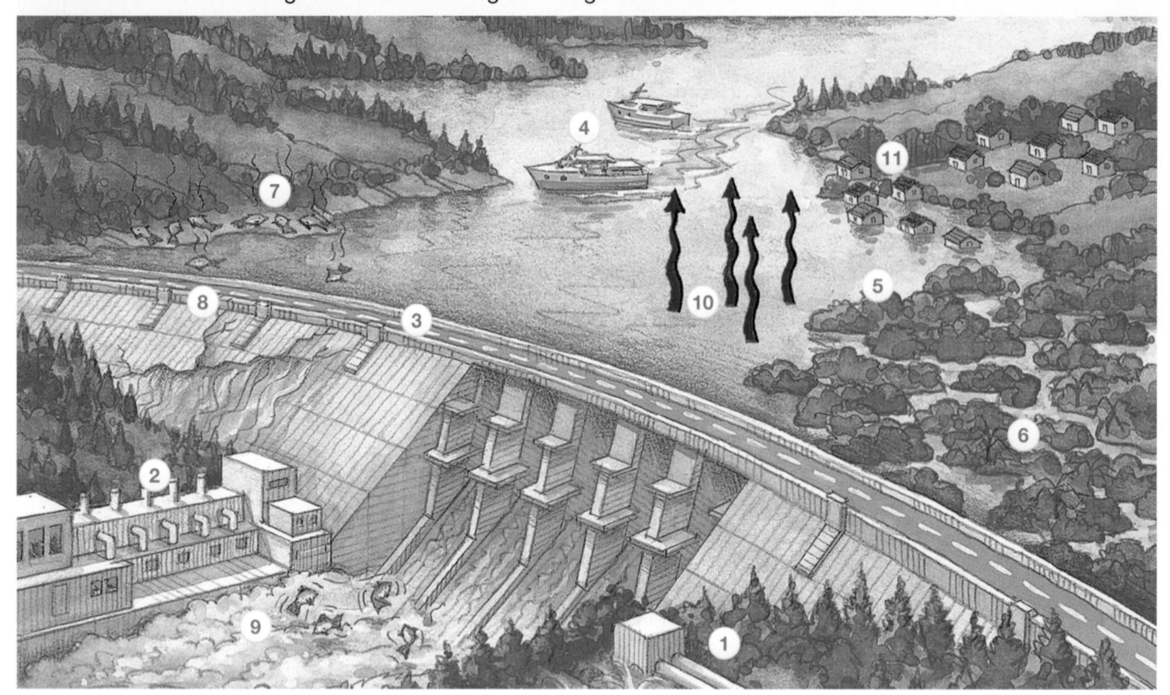

Positive changes

1 A regular water supply allows for irrigation farming. Only 20 per cent of the world's arable land is irrigated, but it produces over 40 per cent of crop output.

2 Released water can generate hydro-electricity, which accounts for 24 per cent of global energy and 90 per cent of renewable energy.

3 Dams can hold back water to reduce flooding and even out seasonal changes in river flow.

4 Income can be generated from tourism, recreation and the sale of electricity, water and agricultural products.

Negative changes

5 Large areas of fertile land upstream become flooded or inundated as water backs up behind the dam wall. Alluvium or silt is deposited in the calm water that previously would have enriched floodplains.

6 Initially, flooded vegetation rots and releases greenhouse gases.

7 The release of cold water from dams creates thermal pollution. Originally the Colorado River had a seasonal fluctuation in temperature of 27 °C. Today, temperatures average 8 °C all year. The water is too cold for native fish reproduction, but is ideal for some introduced species.

8 Some dams are constructed in tectonically unstable areas, which are prone to earthquakes.

9 Dams block the natural migration of fish upstream.

10 Over 7 per cent of the world's fresh water is lost through evaporation from water storages.

11 Between 40 and 80 million people worldwide have been forced from their homes and land due to dams. Often these people are poor or indigenous and receive little compensation.

4.3 Activities

To answer questions online and to receive **immediate feedback** and **sample responses** for every question, go to your learnON title at www.jacplus.com.au. *Note*: Question numbers may vary slightly.

Remember

1. What human activities are responsible for ***changing*** or fragmenting rivers?
2. Using figure 1, describe the location of ***places*** with rivers that are largely unaffected by river fragmentation.
3. Using your atlas, compare a map of world population distribution with figure 1. What do you notice about the ***interconnection*** between population concentration and moderately to highly fragmented rivers?

4. Suggest the ways that native fish can be affected by large dams.

Explain

5. Construct a table with the following headings to classify the impacts of dam building. Use information from figure 2 and include impacts from your own region.

Positive impacts on people	Negative impacts on people
Positive impacts on environment	Negative impacts on environment

6. Use the **Controversial dams** weblink in the Resources tab to view slides of 10 of the world's most controversial dams. Construct a table to (a) name each scheme and its location, (b) list the purposes for each of the dams and (c) list reasons why the schemes are controversial.
7. 'The positive impacts of large dam building projects on people outweigh the negative impacts on the environment.' Do you agree or disagree with this statement? Give reasons for your point of view.
8. Use the **IRO** weblink in the Resources tab to watch a video called 'We all live downstream' by the International Rivers organisation.
   (a) Explain what is meant by the phrase 'Rivers connect landscapes, cultures and livelihoods.'
   (b) What would you consider are the two main messages coming from this video?
   (c) What do you think people can do to reduce the impacts of proposed new dams?

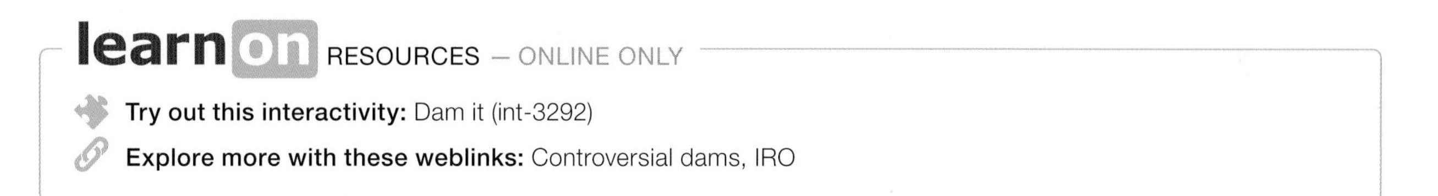

learn on RESOURCES — ONLINE ONLY
Try out this interactivity: Dam it (int-3292)
Explore more with these weblinks: Controversial dams, IRO

4.4 Do we have to dam?

4.4.1 Why should a river flow?

Traditionally, water flowing out to sea was seen as a waste. If it could be stored, then it could be used. Little thought was given to the health of the river and the importance of keeping water in a stream. The benefits of damming rivers for multi-purpose use have always been given priority by governments around the world. But is this the only solution to our growing water needs?

Large-scale or mega dams have always been linked to economic development and improvement in living standards. It has only been in recent times that the real costs of these schemes, environmentally, economically and socially, have been questioned. Figure 1 shows the number of downstream communities in each country that have the potential to be affected by the construction of mega dams.

There is also the concern that multi-purpose dams have conflicting aims. To generate hydro-electricity you need to release a large volume of stored water. To provide **flood mitigation** you need to keep water levels low in a dam, but then you need a large store if you wish to use the water for irrigation. So what do you do?

More than one billion people worldwide lack access to a decent water supply, yet it has been estimated that only 1 per cent of current water use could supply 40 litres of water per person per day, if the water was properly managed. The problem is not so much the quantity or distribution of water resources but the mismanagement of it. During the twentieth century, over $2 trillion has been spent on the construction of more than 50 000 dams. The emphasis now is to switch from *controlling* river flow to *adapting* to river flow. In other words, shifting from a human-centred to an earth-centred approach. This would mean building small-scale projects that promote social and environmental sustainability. In many regions of the world, whole communities are protesting against the need for mega dams in preference to smaller schemes that benefit local people directly (see subtopic 4.6).

FIGURE 1 Distribution of downstream communities affected by large dams

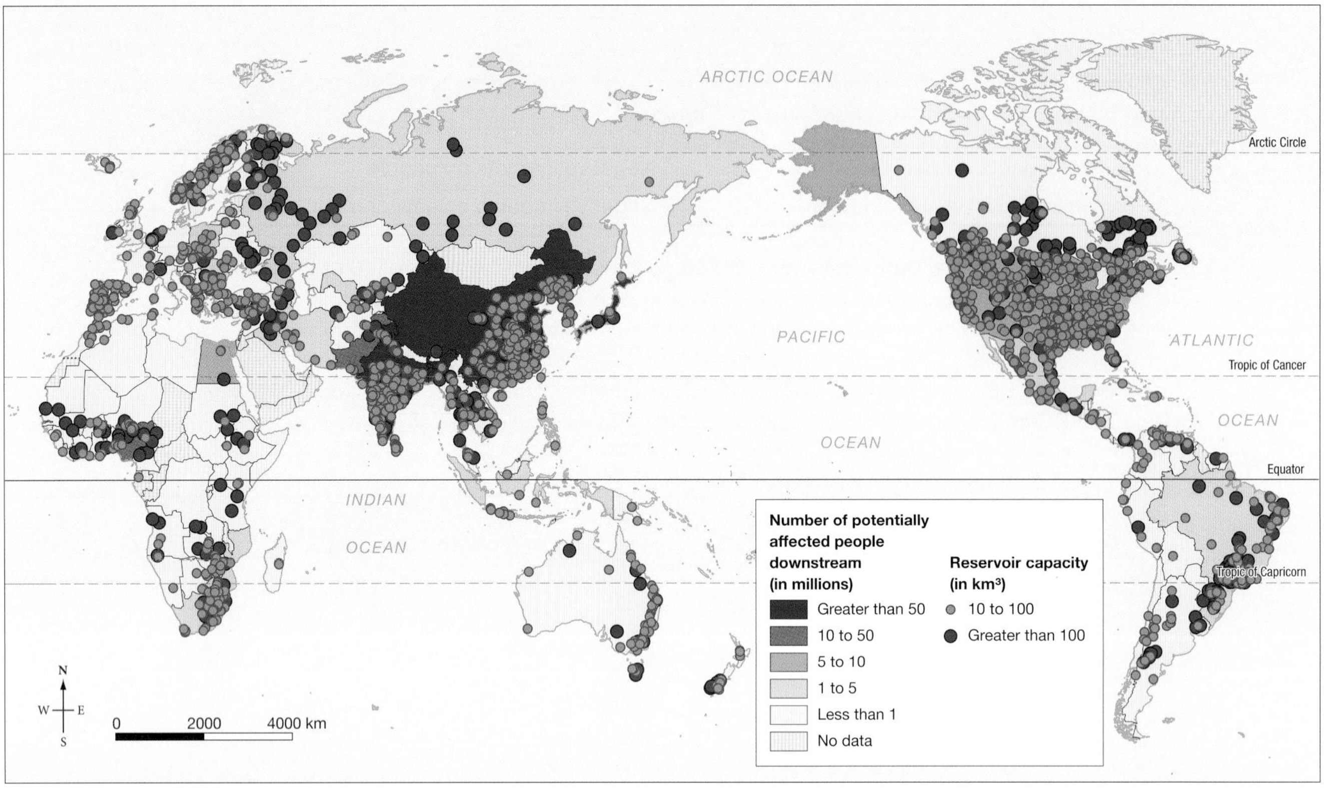

Source: Lehner et al.: High resolution mapping of the world's reservoirs and dams for sustainable river flow management. Frontiers in Ecology and the Environment. GWSP Digital Water Atlas (2008). Map 81: GRanD Database (V1.0). Available online at http://atlas.gwsp.org.

4.4.2 What are the alternatives to dams?

There are viable alternatives to dams that are often cheaper and have fewer social and environmental impacts. The focus has to be, firstly, the reduction in demand for water and, secondly, on being more efficient with the existing water.

How can water savings be made?

Agriculture

Globally, more than 70 per cent of fresh water is used for agriculture. Irrigation is often very inefficient, with over half of the water applied not actually reaching the plants. High rates of evaporation and leaking **infrastructure** waste water. Often governments subsidise and encourage farmers to grow water-thirsty crops, such as cotton, in semi-arid regions. Poorly designed and managed irrigation schemes can become unsustainable if they develop waterlogging and salinity problems.

Vast water savings could be made by improving irrigation methods, switching to less water-consuming crops and taking poor quality land out of production. If the amount of water consumed by irrigation was reduced by 10 per cent, water available for domestic use could double across the globe.

Urban use

It is estimated that as much as 40 per cent of water is wasted in urban areas just through leaking pipes and taps. Savings can be made by:

- reducing leaking pipes and improving water delivery infrastructure
- encouraging the use of water- and energy-efficient appliances and fixtures
- changing the pricing of water to a 'the more you use, the more you pay' system
- offering incentives to industry to reduce water waste and recycle
- harvesting rainwater, collecting rainwater off roofs, recycling domestic wastewater and other efficiency schemes.

Small-scale solutions

It has been estimated that it would cost $9 billion a year between now and 2025 to provide all of the world's people with adequate water and sanitation using small-scale technologies. This amount is only one-third of current spending in developed nations on water and sanitation. It is the equivalent of nine day's defence spending by the United States of America. Rather than one large, expensive dam, smaller projects that benefit local communities can be more desirable. These are often constructed and maintained by people who benefit directly from control over their own resources, at a minimal cost (see figure 2).

FIGURE 2 Research in India has shown that 10 micro dams with one-hectare catchments will store more water than one dam of 10 hectares.

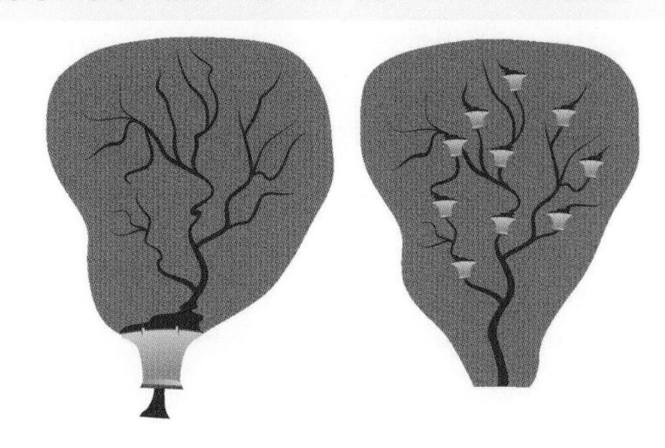

How can we reduce the need for dams?

As many countries are actually running out of suitable places to locate large dams, alternatives need to be found. **Rainwater harvesting** schemes such as illustrated in figure 3 (a) and (b) can be used for storing water. **Micro hydro-dams** (see figure 4) can be used for generating electricity. Both of these schemes are easier and cheaper to build than large dams, and have lower environmental impacts.

FIGURE 3 Two methods for water harvesting: (a) rainwater tank and (b) groundwater recharging

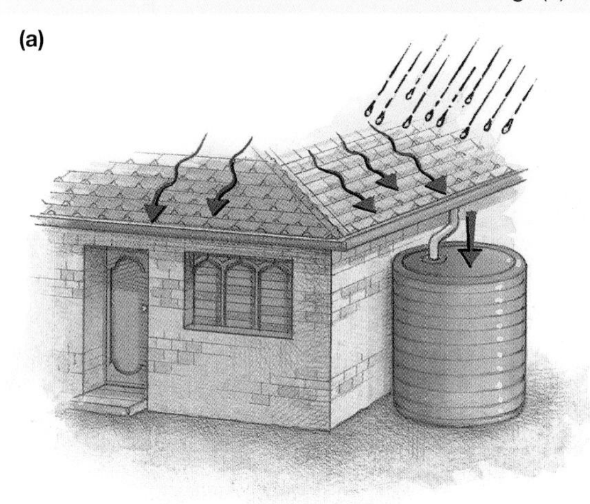

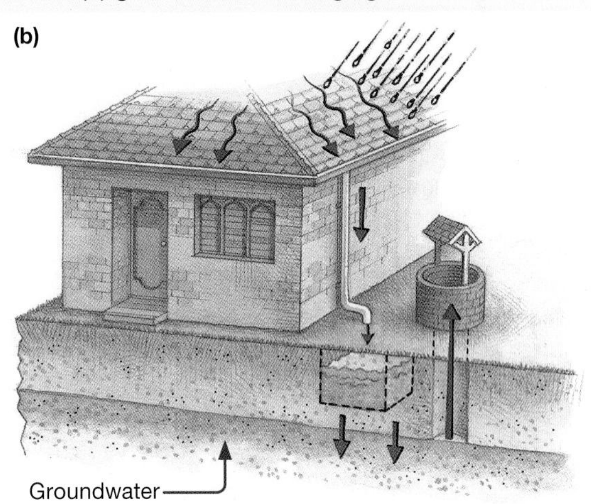

FIGURE 4 Water collected from a stream uphill rushes down the pipe and drives a small turbine in the hut to generate electricity for a local community in the Philippines.

4.4.3 Traditional water harvesting methods

CASE STUDY

Rajasthan, India

The state of Rajasthan is located in the arid north-west of India (see figure 5). The region has only 1 per cent of the country's surface water and a population growth rate of 21 per cent (compared to Australia's 1.5 per cent). The largest state in India faces both water scarcity and frequent droughts. Continual pumping of groundwater has seen underground water supplies dropping.

Traditionally, forests, grasslands and animals were considered property to be shared by all, and were carefully managed by a strict set of rules by local communities. These resources were used sustainably to ensure continual regeneration of plants and trees to enable farming to continue each year. By the mid twentieth century, government initiatives had taken control of local resources and promoted excessive mining and logging in the area. Large-scale deforestation resulted in severe land degradation, which increased the frequency of flash floods and droughts. There was little motivation for villages to maintain traditional water systems, or johads, and so there was a gradual decline in people's economic and social wellbeing.

FIGURE 5 Distribution of rainfall in India. The state of Rajasthan is highlighted.

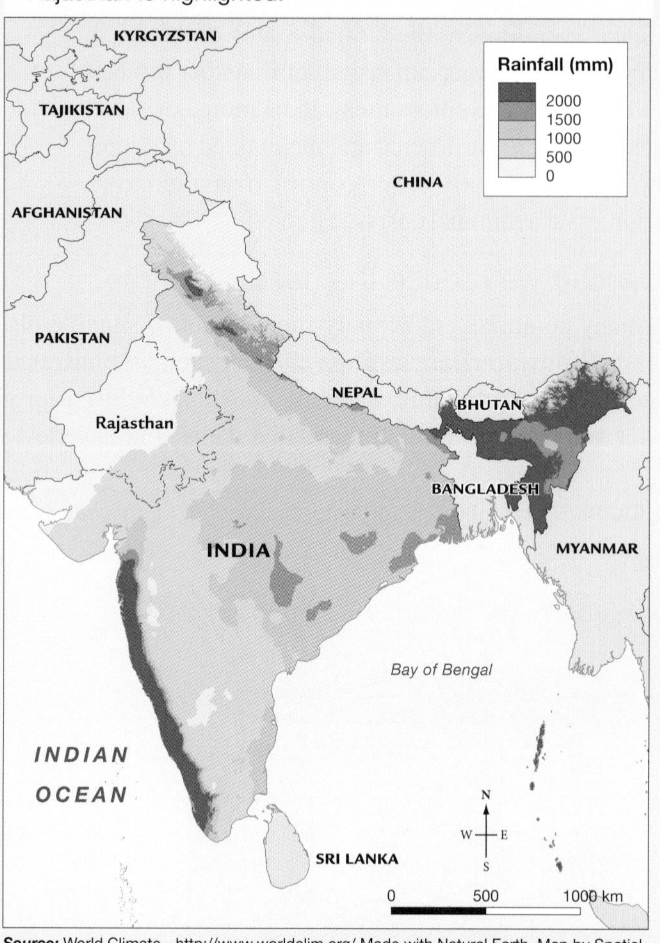

Source: World Climate - http://www.worldclim.org/ Made with Natural Earth. Map by Spatial Vision.

Tarun Bharat Sangh (TBS) is an aid agency that was established in the mid-1980s. It set about trying to re-establish traditional water management practices. It focused its attention on the construction and repair of nearly 10 000 johads in over 1000 villages. Johads are often small, dirt embankments that collect rainwater and allow it to soak into the soil and recharge groundwater **aquifers** (see figure 6).

Another johad design features small concrete dams across gullies that would seasonally flood, trapping the water and allowing it to infiltrate. Water, stored in aquifers, can later be withdrawn when needed via wells. The benefits have been remarkable and the estimated cost calculated to be an average of US$2 or 100 rupees per person. This is compared to over 10 000 rupees per head for water supplied from the Narmada River Dam Project.

FIGURE 6 A johad or traditional small water harvesting dam in India

What have been the benefits?

Environmental benefits

- Groundwater has risen by six metres.
- Five rivers which flowed only after the monsoon season now flow all year (fed by **base flow**).
- Revegetation schemes have increased forest cover by 38 per cent, which helps improve the soil's ability to hold water and reduce evaporation and erosion.

Social benefits

- More than 700 000 people across Rajasthan have benefited from improved access to water for household and farming use.
- There has been a revival of traditional cultural practices in constructing and maintaining johads.
- The role of the village council (Gram Sabha) is promoted for encouraging community participation and social justice.
- With a more reliable water supply communities became more economically viable.

4.4 Activities

To answer questions online and to receive **immediate feedback** and **sample responses** for every question, go to your learnON title at www.jacplus.com.au. *Note*: Question numbers may vary slightly.

Remember

1. (a) Why has water flowing out to sea been considered a waste?
   (b) Is this a human-centred or earth-centred viewpoint?
2. (a) Refer to figure 1. Which countries in the world have the most number of people affected by large dams? Suggest a reason why.
   (b) In what ways would they be affected?
   (c) Would people and ***environments*** upstream of large dams be affected by the dams? Give reasons for your answer.
3. Where are the world's largest (over 100 km^3) dams?

Explain

4. Suggest reasons why large-scale dam projects were seen as indicators of development and progress in countries.
5. Suggest reasons that make a place suitable for a large dam. Consider landforms, climate, soil and rock type.
6. Study the information in figure 5. Explain why Rajasthan has water issues. Use data in your answer.
7. Some ***places*** in India can receive up to 2500 mm of rainfall per year, but this can all fall in 100 hours. Suggest possible repercussions of this for local communities.
8. Have small-scale water management schemes in Rajasthan been successful? Why or why not?

Discover

9. (a) Investigate the different methods of irrigating crops, such as flood, furrow and drip irrigation. What are the advantages and disadvantages of each in terms of water use and waste?
   (b) Which irrigation method would:
      (i) be the most economically viable
      (ii) have the most ***environmental*** benefit?
10. (a) Use the **Water harvest** weblink in the Resources tab.
    (b) Select two locations and describe the different traditional water harvesting schemes in use.
    (c) Compare your two locations and schemes with the johad scheme in Rajasthan, and evaluate how effective they are in terms of economic viability (how affordable they are to construct and manage), ***environmental*** benefit (if they are using the water resources ***sustainably***) and social justice (if the schemes are fair for all people, and if the community benefits).

RESOURCES — ONLINE ONLY

Explore more with this weblink: Water harvest

4.5 SkillBuilder: Creating a fishbone diagram

online only

WHAT IS A FISHBONE DIAGRAM?

A fishbone diagram is a graphic representation of the causes of a particular effect. Fishbone diagrams can also detail the positive and negative impacts of an action or event.

Go online to access:

- a clear step-by-step explanation to help you master the skill
- a model of what you are aiming for
- a checklist of key aspects of the skill
- a series of questions to help you apply the skill and to check your understanding.

FIGURE 1 Fishbone diagram that examines the causes of damage to ecosystems in rural areas

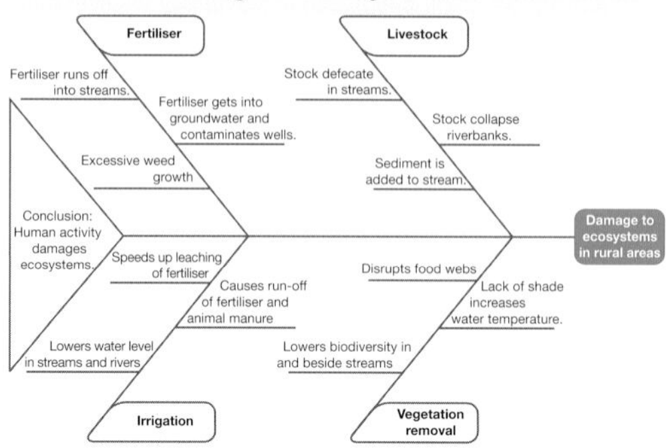

learnON RESOURCES — ONLINE ONLY

Watch this eLesson: Creating a fishbone diagram (eles-1748)

Try out this interactivity: Creating a fishbone diagram (int-3366)

4.6 Is fighting worth a dam?

Access this subtopic at **www.jacplus.com.au**

4.7 What happens when we divert water?

4.7.1 Why is water diverted?

Many of the world's greatest lakes are shrinking, and large rivers such as the Colorado, Rio Grande, Indus, Ganges, Nile and Murray discharge very little water into the sea for months and even years at a time. Up to one-third of the world's major rivers and lakes are drying up, and the groundwater wells for 3 billion people are being affected. The overuse and diversion of water is largely to blame.

Due to the uneven distribution of water and population there is often the need to transfer water, and large-scale **diversions** often require piping or pumping water from one drainage basin to another. For example, water from the Snowy River is diverted into the Murray and Murrumbidgee rivers. Diverting water can alleviate water shortages and allows for the development of irrigation and the production of hydro-electricity. Diversions, however, are not always the most sustainable use of water resources.

4.7.2 A dying Lake Urmia in Iran

The largest lake in the Middle East and one of the largest salt lakes in the world is drying up. Since the 1970s, Lake Urmia in northern Iran has shrunk by nearly 90 per cent, exposing extensive areas of salt flats (see figure 1 (a) and (b)).

FIGURE 1 Lake Urmia (a) in 1998 and (b) in 2011

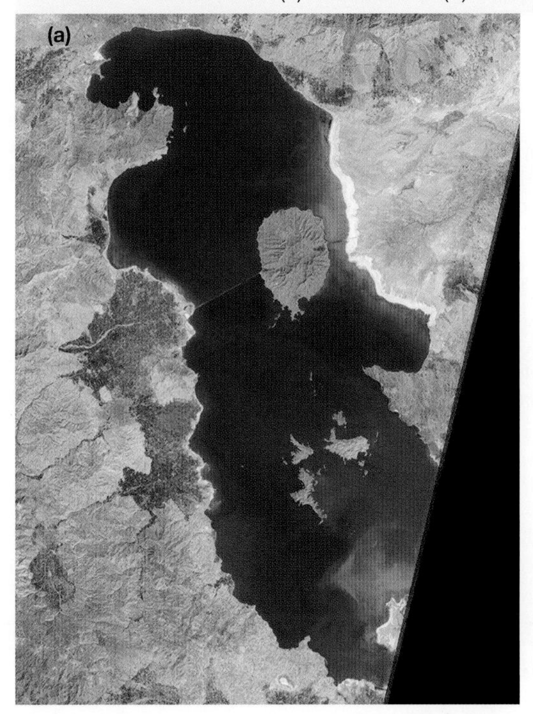

The lake was declared a Wetland of International Importance by the Ramsar Convention in 1971, and a UNESCO Biosphere Reserve in 1976. The lake and its surrounding wetlands serve as a seasonal habitat and feeding ground for migratory birds that feed on the lake's shrimp. This shrimp is the only thing, other than plankton, that can live in the salty water.

Lake Urmia is a **terminal lake**: the rivers, some permanent and some ephemeral, that flow into the lake bring naturally occurring salts. Because of the arid climate, high evaporation causes salt crystals to build up around the shoreline. Figure 2 shows the declining surface area of Lake Urmia.

FIGURE 2 Declining surface area of Lake Urmia

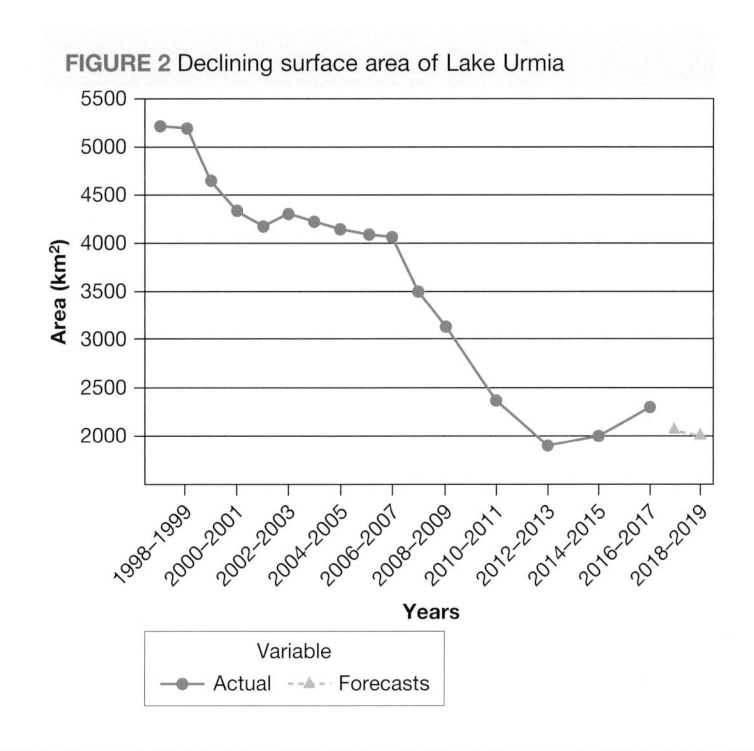

Why is the lake drying up?

A combination of environmental, economic and social factors has been blamed for the large-scale changes in Lake Urmia. Prolonged drought and the illegal withdrawal of water by farmers who do not pay or who take more than their allocation are minor contributors to the problem. Recent research has argued that the scale of shrinkage cannot be explained by fluctuating rainfall; rather, it is directly correlated to water diversions and the increased demand for water in the region (see figure 3). The end result is a form of 'socioeconomic drought' — a man-made drought caused when the demand for water is greater than the available supply.

Impacts of this man-made drought include:

- increased salinity of the shallow lake due to high evaporation and reduced fresh water flowing in via rivers (salt levels have increased from 160 g/litre to 330 g/litre)
- collapse of the lake's ecosystem and food chain (salt levels over 320 g/litre are fatal to the shrimp which form the basis of its food chain)
- loss of habitat as surrounding wetlands dry up, which then reduces tourism to view wetland wildlife
- over 400 km^2 of exposed lakebed around its shores is nothing but salty deserts, unable to support native vegetation or food crops
- salt storms occur as wind blows salt and dust from the exposed, dry lakebed. The storms damage crops and are also a potential health hazard for people.
- less water is available for food production.

FIGURE 3 Reasons for Lake Urmia's decline

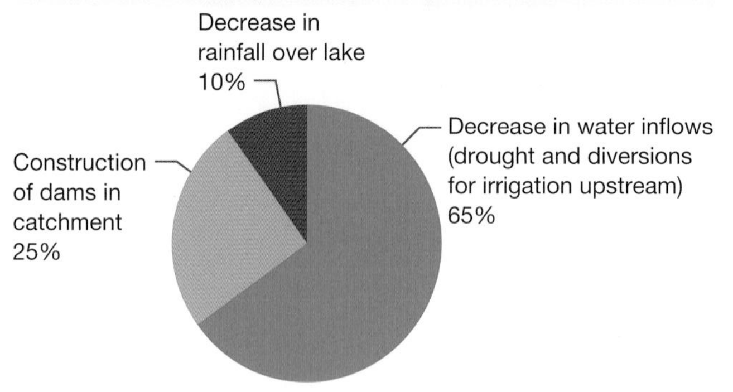

Note: Average rainfall is 235 mm. The last decade has seen this decrease by 40 mm.

FIGURE 4 Distribution of dams, existing and under construction, in the lake's catchment area. This level of diversions is unsustainable.

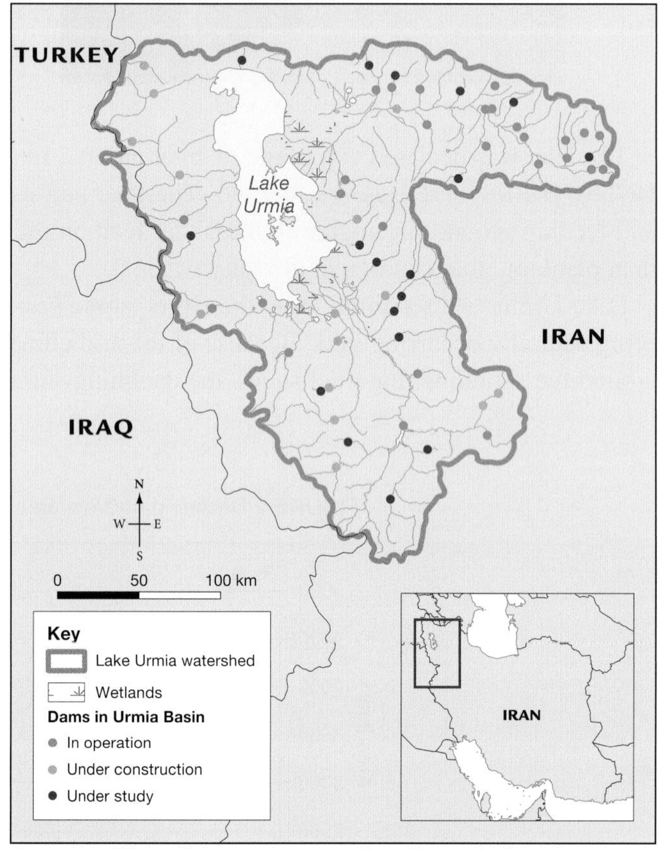

Source: United Nations Environment Programme. Vector Map Level 0 Digital Chart of the World.

Possible actions

Essentially, more water is required to flow into the lake to increase the water level and dilute the salt. This water must come from either reducing water allocated to irrigation and/or transferring water into the basin from the Zab or Aras rivers or even the Caspian Sea, over 300 kilometres away. This would require the cooperation of other countries and the scale of the project would be very expensive. The Iranian Government has pledged $5 billion over 10 years to help revive the lake. Will it be enough to save it?

4.7 Activities

To answer questions online and to receive **immediate feedback** and **sample responses** for every question, go to your learnON title at www.jacplus.com.au. *Note*: Question numbers may vary slightly.

Discover

Conduct some online research and investigate the decline of either Lake Chad in Africa, Owens Lake in the United States or the Aral Sea in Kazakhstan. Use the following points as a guide to writing an investigative report:

- location
- original size and appearance of the lake/sea
- original uses of the lake/sea and surrounding area
- causes and rate of decline
- ***changes*** that have taken place
- impacts on people and the ***environment***
- possible solutions.
  Include annotated images, maps and data where possible.

4.8 Why is groundwater shrinking?

4.8.1 What is groundwater?

Of all the fresh water in the world not locked up in ice sheets and glaciers, less than 1 per cent is available for human use and most of that is groundwater. Groundwater is used by more than two billion people, making it the single most used natural resource in the world. It is also the most reliable of all water sources. Fresh water stored deep underground is essential for life on Earth.

Groundwater is one of the invisible parts of the water cycle as it lies beneath our feet. Rainfall that does not run off the surface or fill rivers, lakes and oceans will gradually seep into the ground. Figure 1 shows where groundwater is stored in porous rock layers called aquifers. Water is able to move through these aquifers and can be stored for thousands of years. Unlike most other natural resources, groundwater is found everywhere throughout the world.

FIGURE 1 Diagram showing groundwater

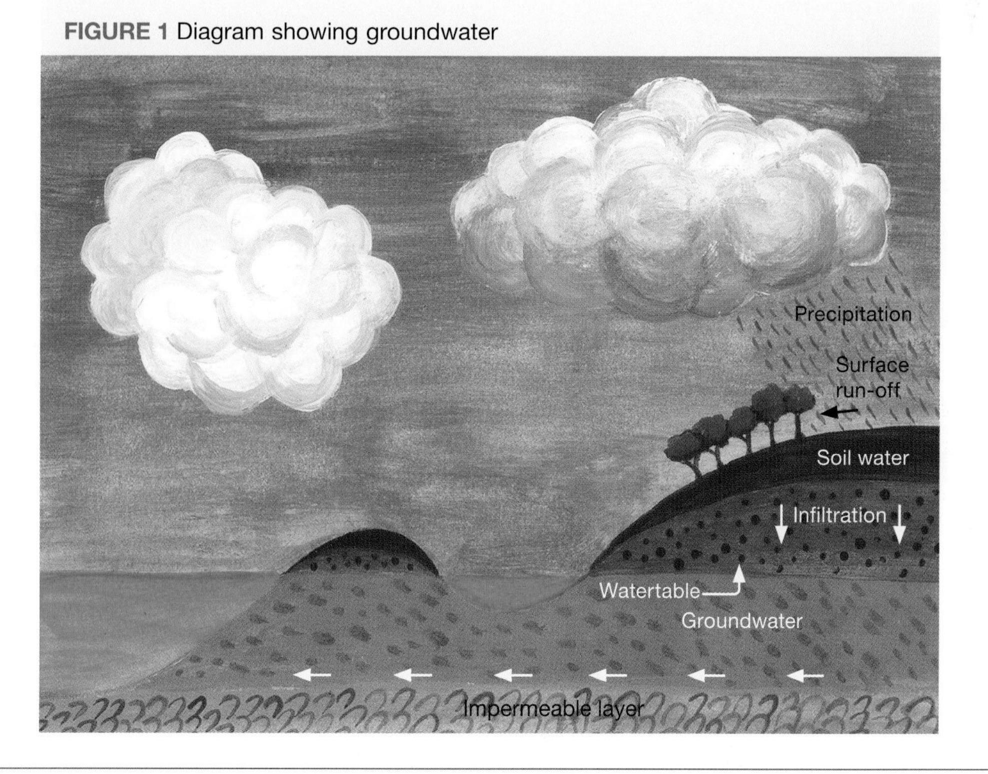

4.8.2 What are the advantages of using groundwater?

Since the mid-twentieth century, advances in drilling and pumping technology have provided people with an alternative to surface water for meeting increasing water demands. Groundwater has many advantages:

- It can be cleaner than surface water.
- It is less subject to seasonal variation and there is less waste through evaporation.
- It requires less and cheaper infrastructure for pumping as opposed to dam construction.
- It has enabled large-scale irrigated farming to take place.
- In arid and semi-arid places groundwater has become a more reliable water supply, which has led to improved water and food security.

If groundwater is removed unsustainably, that is, at a rate that is greater than is being replenished naturally by rainfall, run-off or underground flow, then **watertables** drop and it becomes harder and more expensive to pump. In areas of low rainfall there is very little **recharge** of groundwater so it may take thousands of years to replace. Over-extraction of groundwater can result in wells running dry, less water seeping into rivers and even land **subsidence** or sinking. Figure 2 identifies those places in the world most at risk of groundwater depletion. Many of these are important food bowls for the world.

FIGURE 2 The world's use of groundwater

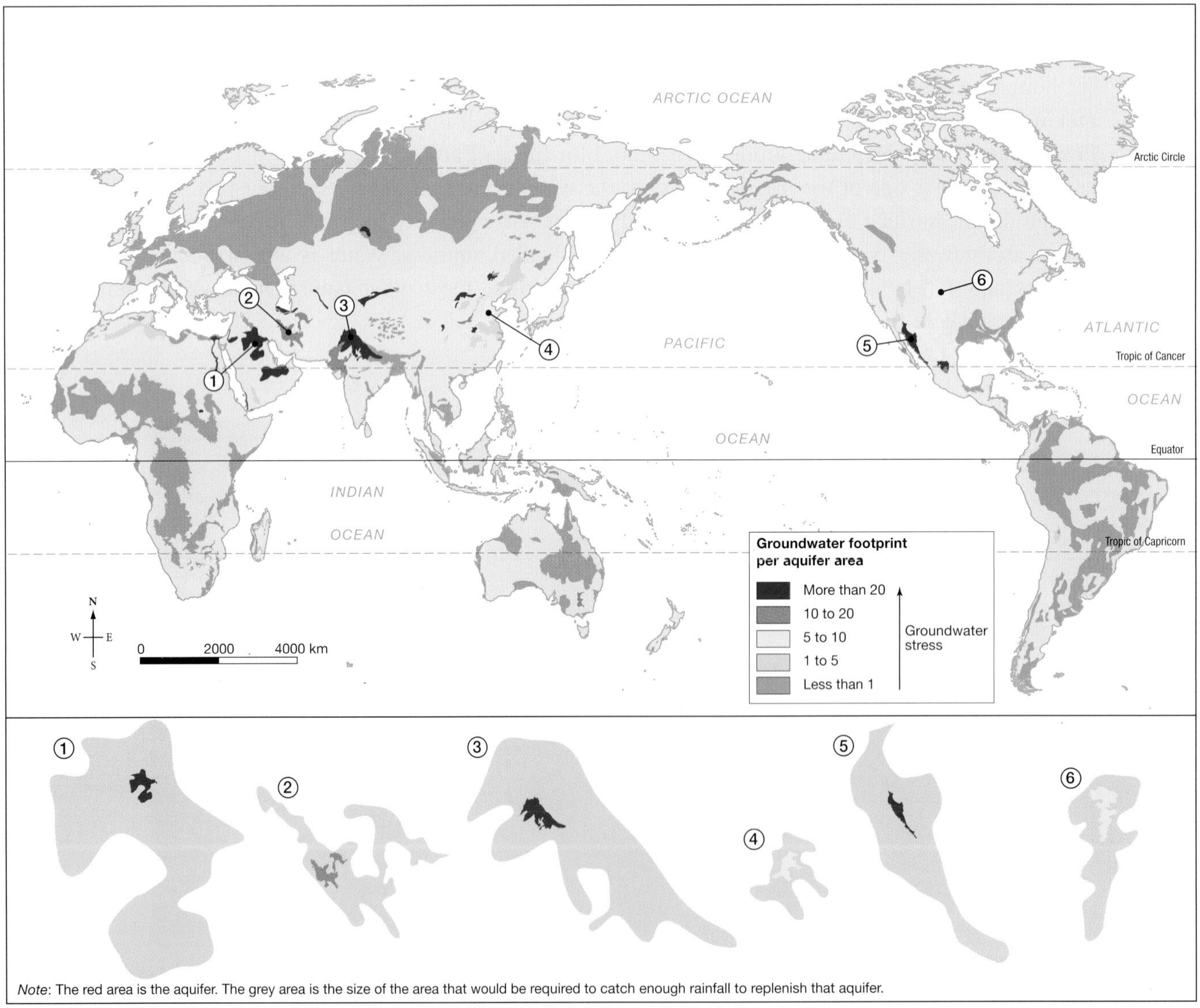

Note: The red area is the aquifer. The grey area is the size of the area that would be required to catch enough rainfall to replenish that aquifer.

Source: BGR & UNESCO 2008: Groundwater Resources of the World 1 : 25 000 000. Hannover, Paris. Map by Spatial Vision

4.8.3 Can we improve our use of groundwater?

In the past we had limited knowledge of the interconnection between groundwater and surface water. As agriculture is the biggest user of groundwater, any improved efficiencies in water use can reduce the demand for pumping more water. Improved irrigation methods and the re-use of treated effluent water are all methods that could reduce our unsustainable use of groundwater. Many countries share aquifers so pumping in one place can affect water supplies in another. There is a need for more international cooperation and management of the aquifer as a single shared resource.

4.8 Activities

To answer questions online and to receive **immediate feedback** and **sample responses** for every question, go to your learnON title at www.jacplus.com.au. *Note*: Question numbers may vary slightly.

Remember

1. Refer to the **That sinking feeling** interactivity (int-3293) and figure 1.
   (a) What is the difference between groundwater and the watertable?
   (b) Describe how water can move vertically and horizontally through the ground.
   (c) What is the ***interconnection*** between atmospheric, surface and groundwater?
2. Refer to figure 2. Describe, with the use of an atlas, the location of ***places*** in the world that have the highest groundwater stress.

Explain

3. What are the advantages and disadvantages of using groundwater for domestic and agricultural purposes?
4. Looking at figure 2, explain the ***scale*** of the area needed to replenish the most stressed aquifers.

Predict

5. Using an atlas, find a map of world food production and compare this with any three ***places*** from figure 2.
   (a) What types of food are produced in those regions of the world where watertables are severely depleted?
   (b) What are the future implications for ***sustainable*** food production in these regions?

Think

6. Who owns groundwater? How can we manage the resource ***sustainably***? Write a paragraph expressing your viewpoint.

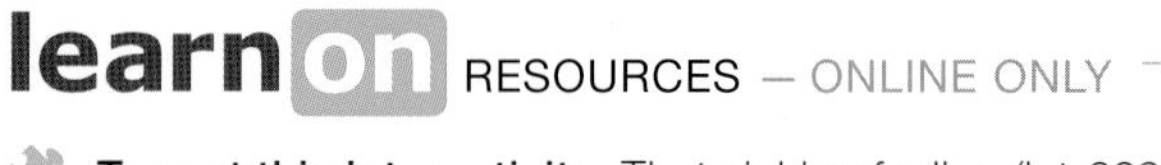

Try out this interactivity: That sinking feeling (int-3293)

4.9 Why is China drying up?

Access this subtopic at **www.jacplus.com.au**

4.10 Why do we drain wetlands?

4.10.1 Why are wetlands important?

Often referred to as the area where 'earth and water meet', **wetlands** are one of the most important and valuable biomes in the world.

What are the threats to wetlands?

FIGURE 1 A wetland in Queensland. What features in this image would be typical for a wetland?

- Dams alter seasonal floods and block supply of sediment and nutrients onto the floodplain and deltas. Often little water and sediment reaches the mouths and deltas of large rivers.
- Agricultural expansion is the largest contributor to wetland loss and degradation globally. Farming often requires the draining of wetlands to create more land. Biodiversity is reduced and water run-off from agriculture is often polluted with fertilisers and pesticides. Increased pumping from aquifers depletes groundwater resources.
- Loss of wetlands affects populations and the migratory patterns of birds and fish. The introduction of invasive species results in changed ecosystems and loss of biodiversity. Seventy per cent of amphibian species are affected by habitat loss.
- Clearing for urban growth, industry, roads and other land uses replaces wetlands with hard **impervious** surfaces, which reduces infiltration and leads to polluted run-off and increased impacts of flooding.
- While wetlands can naturally filter many pollutants, excessive amounts of fertilisers and sewage causes algal blooms and **eutrophication**, depriving aquatic plants and animals of light and oxygen.
- Climate change is expected to increase the rate of wetland degradation and loss.

4.10 Activities

To answer questions online and to receive **immediate feedback** and **sample responses** for every question, go to your learnON title at www.jacplus.com.au. *Note*: Question numbers may vary slightly.

Mapping Activity

1. Use the **Wetlands** weblink in the Resources tab to describe the importance of wetlands.
2. Study figure 2, which shows the Murray River west of Wentworth.
   (a) Identify and give the grid references for two different types of wetlands on this map.
   (b) What is the average elevation of this region?
   (c) Suggest two reasons why this ***place*** is subject to inundation (flooding).
   (d) Use evidence from the map to make a list of the ***changes*** that people have brought to this region.
   (e) If a dam was to be built several kilometres upstream, suggest possible ***changes*** that might occur to the wetlands in the map area.
   (f) Name the features at the following locations and suggest how they are used to manage water.
      (i) 551165
      (ii) 555159
   (g) Locate the state border between NSW and Victoria. Who 'owns' the Murray River?

learnON RESOURCES — ONLINE ONLY

Topographic map of Wentworth, New South Wales (doc-11569)

Try out this interactivity: Wetland wonderlands (int-3294)

Explore more with this weblink: Wetlands

FIGURE 2 Topographic map of Wentworth, New South Wales

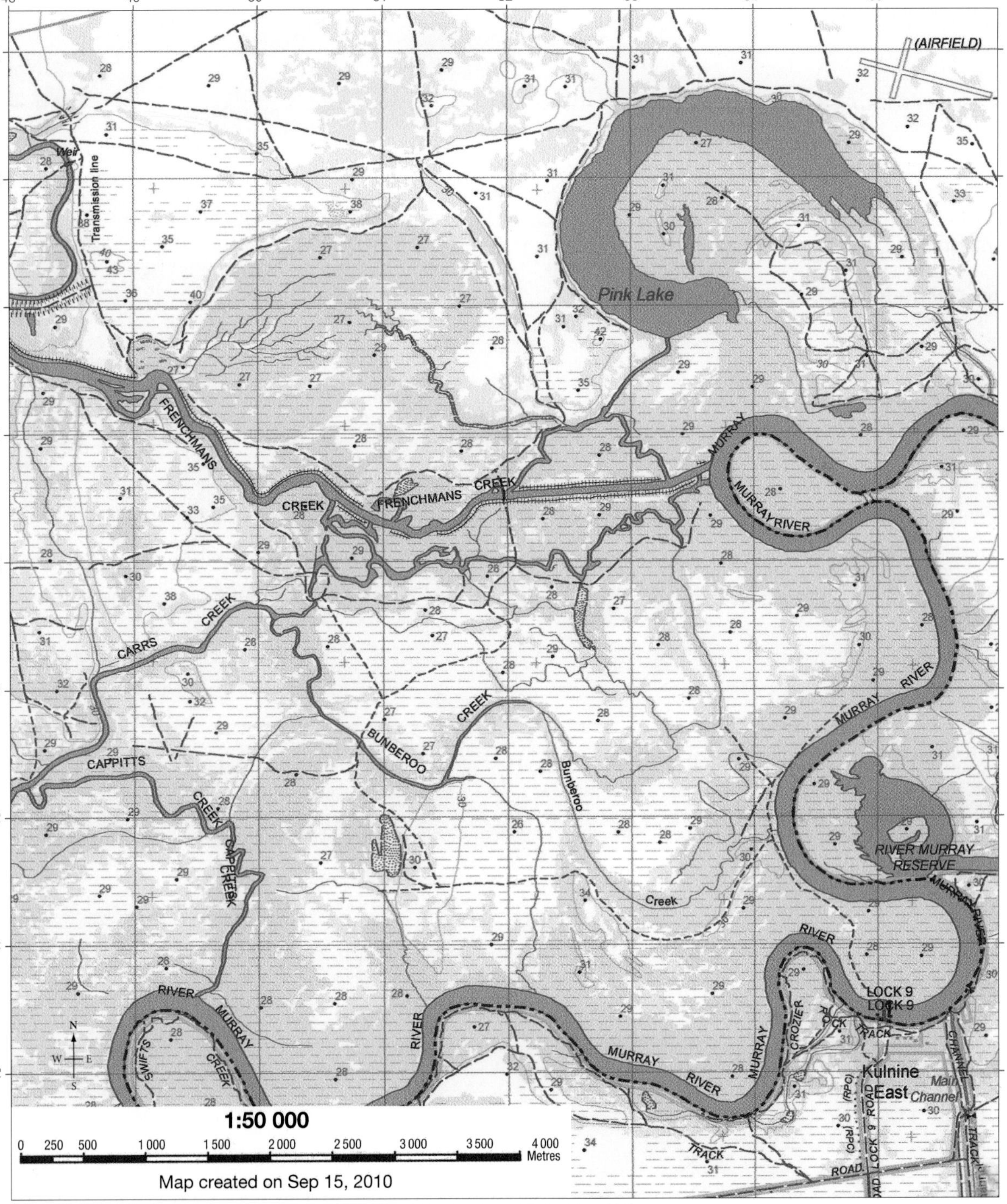
48
49
50
51
52
53
54
55
24
23
22
21
20
19
18
17
16
(AIRFIELD)
Weir
Transmission line
Pink Lake
FRENCHMANS
CREEK
FRENCHMANS
CREEK
MURRAY
MURRAY RIVER
CARRS
CREEK
CAPPITTS
CREEK
CAPPITTS
CREEK
BUNBEROO
CREEK
Bunberoo
Creek
RIVER MURRAY
RESERVE
MURRAY
RIVER
RIVER
RIVER
MURRAY
MURRAY
RIVER
CREEK
SWIFTS
MURRAY
RIVER
CROZIER
LOCK 9
LOCK 9
TRACK
Kulnine
East
Main
Channel
CHANNEL
LOCK 9 ROAD
(RPC)
ROAD
TRACK
N
W
E
S
1:50 000
0 250 500 1 000 1 500 2 000 2 500 3 000 3 500 4 000
Metres
Map created on Sep 15, 2010

Source: Vicmap Topographic © The State of Victoria, Department of Sustainability and Environment, 2010

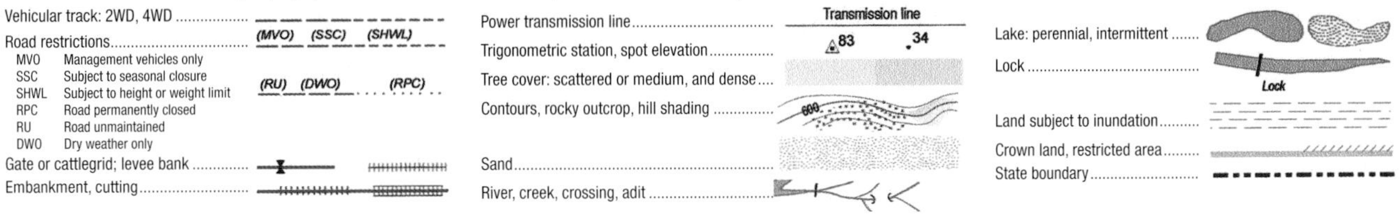
Vehicular track: 2WD, 4WD
Road restrictions
(MVO) (SSC) (SHWL)
MVO Management vehicles only
SSC Subject to seasonal closure
SHWL Subject to height or weight limit
RPC Road permanently closed
RU Road unmaintained
DWO Dry weather only
(RU) (DWO) (RPC)
Gate or cattlegrid; levee bank
Embankment, cutting
Power transmission line
Transmission line
Trigonometric station, spot elevation
83
34
Tree cover: scattered or medium, and dense
Contours, rocky outcrop, hill shading
Sand
River, creek, crossing, adit
Lake: perennial, intermittent
Lock
Lock
Land subject to inundation
Crown land, restricted area
State boundary

4.11 SkillBuilder: Reading topographic maps at an advanced level

WHAT IS READING A TOPOGRAPHIC MAP AT AN ADVANCED LEVEL?

Topographic maps are more than just contour maps showing the height and shape of the land. Reading this information requires more advanced skills such as calculating local relief, gradients and the size of various areas.

Go online to access:

- a clear step-by-step explanation to help you master the skill
- a model of what you are aiming for
- a checklist of key aspects of the skill
- a series of questions to help you apply the skill and to check your understanding.

FIGURE 1 Topographic map of Berri, South Australia

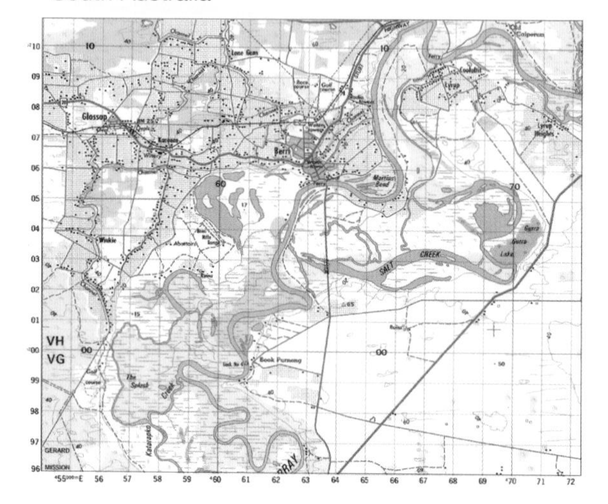

RESOURCES — ONLINE ONLY

Watch this eLesson: Reading topographic maps at an advanced level (eles-1749)

Try out this interactivity: Reading topographic maps at an advanced level (int-3367)

4.12 How can we put water back?

Access this subtopic at **www.jacplus.com.au**

4.13 Review

online only

4.13.1 Review

The Review section contains a range of different questions and activities to help you revise and recall what you have learned, especially prior to a topic test.

4.13.2 Reflect

The Reflect section provides you with an opportunity to apply and extend your learning.

Access this subtopic at **www.jacplus.com.au**

TOPIC 5 Managing change in coastal environments

5.1 Overview

Numerous **videos** and **interactivities** are embedded just where you need them, at the point of learning, in your learnON title at www.jacplus.com.au. They will help you to learn the content and concepts covered in this topic.

5.1.1 Introduction

The coast is home to 80 per cent of the world's population, and it is a popular place to settle for reasons of climate, water resources, land for agriculture and industry, access to transportation systems, and recreation. Hence, it is essential to understand the changes that are occurring to coastal environments, and how they will affect human settlements. The changes are both natural and human-induced. They are sometimes short term (as a result of storms and tsunamis) and sometimes long term (climate change leading to rising sea levels). To cope with these changes, careful planning and management is needed to ensure a sustainable future for human activity at the coast.

Houses along Malibu Beach in California are regularly threatened by severe storms. Is housing the most suitable land use for this area?

Starter questions

1. How do people use coastal ***places***?
2. What ***changes*** have people brought to the coastal area in the image on these pages?
3. What could you do here to help reduce this problem of people versus nature?
4. What ***changes*** to coastal areas have you observed when visiting a beach?

INQUIRY SEQUENCE

5.2 Why must we preserve the coast?

5.2.1 The importance of the coast

Coasts are a dynamic natural system. The forces of nature are constantly at work, either creating new land or wearing it away. Nine out of 10 Australians live within 50 kilometres of the coast. As well as being a favoured place to live, the coast is the most popular destination for tourists and visitors.

All forms of human activities can have impacts on coastal landforms and the **ecosystems** of plant and animal life. Australia's coasts need to be managed to achieve goals of sustainable living for all who share this common environment. In addition, there is a need to balance the diverse viewpoints of human-induced development with conservation principles.

5.2.2 The coastal zone

The coastal zone may be defined as the zone where the land meets the sea (see figure 1). Generally speaking, it includes an area called coastal waters. This area includes the zone between high and low tide and an area of land called the **hinterland**. The Australian coast, which is approximately 37 000 kilometres in length, consists of many different environments such as plains, rivers and lakes, rainforests, wetlands, mangrove areas, estuaries, beaches, coral reefs, seagrass beds and all forms of sea life found on the adjoining continental shelf.

In these varied coastal environments many of Australia's World Heritage sites are found, such as the Great Barrier Reef (see figure 2), Lord Howe Island, Fraser Island and Shark Bay. The coast is also important for human settlement: urban complexes, ports and harbours. Many Aboriginal and Torres Strait Islander people lived and continue to live in coastal communities. Historical evidence of middens, art sites, fish traps, stone and ochre quarries, and burial and religious sites show the long history of occupation of the Australian coast. Therefore, we need to manage the coast in a much more sustainable way.

FIGURE 1 Typical Australian coastal scenario, Point Danger, Tweed Heads, New South Wales

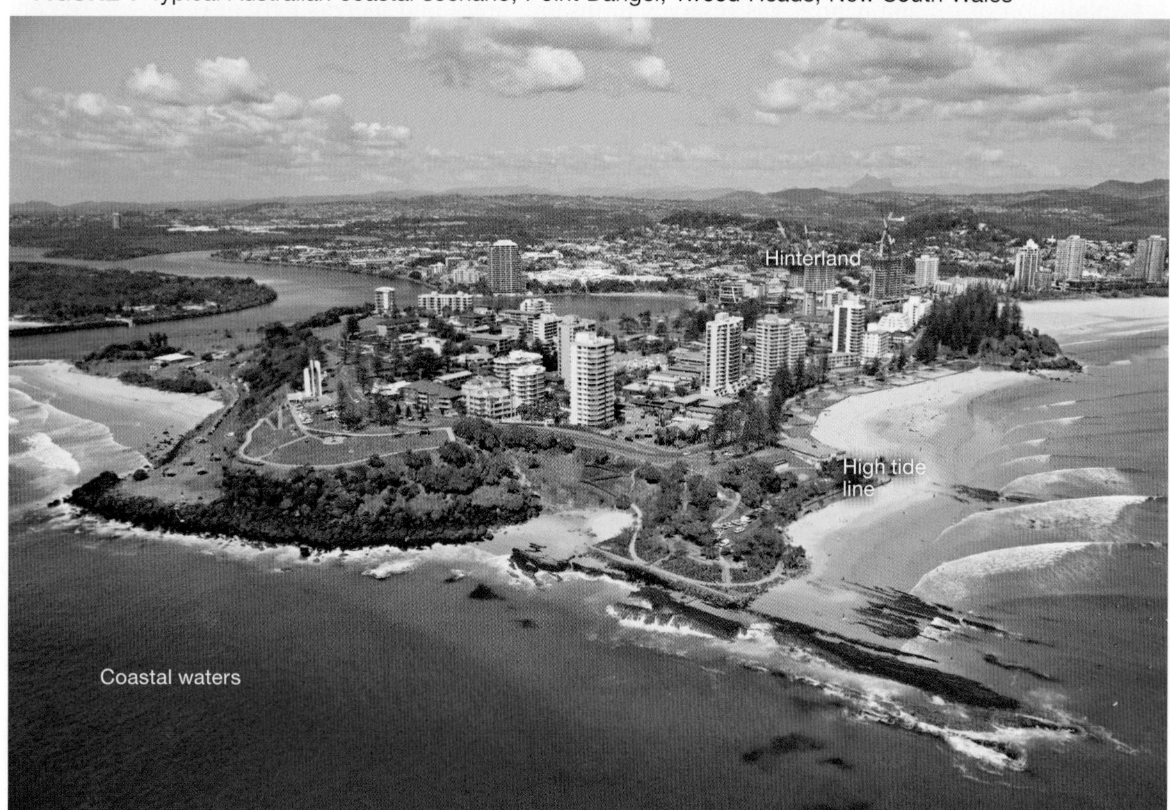

5.2.3 What types of human activities affect coasts?

Human impacts on coastlines include the construction of ports, boat marinas and sea walls; changes in land use (for example, from a natural environment to agricultural or urban environments); and the disposal of waste from coastal and other settlements.

The 2015 *World Ocean Review* indicates that for oceans and coasts to be sustainably managed into the future new environmental policies must be implemented. The issues identified in the review that need to be addressed include:

FIGURE 2 The Great Barrier Reef was placed on the World Heritage List in 1981.

- **Marine pollution**
  - Toxic substances and heavy metals from industrial plants (liquid effluent and gaseous emissions)
  - Nutrients, in particular phosphate and nitrogen, from agricultural sources and untreated wastewater (eutrophication of coastal waters)
  - Ocean noise pollution from shipping and from growing offshore industry (exploitation of oil and natural gas reserves, construction of wind turbines, future mineral extraction)

- **Growing demand for resources**
  - Exploitation of oil and natural gas reserves in inshore areas and increasingly also in deep-sea areas, resulting in smaller or greater amounts of oil being released into the sea
  - Sand, gravel and rock for construction purposes
  - For the development of new pharmaceuticals: extraction of genetic resources from marine life such as bacteria, sponges and other life forms, the removal of which may result in damage to sea floor habitats
  - Future ocean mining (ore mining at the sea floor) which may damage deep sea habitats
  - Aquaculture (release of nutrients, pharmaceuticals and pathogens)
- **Overfishing**
  - Industrial-scale fishing and overexploitation of fish stocks; illegal fishing
- **Habitat destruction**
  - Building projects such as port extensions or hotels
  - Clear-felling of mangrove forests
  - Destruction of coral reefs as a result of fishing or tourism
- **Bioinvasion**
  - Inward movement of non-indigenous species as a result of shipping transport or shellfish farming; changes in characteristic habitats
- **Climate change**
  - Ocean warming
  - Sea-level rise
  - Ocean acidification.

Source: World Ocean Review bronze globe sculpture 2015

The greatest threat to coasts today is rising sea levels. It is recognised that global warming is a result of **enhanced greenhouse gas emissions**, which is human-induced. This is leading to the melting of polar ice caps and glaciers. Some of the changes to coastal environments that will result due to global warming include:

- increases in intensity and frequency of storm surges and coastal flooding
- increased salinity of rivers and groundwaters resulting from salt intrusion
- increased coastal erosion
- inundation of low-lying coastal communities and critical infrastructure
- loss of important mangroves and other wetlands
- impacts on marine ecosystems such as coral reefs.

FIGURE 3 To bring attention to issues of global warming, a meeting was held on the sea floor by government representatives of the Republic of the Maldives.

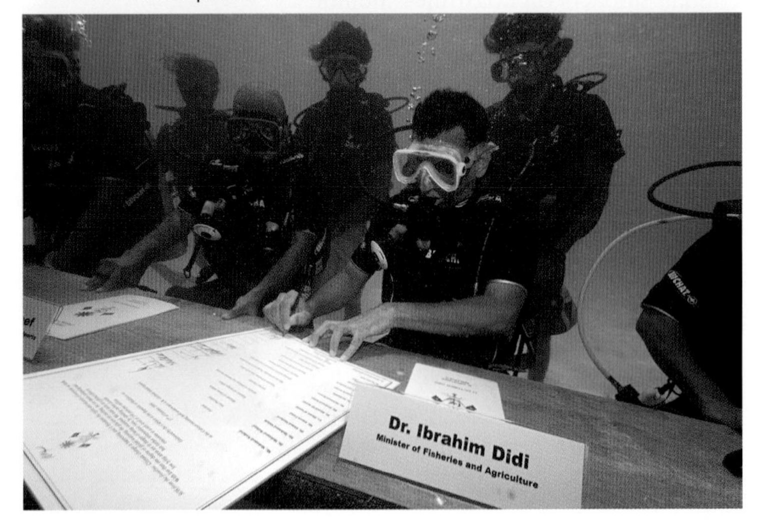

Coastal environments have not always been managed sustainably. In the past, decision-makers had limited knowledge about the fragile nature of many coastal ecosystems, and they had limited environmental worldviews about the use of coastal areas. Their aim was to develop coastal areas for short-term economic gains. This was based on the belief that nature's resources were limitless. Building high-rise apartments and tourist resorts on sand dunes seemed a good idea — until they fell into the sea when storms eroded the shoreline.

Over time, people have realised that coastal management requires an understanding of the processes that affect coastal environments. To manage the coast sustainably we need to understand:

- the coastal environment and the effect of physical processes
- the effect of human activities within the coastal zone
- the different perspectives of coastal users
- how to achieve a balance between conservation and development
- how decisions are made about the ways in which coasts will be used
- how to evaluate the success of individuals, groups and the levels of governments in managing coastal issues.

5.2 Activities

To answer questions online and to receive **immediate feedback** and **sample responses** for every question, go to your learnON title at www.jacplus.com.au. *Note*: Question numbers may vary slightly.

Remember

1. Why are coasts important to people?
2. What are some impacts that people have on coastal areas?
3. Refer to figure 1. What land use is found at Point Danger? How might this affect the dynamic nature of the coastal zone?

Explain

4. What is meant by a World Heritage site and why is it important to preserve the Australian sites identified in this section?
5. Select one of the impacts of rising sea levels on coasts identified in this section and explain why this would be a problem to a selected coastal settlement in Australia.
6. Explain how activities in coastal hinterlands can have an impact on the coast.

Discover

7. Find out more about the location of World Heritage sites in the coastal zone and show them on a map of the world.
8. The Gold Coast is a popular tourist destination in Queensland. Research past and ongoing developments along its coastal zone and display these in the form of a photographic essay.

Predict

9. Predict what might happen to the Great Barrier Reef if it wasn't listed on the World Heritage List.
10. Predict the consequences for the Great Barrier Reef with rising sea levels.

Think

11. Explain why ***environmental***, social and economic criteria must be applied to manage a coastal area such as in figure 1.

5.3 How are coasts built up and worn away?

Access this subtopic at **www.jacplus.com.au**

5.4 SkillBuilder: Comparing aerial photographs to investigate spatial change over time

online only

WHY IS IT USEFUL TO COMPARE AERIAL PHOTOGRAPHS?

Aerial photos are images taken above the Earth from an aircraft or satellite. Two images taken at different times, from the same angle, and placed side by side, show change that has occurred over time. Comparing aerial photographs is useful because each photograph captures details about a specific place at a particular time.

FIGURE 1 Lake Urmia (a) in 1988 and (b) in 2011

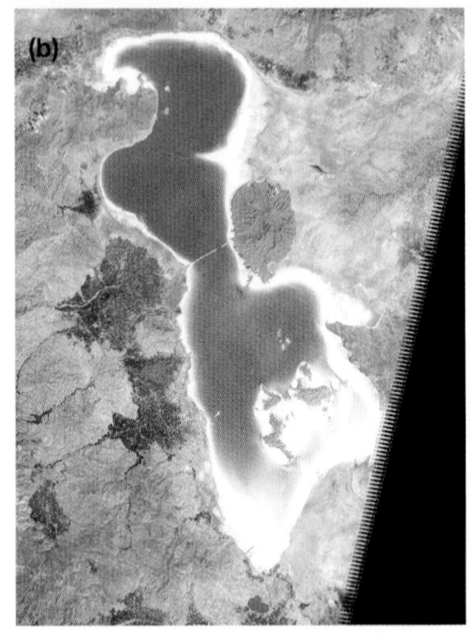

Go online to access:

- a clear step-by-step explanation to help you master the skill
- a model of what you are aiming for
- a checklist of key aspects of the skill
- a series of questions to help you apply the skill and to check your understanding.

learnon RESOURCES — ONLINE ONLY

Watch this eLesson: Comparing aerial photographs to investigate spatial change over time (eles-1750)

Try out this interactivity: Comparing aerial photographs to investigate spatial change over time (int-3368)

5.5 How do coastal areas change?

5.5.1 Introduction

The main pressures on many coastal systems relate to the development of towns and tourist facilities. Careful management can enable growth of urban areas while at the same time protecting the natural coastal features.

5.5.2 Merimbula, New South Wales

The 'Sapphire Coast' in south-east New South Wales is a popular tourist destination because of its array of beautiful beaches, stunning scenery and mild, sunny weather. Merimbula is a coastal resort town in this Sapphire Coast region. Similar to any other popular coastal location, it experiences natural changes as well as the pressures relating to development.

The natural landform features along this coastline include a series of headlands separated by bay head beaches. Merimbula Lake has formed from a slow and gradual build up of a sand barrier, leaving only a narrow channel for salt water to enter and fresh water to exit. The shallow and sheltered waters of the lake provide an ideal environment for oyster farming and recreation. Figure 1 shows the narrow entrance to Merimbula Lake.

FIGURE 1 Topographic map extract of Merimbula

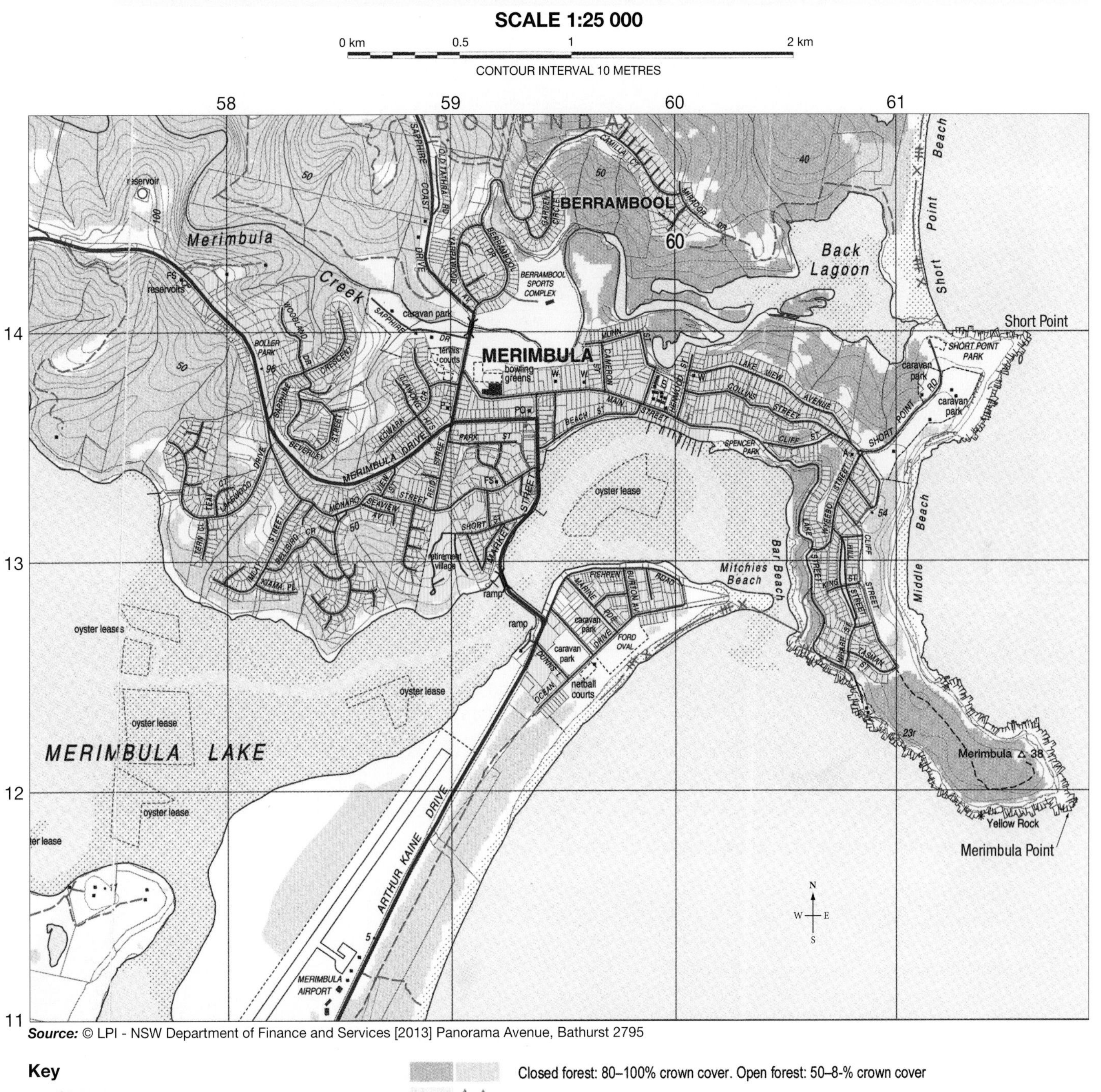
SCALE 1:25 000
0 km
0.5
1
2 km
CONTOUR INTERVAL 10 METRES
58
59
60
61
14
13
12
11
BOURNDA
BERRAMBOOL
Merimbula
Creek
reservoir
reservoirs
caravan park
BERRAMBOOL SPORTS COMPLEX
Back Lagoon
Short Point Beach
Short Point
SHORT POINT PARK
caravan park
MERIMBULA
bowling greens
tennis courts
BOLLER PARK
MERIMBULA DRIVE
ARTHUR KAINE DRIVE
MARKET STREET
MAIN STREET
BEACH ST
LAKE VIEW AVENUE
COLLINS STREET
CLIFF ST
SHORT POINT RD
SPENCER PARK
oyster lease
retirement village
ramp
caravan park
netball courts
FORD OVAL
Mitchies Beach
Bar Beach
Middle Beach
MERIMBULA LAKE
Merimbula △ 38
Yellow Rock
Merimbula Point
MERIMBULA AIRPORT
N
W
E
S

Source: © LPI - NSW Department of Finance and Services [2013] Panorama Avenue, Bathurst 2795

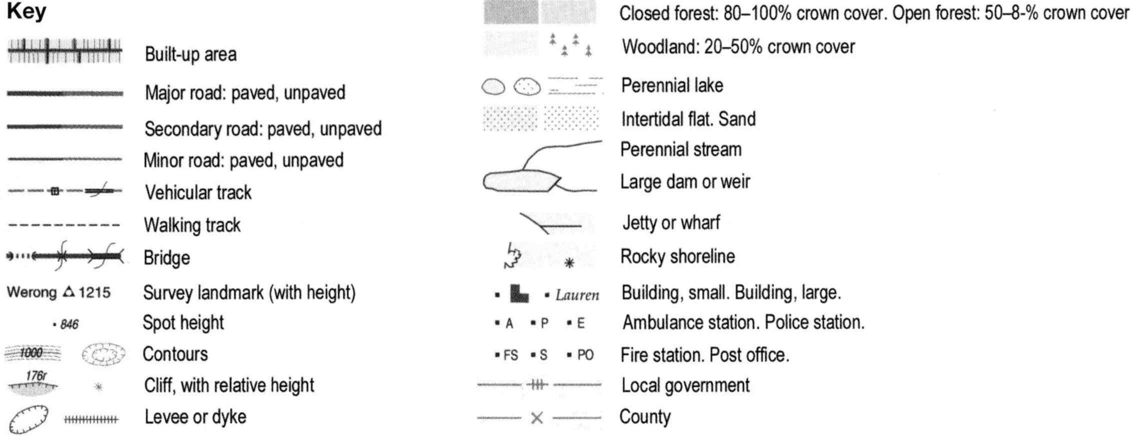
Key
Built-up area
Major road: paved, unpaved
Secondary road: paved, unpaved
Minor road: paved, unpaved
Vehicular track
Walking track
Bridge
Werong △ 1215
Survey landmark (with height)
• 846
Spot height
1000
Contours
176r
Cliff, with relative height
Levee or dyke
Closed forest: 80–100% crown cover. Open forest: 50–8-% crown cover
Woodland: 20–50% crown cover
Perennial lake
Intertidal flat. Sand
Perennial stream
Large dam or weir
Jetty or wharf
Rocky shoreline
• Lauren
Building, small. Building, large.
• A • P • E
Ambulance station. Police station.
• FS • S • PO
Fire station. Post office.
Local government
County

FIGURE 2 Aerial view over Merimbula Lake

5.5 Activities

To answer questions online and to receive **immediate feedback** and **sample responses** for every question, go to your learnON title at www.jacplus.com.au. *Note*: Question numbers may vary slightly.

Refer to the topographic map of Merimbula (figure 1) to answer these questions. You might also like to use Google Earth for a more recent image of the location.

Think

1. Make a tracing of the coastline. Show, shade and label the following natural features:
   - beaches
   - rocky areas
   - Merimbula Lake
   - Back Lagoon
   - Merimbula Creek.
2. Create an overlay map to show the distribution of built-up (urban) areas. Finish your map with BOLTSS.
3. (a) Look closely at the contour lines. What is the relationship between elevation and the built-up areas?
   (b) Approximately what percentage of built-up areas would be on land higher than 20 metres above sea level? Are there any exceptions to this rule? Where?
   (c) Suggest a reason for your observations.
4. Mark on your map the area(s) where you would expect wave action to be the most powerful. Include a symbol for this in your legend.
5. Give a reason why sand has built up to form a beach at Middle Beach and not at Merimbula Point.
6. Would you expect the water in Back Lagoon to be fresh or salty? Use evidence from the map.
7. In what ways have people ***changed*** this coastal ***environment***? List and describe how these ***changes*** might influence the natural processes along the coast.

Predict

8. If, in the future, the sea level was to rise by 10 metres, which of the following features would be safe from the rising sea? Why or why not?
   (a) The caravan park, located at GR613137
   (b) Merimbula Airport, located at GR585113

9. In what direction(s) is Merimbula likely to expand in the future? Justify your decision.
10. Imagine that a series of storms erodes the sand off Merimbula Beach. The local council then decides to build a series of groynes along the beach to trap sand that moves north in a longshore drift current.
    (a) On your map, mark in six groynes approximately 50 metres long and 500 metres apart. How might the beach change in appearance after the groynes are built?
    (b) Draw the new shape of the beach on your map using a black dotted line. Include this symbol in your legend.
    (c) Will sand continue to accumulate at the mouth of Merimbula Lake after the groynes are built? Explain your answer.

learnon RESOURCES — ONLINE ONLY

Topographic map extract of Merimbula, New South Wales (doc-11571)

Try out this interactivity: Predict changes around Merimbula (int-3296)

5.6 SkillBuilder: Comparing an aerial photograph and a topographic map

online only

WHAT COMPARISONS CAN BE MADE BETWEEN AERIAL PHOTOGRAPHS AND TOPOGRAPHIC MAPS?

Comparing an aerial photograph with a topographic map enables us to see what is happening in one place. Each format shows different information. A photograph can provide a clear impression of the activities taking place on the land or under the water at a particular moment in time. Topographic maps allow the cartographer to add information that cannot be identified from the air, such as place names and building names.

Go online to access:

- a clear step-by-step explanation to help you master the skill
- a model of what you are aiming for
- a checklist of key aspects of the skill
- a series of questions to help you apply the skill and to check your understanding.

FIGURE 1 Topographic map of Mossman and the Daintree River National Park, Queensland, 1982

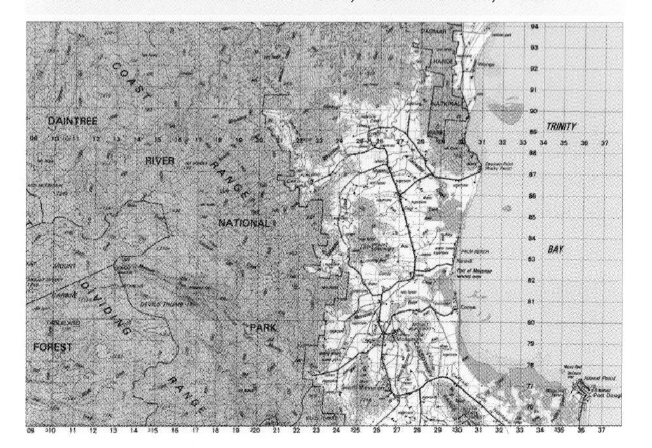

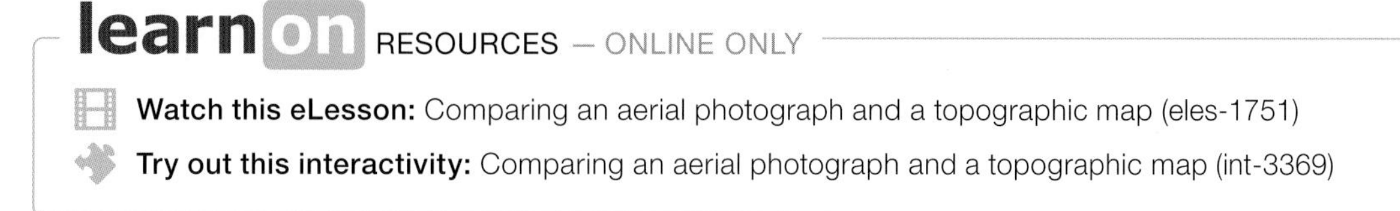

learnon RESOURCES — ONLINE ONLY

Watch this eLesson: Comparing an aerial photograph and a topographic map (eles-1751)

Try out this interactivity: Comparing an aerial photograph and a topographic map (int-3369)

5.7 Why are low-lying islands disappearing?

5.7.1 The impact of climate change

As a result of climate change, many low-lying islands will be flooded by the sea. Thermal expansion of oceans and melting ice leads to rising sea levels, threatening many coastal communities. Many island groups in the Pacific and Indian Oceans will be almost completely inundated by 2050.

Coastal storms, **tsunamis**, flooding, inundation, erosion, deposition and saltwater intrusion into freshwater supplies present a combined threat to coastal regions. With stronger windstorms possible, many low-lying communities will be at risk from storm surges (see figure 1 and table 1).

People living on low-lying islands will be among the first wave of 'climate refugees'. Due to environmental change, mainly through rising sea levels, some people have already had to move, and many more could be without a home in our lifetime.

FIGURE 1 Low-lying islands in the Pacific under threat of disappearing due to climate change

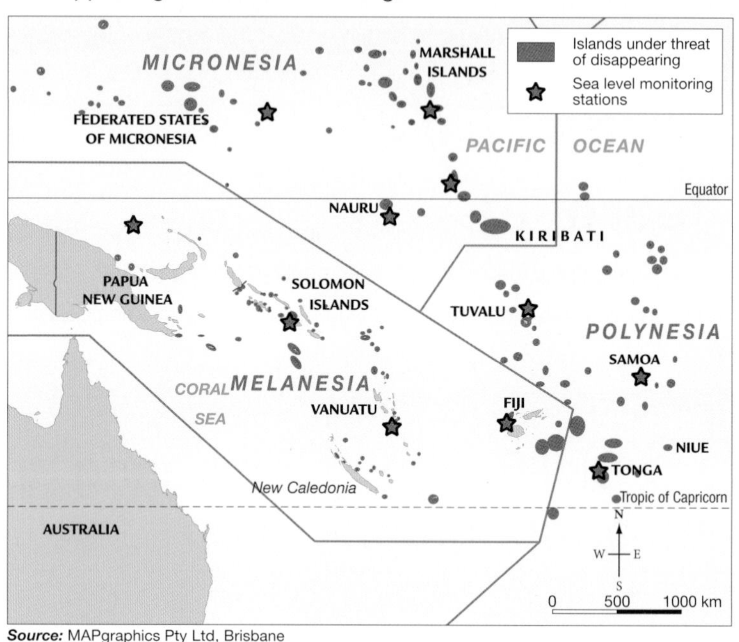

Source: MAPgraphics Pty Ltd, Brisbane

TABLE 1 Selected Pacific Island nations, area and population

Island	Land area (km²)	National extent (km²)	Population	Gross income per person per year (US$)	Highest elevation (metres above sea level)
Kiribati	717	3550000	113400	1690	80
Marshall Islands	181	2131000	54880	3346	10
Tuvalu	26	900000	10640	3286	5
Australia	7690000	7690000	24000000	46550	2229

5.7.2 Rising sea levels in the Pacific

Many of the Pacific Islands are small and can in some cases be described as an **atoll**. Their national boundaries, which include the waters and economic zones they control, extend over vast distances. Hence, islands such as Kiribati, Tuvalu and the Marshall Islands in the south-west Pacific, which are only a few metres above sea level, are particularly vulnerable to rising sea levels and associated severe storm activity due to climate change (see figure 1).

The economies of these Pacific Islands are small-scale, and earnings are not high, with a reliance on what limited natural resources occur on the islands and in the surrounding ocean waters. Due to sandy soils and low altitudes, although rainfalls can be plentiful, little can be retained as streams are few and groundwater is scarce. Hence any incursion by sea water can be devastating for agricultural produce (see figure 2), the urban environment and tourism, which has more recently become a money earner for these islands.

Apart from the predicted rise in sea levels due to global warming, a secondary impact on the life of Islanders will be increases in the temperature of the sea, which will affect coral reefs and fish stocks that live in that environment. As for the Great Barrier Reef in Australia, bleaching and death of coral reefs can lead to the destruction of the whole aquatic ecosystem, and this will have devastating impacts on the Islanders' main diet, which is fish and other forms of seafood.

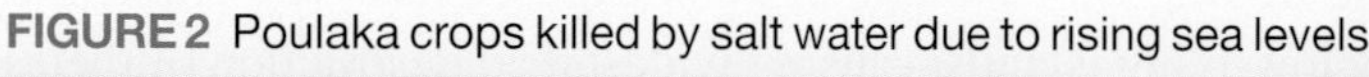

FIGURE 2 Poulaka crops killed by salt water due to rising sea levels

What can the Islanders do?

If food crops are destroyed by rising sea levels, **storm surges** and saltwater pollution, the Pacific Islanders do not have much scope for importing food due to their remoteness, high transport costs and low earnings of individuals. Combined with loss of seafood stocks, the Islanders will need to move to other islands to find a new home and livelihood.

The Pacific Islanders are strong advocates for the policies of the **Kyoto Protocol** on climate change. The Protocol has set up a range of measures to reduce the impact of greenhouse gas emissions by introducing carbon trading schemes and energy-efficient forms of technology such as wind and solar power. The leaders of the Pacific nations have spoken at the United Nations and many international climate change forums to make others aware of their delicate situation and vulnerability to rising sea levels. They have also approached nations such as Australia and New Zealand to see whether they might be able to establish a migration policy into the future.

5.7.3 Rising sea levels in the Maldives

The Maldive Islands are located in the Indian Ocean, to the south-west of India (see figure 3). There are about 1200 coral islands, grouped into 26 atolls, most of which average no more than one metre above sea level (the highest point in the island group is just 2.4 metres above sea level). Economically, the nation depends on tourism and the continuing appeal of its beautiful beaches.

FIGURE 3 Location of the Maldive Islands

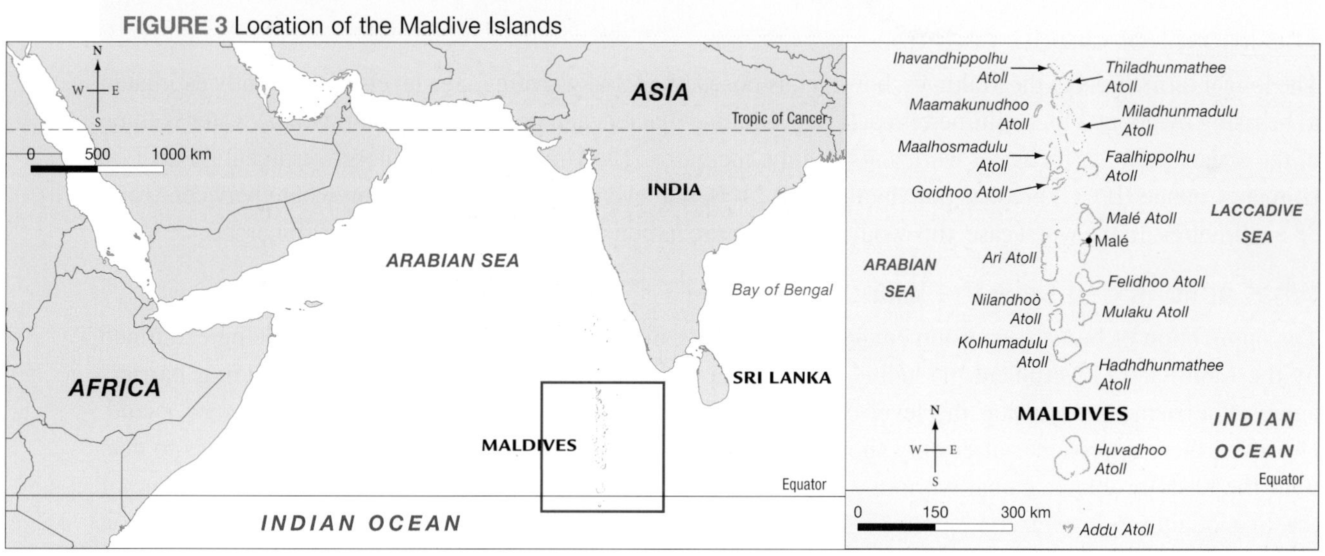

Source: Map by Spatial Vision

The Boxing Day tsunami of 2004 exposed how vulnerable the Maldives are, when the wave swept across many low-lying islands, causing widespread destruction of their fruit plantations. The relatively low number of deaths was due to the fact that most of the population lives in Malé, which is protected by a huge sea wall (see figure 4).

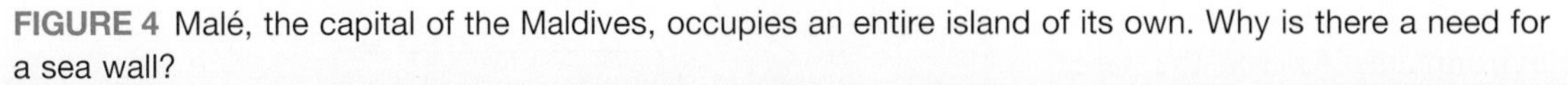

FIGURE 4 Malé, the capital of the Maldives, occupies an entire island of its own. Why is there a need for a sea wall?

Only nine islands were reported to have escaped any flooding, while 57 islands faced serious damage to critical infrastructure, 14 islands had to be totally evacuated, and six islands were destroyed. A further 21 resort islands were forced to close because of serious damage. The total damage was estimated to be more than US$400 million, or some 62 per cent of the GDP. One hundred and two Maldivians and six foreigners reportedly died in the tsunami.

The impact of climate change

The longer term threat to the Maldives, however, is posed by global warming. Sea levels are currently estimated to be rising by about 2 to 3 millimetres each year. Melting glaciers and polar ice are adding to the water volume of the oceans; also, as the water warms, its volume increases. The United Nations Intergovernmental Panel on Climate Change (IPCC) predicts that, by the year 2100, sea levels will have risen by anywhere between 9 and 88 centimetres. In the worst case, this would see the entire nation of the Maldives virtually submerged.

What actions can save the islands?

The application of human–environment systems thinking in the form of various schemes is being examined by the Maldivian Government, including moving populations from islands more at risk, building barriers against the rising sea, raising the level of some key islands and even building a completely new island. However, these approaches offer only short-term solutions. The longer term sustainable challenge is to deal with the basic problem: global warming itself. It is perhaps understandable that the Maldives was one of the first countries to sign the Kyoto Protocol, which sought international agreement to cut back carbon dioxide emissions.

Unless the international community agrees to an environmental worldview that incorporates changes to make large cuts in emissions, the problems facing the Islanders will get worse. Numerous people will have to seek refuge in other countries. Without global action, eventually the Islanders will lose their countries.

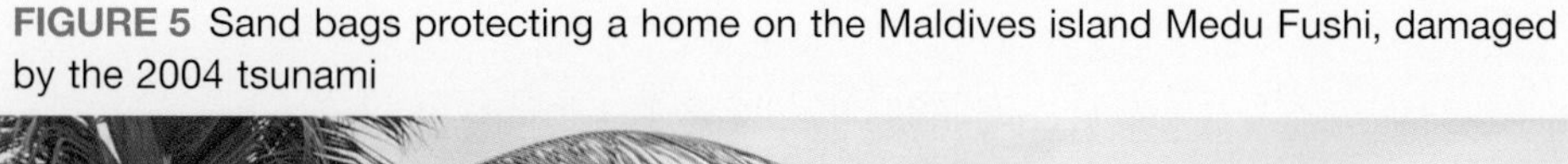

FIGURE 5 Sand bags protecting a home on the Maldives island Medu Fushi, damaged by the 2004 tsunami

5.7 Activities

To answer questions online and to receive **immediate feedback** and **sample responses** for every question, go to your learnON title at www.jacplus.com.au. *Note*: Question numbers may vary slightly.

Remember

1. Many islands in the Pacific and Indian Oceans are under threat from rising sea levels. Why is this so?
2. What can the governments and Pacific Island peoples do, both in the short term and long term, to solve the problems they will face due to climate change?
3. How is climate ***change*** threatening water supplies and affecting food resources?

Discover

4. Use the **Maldives** weblink in the Resources tab to watch the news article on the Maldivian island of Maduwaree. Summarise what is contributing to the receding coastline. Detail what ***environmental***, social and economic impacts the ***change*** is having on the inhabitants, and outline what management solutions are proposed.

FIGURE 6 SWOT analysis

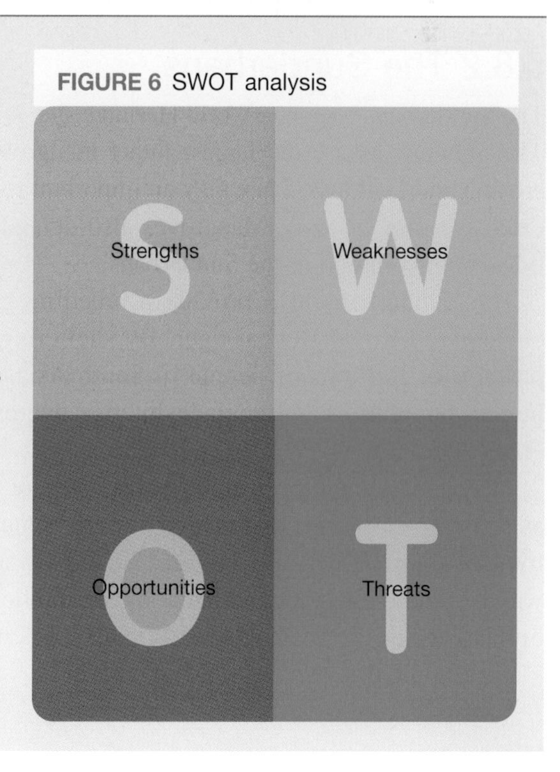

Predict

5. As a class, discuss other parts of the world you can think of that might also be threatened if sea levels were to rise by about one metre over the next 100 years. Justify your views.

Think

6. In groups, discuss the following statements concerning the impacts of climate ***change*** in the Pacific and Indian Oceans using ***environmental***, economic and social criteria. Alternatively you may wish to use a SWOT analysis (see figure 6) to help in evaluating each statement.
   (a) Australia and New Zealand should be prepared to re-settle the 100 000 people of the islands of Kiribati if sea level rises create a 'climate refugee' problem.
   (b) Increasing the height of the sea wall around the island of Malé will solve the threats of rising sea levels in the Maldives and ensure tourism into the future.

RESOURCES — ONLINE ONLY

Explore more with this weblink: Maldives

5.8 How do inland activities affect coasts?

5.8.1 Flood-prone Bangladesh

The country of Bangladesh is a large **alluvial plain** crossed by three rivers: the Ganges, Brahmaputra and Meghna. Each river carries massive volumes of water from its source in the Himalayas, spreads out along the **deltaic plain**, and empties into the world's biggest delta, the Bay of Bengal. This makes Bangladesh's coastline one of the most flood-prone in the world.

Apart from flooding by rivers in the delta, sea level rises caused by global warming will lead to the expansion of ocean waters and additional inflows from melting Himalayan snow. Scientists predict a one-metre sea level rise by 2100 if global warming continues at the current rate. The IPCC predicts rising sea levels will overtake 17 per cent of Bangladesh by 2050, displacing at least 20 million people.

5.8.2 The Sundarbans

The Sundarbans region, a World Heritage site, is just one area of Bangladesh at risk from increased flooding. The Sundarbans are the largest intact mangrove forests in the world. Mangroves protect against coastal erosion and land loss. They play an important role in flood minimisation because they trap sediment in their extensive root systems. Mangroves also defend against storm surges caused by tropical cyclones or king tides, both common in the Sundarbans.

The Sundarbans also provide a breeding ground for birds and fish, as well as being home to the endangered Royal Bengal tiger. By sheltering juvenile fish, the mangrove forest provides a source of protein for millions of people in South Asia. Recently, the Sundarbans have also attracted a growing human population as Bangladeshis flee overcrowding in the capital city, Dhaka, or flooding and poverty in rural areas.

Increasing human occupation poses a severe threat to the Sundarbans. Most Bangladeshis rely on wood as a source of energy, and mangroves are being cleared to make charcoal for cooking. Aquaculture industries also have a negative impact. Mangroves are cleared to accommodate huge ponds for fish breeding, which quickly become polluted by antibiotics, waste products and toxic algae. This damage to the Sundarbans destroys Bangladesh's natural defence against flooding.

FIGURE 1 Flooding in Bangladesh — some causes

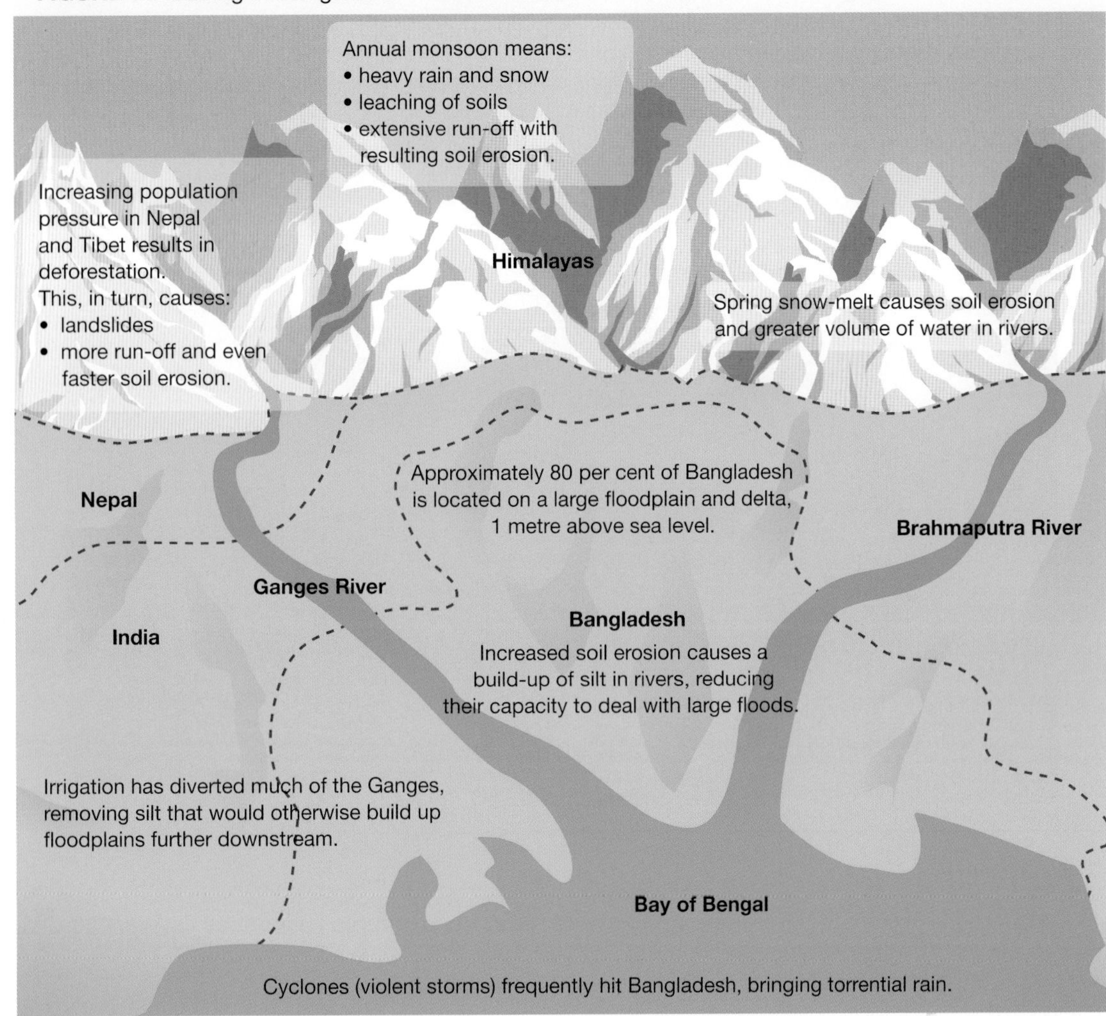

Annual monsoon means:
• heavy rain and snow
• leaching of soils
• extensive run-off with resulting soil erosion.
Increasing population pressure in Nepal and Tibet results in deforestation.
This, in turn, causes:
• landslides
• more run-off and even faster soil erosion.
Himalayas
Spring snow-melt causes soil erosion and greater volume of water in rivers.
Nepal
Approximately 80 per cent of Bangladesh is located on a large floodplain and delta, 1 metre above sea level.
Brahmaputra River
Ganges River
India
Bangladesh
Increased soil erosion causes a build-up of silt in rivers, reducing their capacity to deal with large floods.
Irrigation has diverted much of the Ganges, removing silt that would otherwise build up floodplains further downstream.
Bay of Bengal
Cyclones (violent storms) frequently hit Bangladesh, bringing torrential rain.

5.8.3 The impact of flooding

The increase in temperature which has led to an increased melting of glaciers and snow inland in the Himalayas will exacerbate the existing problems of flooding in Bangladesh. Climate change also causes shifts in weather patterns. If the **monsoon** season (from June to October) coincided with an unseasonal snowmelt, flooding would occur on a scale never before seen, especially with the event of tropical cyclones. Land will be lost and people displaced. Many islands fringing the Bay of Bengal are already under water, producing 'climate refugees', people who have literally nowhere to go.

The 1991 Bangladesh cyclone was among the deadliest tropical cyclones on record (see figure 2). The cyclone struck the Chittagong district of south-eastern Bangladesh with winds of around 250 km/h. The storm forced a six-metre storm surge inland over a wide area, killing at least 138 000 people and leaving as many as 10 million homeless.

Because of these risks, Bangladesh needs to plan and implement management strategies based on understandings of the reasons behind the changes and consideration of interactions between environmental, economic and social factors operating in the region. The government encourages farming methods that avoid deforestation, and a ban is proposed on heavy-polluting vehicles. A proposed economic solution is ecotourism, as it attracts foreign currency while preserving the natural ecosystems and promoting sustainable development.

FIGURE 2 A village on one of Bangladesh's coastal islands was devastated by a cyclone in 1991. Although people in areas such as these are aware of the risk, overcrowding often prevents them from moving to safer regions.

5.8 Activities

To answer questions online and to receive **immediate feedback** and **sample responses** for every question, go to your learnON title at www.jacplus.com.au. *Note*: Question numbers may vary slightly.

Remember

1. How do mangroves minimise the impact of floods and coastal erosion?
2. Name two reasons mangroves are being cleared in the Sundarbans.
3. What are 'climate refugees'?

Explain

4. Refer to figure 1. Explain how the geography of Bangladesh makes it so vulnerable to the threat posed by climate ***change***.
5. How can ecotourism play a role in preserving Bangladesh's ecosystems?

Discover

6. List the factors that are displacing Bangladeshis and forcing them to move to the Sundarbans.

Think

7. Refer to figure 1.
   (a) Describe how cyclones can contribute towards flooding in Bangladesh.
   (b) List some short-term and long-term actions that neighbouring nations Tibet, India and Nepal could implement to lessen the impact of flooding in ***places*** like Bangladesh.
   (c) Divide a table into three columns with the headings 'Food production', 'Transport' and 'Settlement', and list the consequences of flooding for each category.
8. Discuss in small groups to what extent you think economic goals and objectives are important with respect to ***environmental*** goals such as reducing greenhouse gas emissions and societal ***change***. Listen respectfully to one another's views. Decide as a group what policy direction you would push if you were in a position of influence in government, and present this view jointly to the class.

learnon RESOURCES — ONLINE ONLY

Try out this interactivity: Bangladesh awash (int-3297)

5.9 Who shifted the sand?

Access this subtopic at **www.jacplus.com.au**

5.10 How do we manage coastal change?

5.10.1 Changing coastlines

Coastal areas are not static or fixed places and as such they are subject to two main agents of change. These can be defined as natural environmental processes and human-induced processes. In terms of natural environmental processes, where deposition processes dominate, coasts have been growing. However, where erosion processes dominate, coastal land is lost to the sea. Wherever people have imposed their structures, in the form of housing, harbour works and the like on coasts, there is a need to manage or at least moderate the processes of coastal change in a sustainable manner.

Of the world's population, 41 per cent, or some 2.2 billion people live within 100 kilometres of the coast, and most of the world's megacities are located on the coast. According to predictions made by the World Ocean Review in 2010 and more recently in 2015, at least one billion people who live in low lying coastal areas could experience inundation and/or erosion of their lands into the future (see table 1). This change to coasts is seen as stemming essentially from climate change which, as a human-induced event, is leading to rising seas and more frequent severe storm events. A consequence will mean an increase in what are known as 'climate refugees', people who will have to relocate due to coastal changes.

TABLE 1 Nations with the largest populations and the highest proportions of population living in low-lying coastal areas

Top 10 nations classified by population in low-lying coastal regions			Top 10 nations classified by proportion of population in low-lying coastal areas		
Nation	Population in low-lying coastal regions (10^3)	% of population in low-lying coastal regions	Nation	Population in low-lying coastal regions (10^3)	% of population in low-lying coastal regions
1. China	127 038	10%	1. Maldives	291	100%
2. India	63 341	6%	2. Bahamas	267	88%
3. Bangladesh	53 111	39%	3. Bahrain	501	78%
4. Indonesia	41 807	20%	4. Suriname	325	78%
5. Vietnam	41 439	53%	5. Netherlands	9590	60%
6. Japan	30 827	24%	6. Macau	264	59%
7. Egypt	24 411	36%	7. Guyana	419	55%
8. United States	23 279	8%	8. Vietnam	41 439	53%
9. Thailand	15 689	25%	9. Djibouti	250	40%
10. Philippines	15 122	20%	10. Bangladesh	53 111	39%

5.10.2 Protecting the coast

The protection of the coast through management programs is a costly business which aims to overcome problems associated with land loss, waterlogging and incursions of **groundwater salinity**.

The Netherlands and Germany together spend 250 million euros on coastal works each year. The Netherlands, a country with two-thirds of its land below sea level, has proven that protecting the coastline is possible through a large investment of capital. The most common form of coastal protection in the Netherlands are **dykes** to hold back the sea; however, a recent addition is **floating settlements** that can rise and fall as sea levels change (see figure 1).

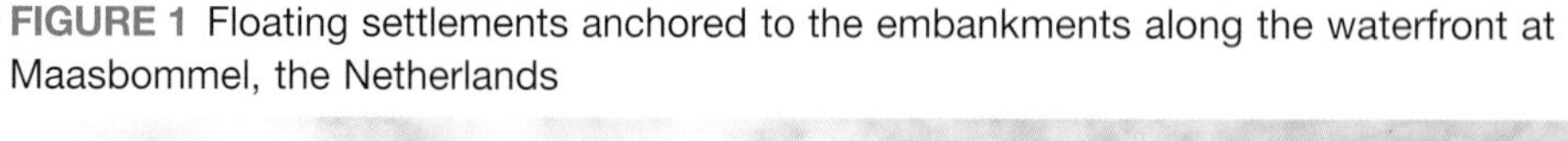

FIGURE 1 Floating settlements anchored to the embankments along the waterfront at Maasbommel, the Netherlands

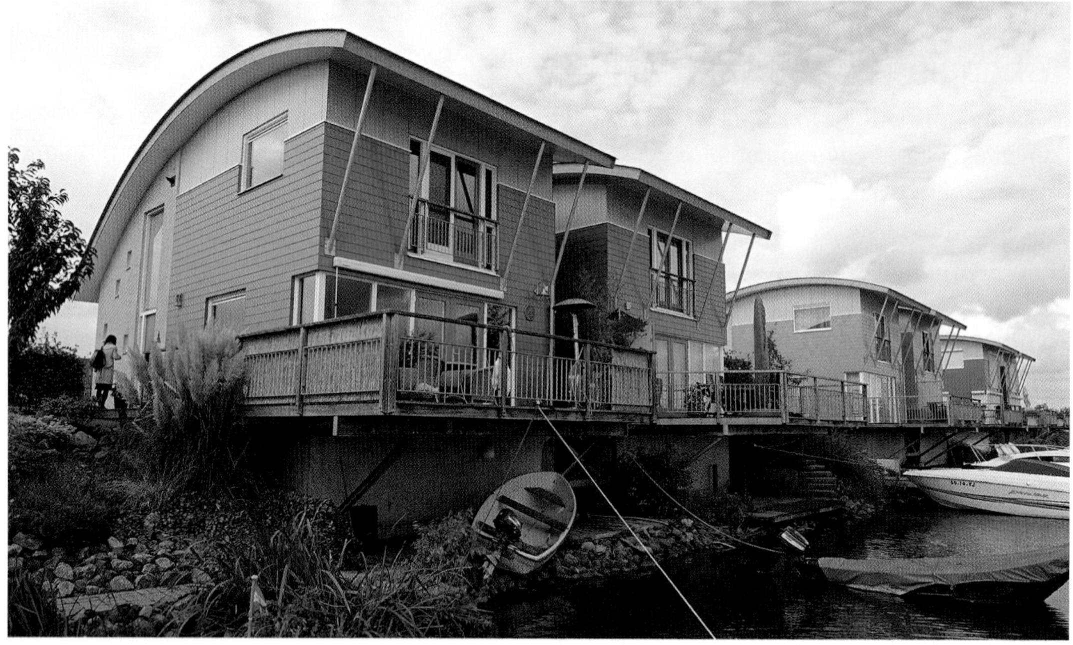

5.10.3 Coastal management in Australia

If coastlines are to be protected, a wide range of strategies must be employed to combat changes to the coastline and, in particular, flooding of low-lying areas and increased erosion of beaches and bluffs. The techniques shown in table 2 are used in Australia.

TABLE 2 Possible management solutions to reduce the impacts of sea level rise and erosion

Solution	Description	Diagram	Advantages	Disadvantages
Beach nourishment	The artificial placement of sand on a beach. This is then spread along the beach by natural processes.	Established vegetation – shrubs and sand grasses Initial nourishment designed for 10 years Fencing Sea level Existing profile	Sand is used that best matches the natural beach material. Low environmental impact at the beach	The sand must come from another beach and may have an environmental impact in that location. Must be carried out on a continuous basis and therefore requires continuous funds

(Continued)

TABLE 2 Possible management solutions to reduce the impacts of sea level rise and erosion (*Continued*)

Solution	Description	Diagram	Advantages	Disadvantages
Groyne	An artificial structure designed to trap sand being moved by longshore drift, therefore protecting the beach. Groynes can be built using timber, concrete, steel pilings and rock.		Traps sand and maintains the beach	Groynes do not stop sand movement that occurs directly offshore. Visual eyesore
Sea wall	A structure placed parallel to the shoreline to separate the land area from the water		Prevents further erosion of the dune area and protects buildings	The base of the sea wall will be undermined over time. Visual eyesore Will need a sand nourishment program as well High initial cost Ongoing maintenance and cost
Offshore breakwater	A structure parallel to the shore and placed in a water depth of about 10 metres		Waves break in the deeper water, reducing their energy at the shore.	Destroys surfing amenity of the coast Requires large boulders in large quantities Cost would be extremely high
Purchase property	Buy the buildings and remove structures that are threatened by erosion		Allows easier management of the dune area Allows natural beach processes to continue Increases public access to the beach	Loss of revenue to the local council Possible social problems with residents who must move Exposes the back dune area, which will need protection Cost would be extremely high Does not solve sand loss

5.10 Activities

To answer questions online and to receive **immediate feedback** and **sample responses** for every question, go to your learnON title at www.jacplus.com.au. *Note*: Question numbers may vary slightly.

Remember

1. Refer to table 1. Which country is most susceptible to ***changing*** coastlines in terms of absolute population numbers?
2. Why does the Netherlands spend money on coastal protection?

Discover

3. Explain how a coastal defence system such as a dyke works.

Predict

4. What impact would sea level rise and erosion have on future food security?

Think

5. What might be the impact of sea level rise and coastal erosion on the tourist industries of the Gold Coast area of Australia? What strategies of coastal protection, mentioned in this topic, could help solve the problems, and how might they work?
6. Evaluate the strengths and weaknesses of two of the management strategies shown in table 2.
   (a) Which strategy would have the least ***environmental*** impact?
   (b) Which strategy would have the greatest economic impact or be the most costly to maintain?
   (c) Which strategies could improve social amenities such as tourism and recreation in coastal areas? Give reasons for your answer.

5.11 Review

5.11.1 Review

The Review section contains a range of different questions and activities to help you revise and recall what you have learned, especially prior to a topic test.

5.11.2 Reflect

The Reflect section provides you with an opportunity to apply and extend your learning.

Access this subtopic at **www.jacplus.com.au**

TOPIC 6
Marine environments — are we trashing our oceans?

6.1 Overview

Numerous **videos** and **interactivities** are embedded just where you need them, at the point of learning, in your learnON title at www.jacplus.com.au. They will help you to learn the content and concepts covered in this topic.

6.1.1 Introduction

Imagine you are on a beach. You are looking out to sea at the endless, constantly moving mass of water that stretches to the horizon. Why does it move, how does it move, what lies beneath?

Life on Earth would not be possible without our oceans. Humans are interconnected to the oceans, which provide or regulate our water, oxygen, weather, food, minerals and resources. Oceans also create a surface for transport and trade and provide a habitat for 80 per cent of all life on Earth. Our oceans are under threat as we use them to extract resources, dump waste and destroy them. It has been very much a case of 'out of sight, out of mind'. Let's now look at this problem in more detail.

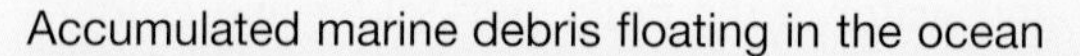

Accumulated marine debris floating in the ocean

Starter questions

1. What are your first thoughts when you view this photograph?
2. Suggest items that might be floating in this rubbish.
3. Where do you think this waste has come from, and how did it get here?
4. What waste does your family generate, and what happens to it?

INQUIRY SEQUENCE

6.2 Why is there motion in the ocean?

6.2.1 What are ocean currents?

In January 1992, a ship sailing from Hong Kong to the United States lost a shipping crate containing 28 000 plastic bath toys at sea during a storm. The toys drifted off in the currents, the first ones eventually reaching the Alaskan coast in November of that year. More than 20 years later, many are still floating! The tracking of these toys has enabled scientists to improve their understanding of ocean currents.

Why doesn't water at the equator get hotter and hotter and water at the poles get colder and colder? The answer is ocean currents. Currents are movements of water from one region to another, often over long distances and time periods. Currents effectively interconnect the world's oceans and seas. They are critically important for 'stirring' the waters and transporting heat, oxygen, carbon dioxide, salts, nutrients, sediments and marine creatures.

A knowledge of currents is vital for navigation, shipping, search and rescue and the dispersal of pollutants. The direction that currents take is influenced by a number of factors, including the Earth's rotation, the shape of the sea floor, water temperature, salinity levels and the wind.

6.2.2 What are the different types of ocean currents?

Surface currents

The action of winds blowing over the surface of the water sets up the movement of water in the top 400 metres of the ocean, creating surface currents. These currents flow in a regular pattern, but they can vary in depth, width and speed. Due to the rotation of the Earth, the **Coriolis force** deflects currents into large circular patterns called **gyres**, which flow clockwise in the Northern Hemisphere and anticlockwise in the Southern Hemisphere (see figure 1).

Deep water currents

Deep water currents are powered by **thermoline circulation** and make up about 90 per cent of water movements in the ocean. Surface currents make up the remaining 10 per cent.

FIGURE 1 The Global Ocean Conveyor Belt and the five main ocean gyres

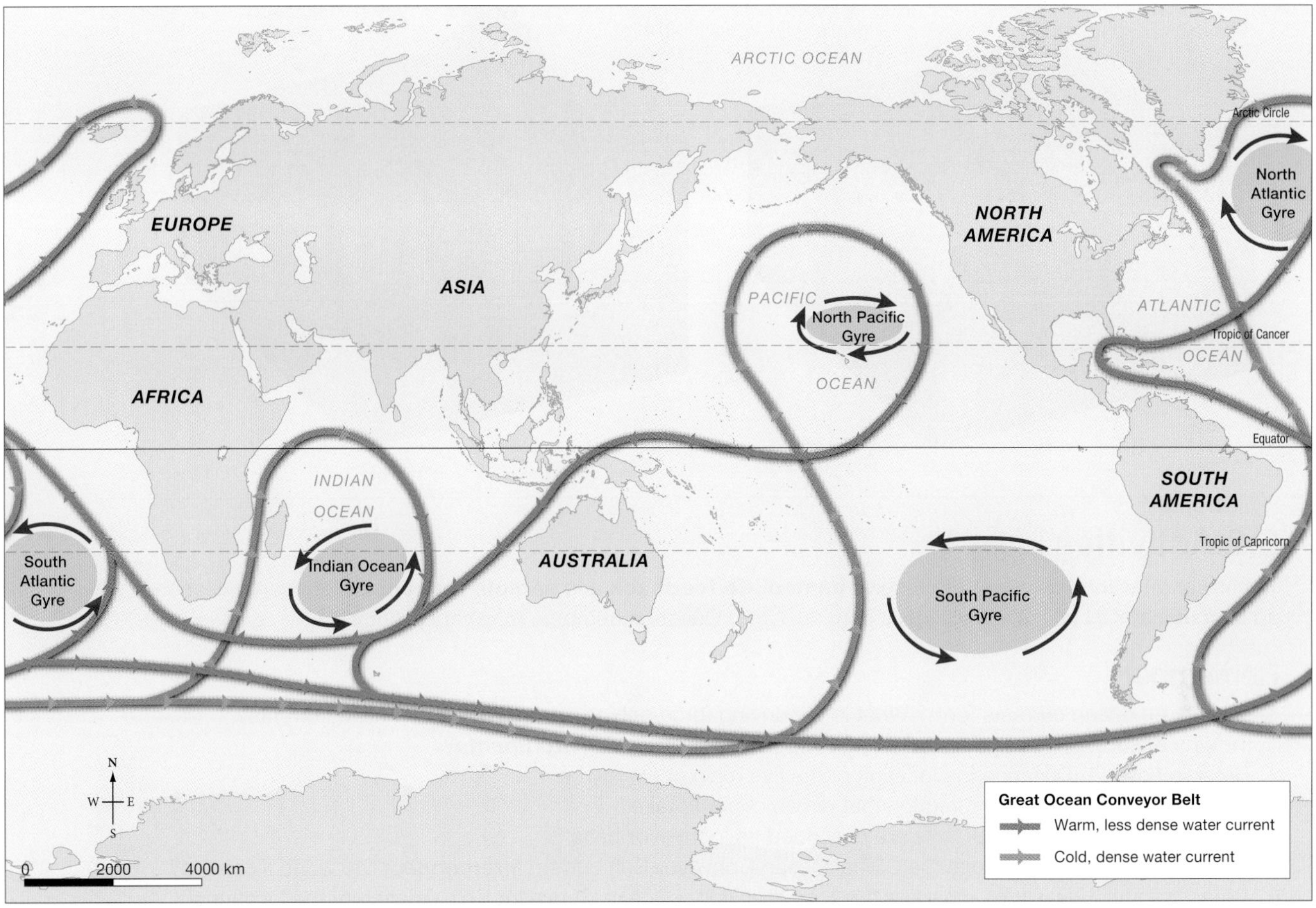

Source: Map by Spatial Vision

Global Ocean Conveyor Belt

The Global Ocean Conveyor Belt is the largest of the thermoline-driven ocean currents (see figure 1). Warm water, which holds less salt and is less dense than cold water, travels from the equator near the surface into higher latitudes. There it loses some of its heat to the atmosphere. The current mixes with colder Arctic waters and this cold, salty water becomes more dense and sinks, flowing as a deep ocean current. This creates a continual looping current which moves at a rate of 10 cm/s and may take up to 1000 years to complete one loop. The quantity of water moved in the Global Ocean Conveyor Belt is more than 16 times the water volume of all the world's rivers.

Upwellings and downwellings

The movement of cold water currents from the deep sea to the surface is called an upwelling. This is shown in figure 2(a). Regions where these occur are very productive fishing grounds as the upwellings bring nutrients from the seabed, which provide food for the growth of plankton, often the start of marine food chains. Over 50 per cent of the world's fish are caught in these areas.

Downwellings, shown in figure 2(b), occur when currents sink, taking with them oxygen and carbon dioxide from the atmosphere. These currents essentially 'stir up' the water and help distribute heat, gases and nutrients.

FIGURE 2 (a) Upwelling and (b) downwelling

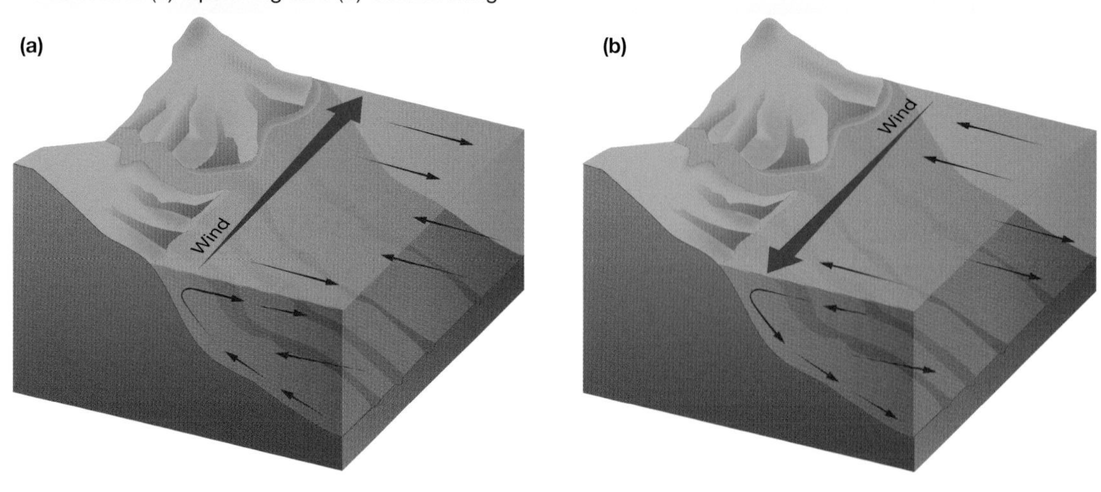

6.2 Activities

To answer questions online and to receive **immediate feedback** and **sample responses** for every question, go to your learnON title at www.jacplus.com.au. *Note*: Question numbers may vary slightly.

Remember

1. (a) Why do ocean currents form? What is the driving force behind surface and thermoline currents?
   (b) Why are upwellings and downwellings important for marine ***environments***?
2. Refer to figure 1. Describe the location of the five main ocean gyres.
3. What factors influence the direction that ocean currents take?
4. Why do you think ocean currents are described as 'conveyor belts'?
5. Looking at figure 1, how does the Global Ocean Conveyor Belt current ***interconnect*** the world's oceans?
6. Refer to figure 1. Describe the route taken by the Global Ocean Conveyor Belt. At each of the locations marked A–E, name the ocean, the direction the current is taking, the continent it is passing, and its thermoline features (warm, cold, higher salt content, lower salt content).

Discover

7. Research the ***interconnection*** between the Humboldt current (cold upwelling) on the west coast of South America and El Niño events.

Predict

8. Suggest what ***changes*** might happen to the Global Ocean Conveyor Belt if there was a significant melting of the polar ice caps.

Think

9. Why doesn't water at the equator keep getting hotter and water at the poles keep getting colder? Use your knowledge of currents to write an explanation for a younger student.

Try out this interactivity: Motion in the ocean (int-3298)

6.3 Where does trash travel?

6.3.1 What is marine pollution?

What happens to that empty drink can or plastic bag that misses the bin? There's a good chance it might wash down the gutter, into the drain and out to sea, never to be seen again. The world's largest rubbish dump is not on land, it is in the ocean. Accidentally or deliberately, the oceans receive millions of tonnes of man-made pollutants each year, which are collected in currents and swirled around the oceans.

Marine pollution is any harmful substance or product that enters the ocean. Most are human pollutants including fertilisers, chemicals, sewage, plastics and other solids, including over 1000 shipping containers per year.

Close to 80 per cent of marine pollutants start off on land and are either washed or deposited into rivers, from where they make their way to the coast (see figure 1). Even industrial air pollution can be returned to Earth's surface via rainfall (see figure 2).

FIGURE 1 Most marine debris starts off on land. Much of the litter in this creek in the Philippines will end up in the sea.

FIGURE 2 The sources of marine pollution

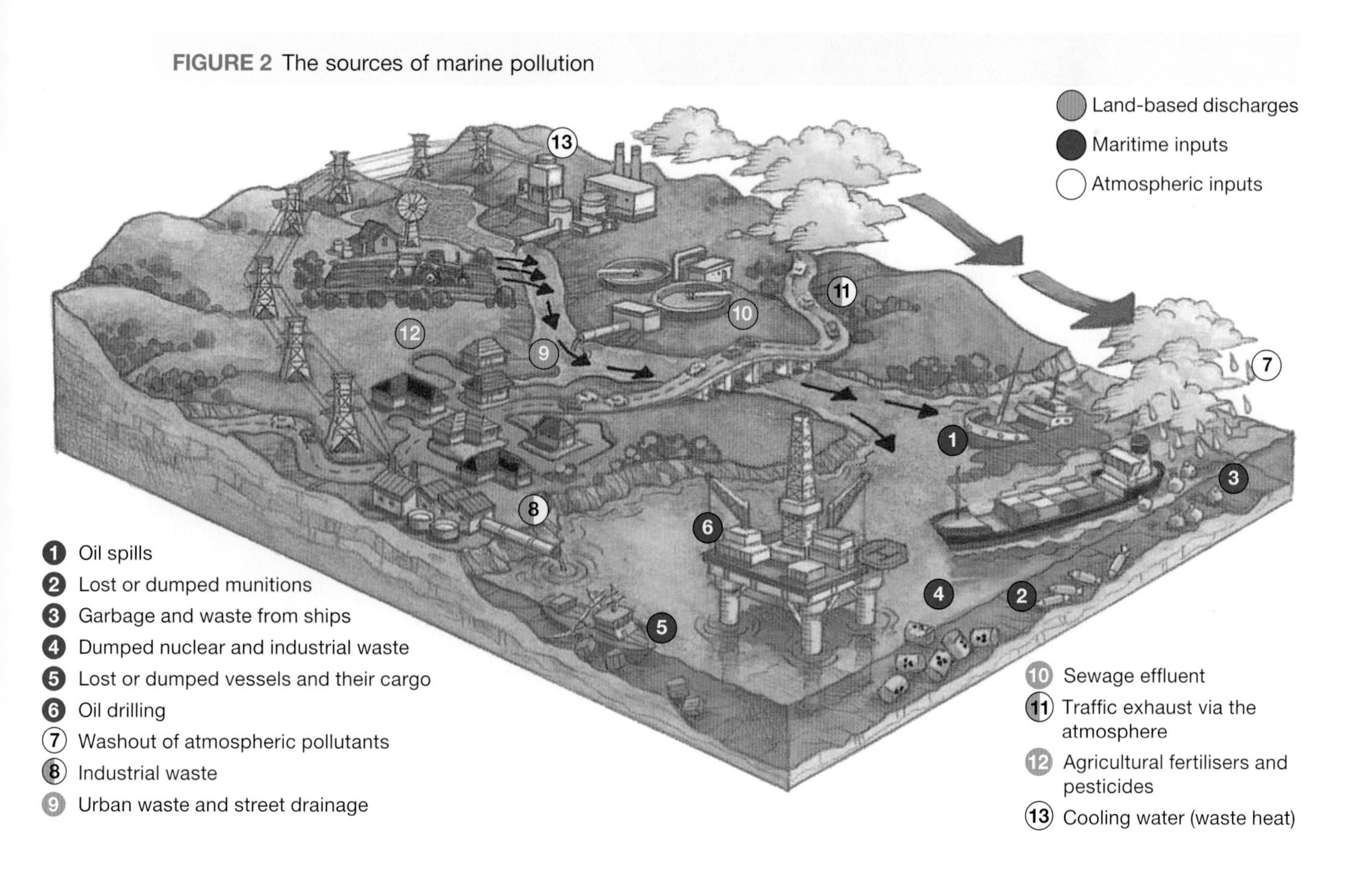

6.3.2 What is marine debris?

Marine debris is litter and other solid material that washes or is dumped into the oceans, much of which is plastic (see figure 3). The special features of plastic that make it such a useful product — it is light, cheap to produce and disposable — also make it a major problem for the ocean (see figure 4). Over 270 million tonnes of plastic are produced each year, approximately 55 per cent of which is recovered, recycled or sent to landfill. The rest is unaccounted for, lost in the environment and eventually washed out to sea, often ending up in the gut or wrapped around the neck of marine creatures, or even buried in Arctic ice. Data collected by scientists from the US, France, Chile, Australia and New Zealand calculated that there were more than 5 trillion pieces of plastic, weighing 269 000 tonnes, floating in the world's oceans. A survey of Australia's coastline found that plastics made up 74 per cent of marine litter. Surface currents and wind can also move debris back on to the coast, where it can become buried in sand or swept back out to sea again.

FIGURE 3 Top 10 marine debris items

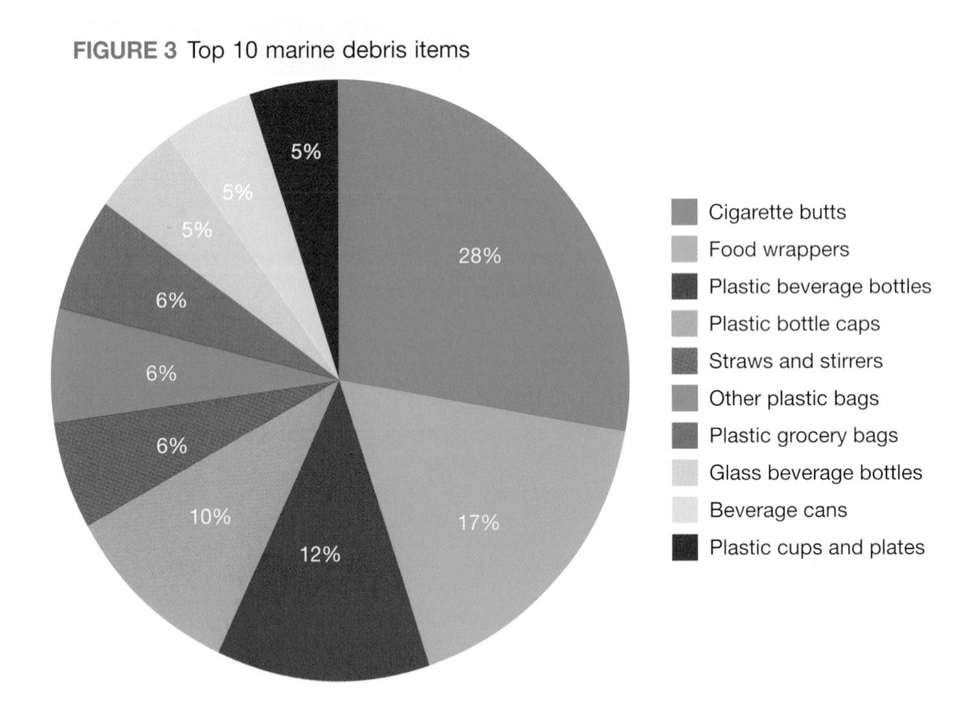

Note: Data is the result of 25 years of surveying debris collected by volunteers in annual debris clean-ups in over 100 countries.

Unlike most other litter, plastics generally are not **biodegradable**. The technological features of plastic mean that when it is exposed to constant wind, waves, salt and sunlight, it breaks down into tiny fragments known as microplastics (20–50 microns in diameter, thinner than a human hair), which can float or sink to the seabed. Samples taken from selected sites in the Mediterranean Sea and the Atlantic and Indian Oceans have shown microplastics as deep as 3000 metres and in concentrations 1000 times higher than those found floating on the surface.

FIGURE 4 Discarded plastic bags resembling jellyfish, floating in the ocean

6.3.3 Where do we find the most marine debris?

The worst-affected places for marine debris tend to be heavily populated coastal places and popular tourist destinations; for example, the Caribbean Sea. Research in 2014 identified the 10 countries that generate the most marine plastic debris, with more than 50 per cent coming from just five countries: China, Indonesia, the Philippines, Vietnam and Sri Lanka (see figure 5).

FIGURE 5 Top 10 sources of marine plastic waste

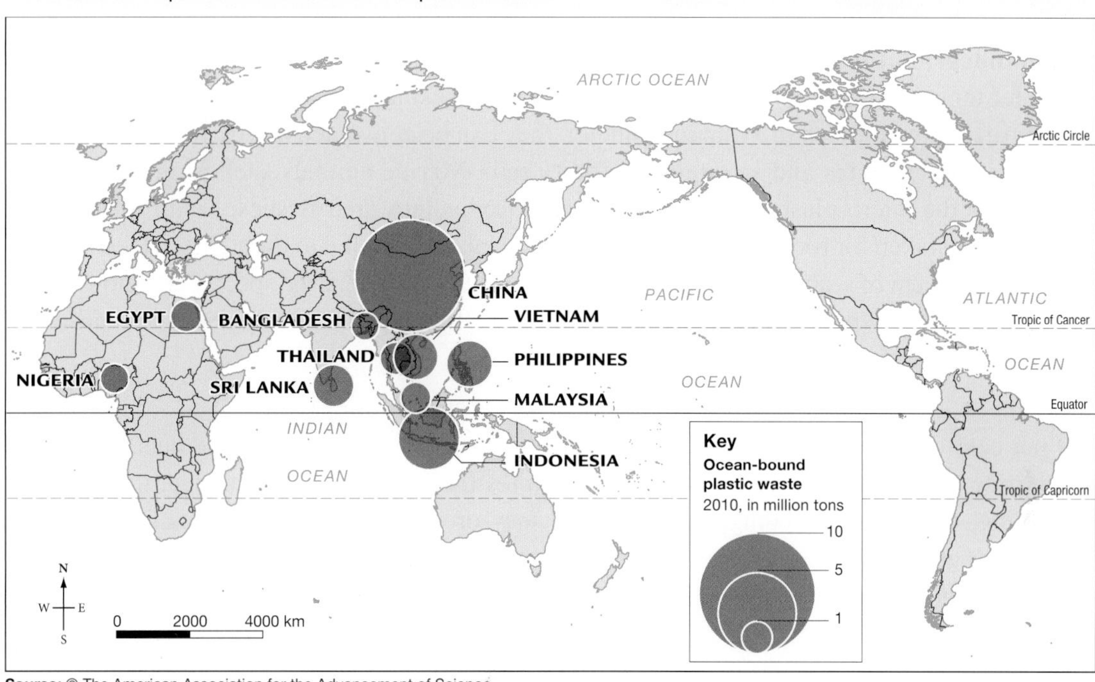

Source: © The American Association for the Advancement of Science

Many of these countries have growing economies and demand for plastic products, but as yet do not have the infrastructure to collect, recycle and dispose of plastic waste before it enters the sea. By comparison, developed countries tend to have systems to trap and collect this waste.

What is the Great Pacific Ocean Garbage Patch?

A swirling sea of plastic bottles, garbage bags and other rubbish is growing in the middle of the North Pacific Ocean, thousands of kilometres from the nearest coastline. Why is it there and how did it get there? Discarded waste from the east coast of Japan and west coast of the United States gets swept up in the North Pacific gyre. Within the marine environment, the slow-moving currents and winds push material into the calmer centre of the gyre, where much of it stays and accumulates. It can take a year for material to reach the centre of the gyre from Japan and five years from the United States. The accumulation of debris has

FIGURE 6 Location of the Great Pacific Ocean Garbage Patch

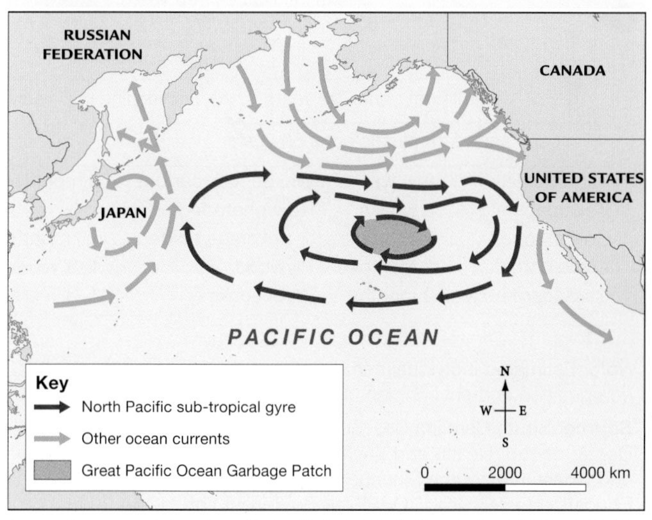

Source: Greenpeace International Made with Natural Earth. Map by Spatial Vision (GAT-22)

Note: The Great Pacific Ocean Garbage Patch floats between Japan and the United States, just north of the Hawaiian Islands. The rotational current caused by the North Pacific gyre draws in garbage from neighbouring coastlines, where it becomes trapped in large quantities in the calmer waters of the gyre's centre.

earned this region the name the 'Great Pacific Ocean Garbage Patch' (see figure 6). Very little garbage is visible on the surface; rather, it is a thick soupy mass of minute pieces of plastic with an average depth of 10 metres. The size of the patch is estimated to be anywhere from 700 000 to more than 15 million square kilometres. Scientists have detected up to 1 million plastic particles per square kilometre in the patch. Another large garbage patch is located in the Atlantic Ocean.

6.3.4 What are the environmental impacts of marine debris?

Figure 7 gives estimates for the length of time some marine debris takes to decompose. Most plastics undergo **photodegradation**, which is much slower in water than on land due to reduced exposure to the sun and cooler temperatures. As the particles break down into smaller particles, they 'thicken' the water and can release toxins. If less than 5 mm in diameter, they can be consumed by sea creatures, which in turn are eaten by bigger creatures and so on up the food chain. Marine animals such as mussels which filter seawater take up the micro-plastics, which can release toxins into their tissues. Small floating pieces of debris are often mistaken for food and are scooped up by seabirds and fed to their chicks (see figure 8).

More than 44 per cent of seabirds are known to eat plastic, while 267 marine species are known to swallow plastic bags, mistaking them for jellyfish (see figure 4). An estimated 100 000 marine mammals and up to 1 million seabirds die each year after ingesting plastic.

Ghost nets

Up to 10 per cent of marine debris is made up of abandoned and discarded fishing nets, known as ghost nets, which pose a very common threat to marine creatures (see table 1). Once tangled, they are prevented from swimming, fishing and breeding, and ultimately they drown. Over time, the nets fill with debris and form rafts which grow to hundreds of metres in diameter. These can drag across reefs or scrape along the seabed, causing considerable damage (see section 6.5).

FIGURE 7 Time periods for the decomposition of marine litter

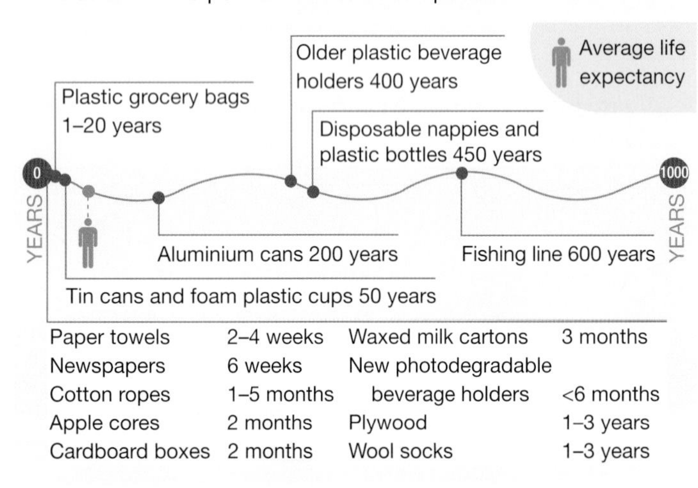

Note: Estimated individual item timelines depend on product composition and environmental conditions.

Source: South Carolina Sea Grant Consortium, South Carolina Department of Health and Envrionmental Control (DFHC) — Ocean and Coastal Resource Management, Centers for Ocean Sciences Education Excellence (COSEE) — Southeast and NOAA 2008

FIGURE 8 Foreign objects found in the stomach of a seabird. How many different items can you identify?

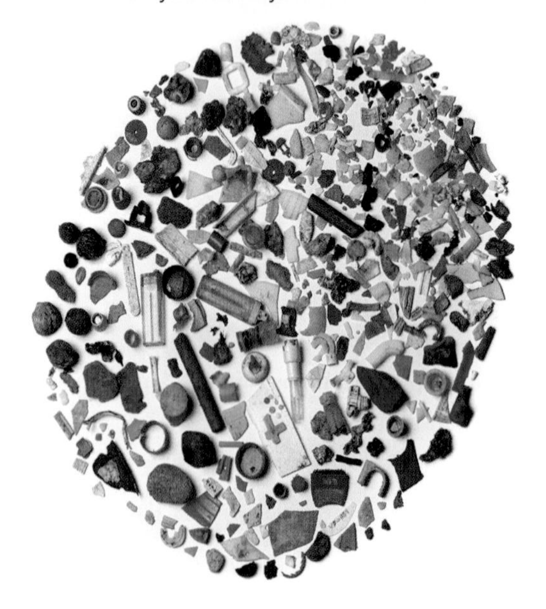

TABLE 1 Numbers of wildlife found entangled in marine debris, from 25 years of data

	Amphibians	Birds	Corals/ Sponges	Fish	Invertebrates	Mammals	Reptiles	Total
Beverage bottles	3	8	0	27	47	13	2	100
Beverage cans	1	2	0	15	17	1	0	36
Crab/ Lobster/ Fish traps	1	11	1	48	106	3	3	173
Fishing hooks	2	76	0	54	10	3	6	151
Fishing line	9	722	14	553	237	46	55	1636
Fishing nets	3	153	1	249	207	29	30	672
Bags (plastic)	13	102	0	142	91	33	23	404
Ribbon/ String	0	91	0	37	29	7	2	166
Rope	4	160	0	114	53	71	24	426
6-Pack holders	2	63	0	52	21	3	5	146
Plastic straps	2	30		34	12	5	5	88
Wire	1	31	1	16	13	7	6	75
Total	**41**	**1449**	**17**	**1341**	**843**	**221**	**161**	**4073**

Hitchhikers

Small marine creatures, such as barnacles, that normally spend their lives attached to rock, coral or coconut shells, can 'hitch a ride' on marine debris. The arrival of pest species in new locations can seriously affect ecosystems as they compete with native species for food or habitat.

Fishing industry

While the fishing industry contributes to marine debris, the industry itself is also affected by the litter. A survey in northern Scotland found that 92 per cent of fishermen had continual problems with marine debris in their nets, snagging nets on rubbish, and that some fishing grounds were avoided due to high litter concentrations.

People

Due to the action of currents, garbage discarded in one country can end up on the beaches of another country thousands of kilometres away. Thus the impacts of marine litter on people are mostly found in coastal regions. Impacts include the rising cost of clearing debris from beaches, loss of tourism revenue, and debris interfering with boating and **aquaculture**. At the extreme end, humans eat fish that might have ingested toxic substances.

6.3 Activities

To answer questions online and to receive **immediate feedback** and **sample responses** for every question, go to your learnON title at www.jacplus.com.au. *Note*: Question numbers may vary slightly.

Remember

1. What are the two biggest contributors to marine pollution across the world's ocean ***space***?
2. (a) Refer to figure 2. Give an example of a pollutant from each of the following sources of marine pollution: (i) atmospheric-based, (ii) land-based, (iii) marine-based.
   (b) Which of the three sources makes up the largest component of marine pollution?

Explain

3. Explain how a plastic bag discarded after a picnic in Los Angeles can end up in the middle of the Pacific Ocean.
4. Refer to figure 3. How would these items compare to a survey of marine litter conducted 50 years ago? What do you think has ***changed*** the most?
5. Refer to figure 5.
   (a) Using the scale provided, estimate the amount of plastic waste that comes from the top three polluters.
   (b) Suggest two reasons why Australia is not on the map.
6. Refer to figure 7. Compare the decomposition ***changes*** for natural materials and man-made materials as seen in this timeline. What does this indicate to the packaging industry and consumers?
7. Refer to table 1. What three items create the most problems for marine wildlife? Suggest reasons why.
8. Is our use of plastic a ***sustainable*** practice? Justify your answer.
9. What are the ***environmental***, economic and technological factors that have created the Great Pacific Ocean Garbage Patch?
10. What are the ***environmental changes*** that rubbish brings to oceans?

Discover

11. Use the **Plastiki Expedition** weblink in the Resources tab to learn about this project. Plastiki is a catamaran that was built by a team led by David De Rothschild. It is made totally out of recycled plastic bottles, and was used to sail the Pacific Ocean to demonstrate the impacts of plastic on the ***environment***. Write a newspaper report of the journey and the team's observations, and map the route they took.

Try out this interactivity: Garbage patch (int-3299)

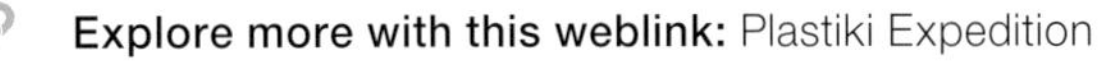

Explore more with this weblink: Plastiki Expedition

6.4 How can we best clean up ocean debris?

6.4.1 What can be done?

The way we consume and discard our resources has created one of the biggest environmental challenges in the world. Our throwaway society has literally thrown all our waste into the oceans! 'The water in our oceans is like blood for our planet. If we continue to fill it with toxic materials such as plastic, it will be to the detriment of life on Earth' (D. Woodring, Project Kaisei director).

Marine debris might start as a local problem, but it also creates a global problem as it often travels a great distance from its original source, crossing both geographic and political boundaries. Marine debris will be reduced only if land-based sources can be controlled. Communities and governments need to develop effective waste reduction schemes if we want to manage our oceans sustainably (see figure 1). If no action is taken, by 2025 we could end up with one tonne of plastic for every three tonnes of fish in our oceans.

Can't we just scoop it up?

Scooping up marine debris is not as easy as it sounds. Firstly, debris like the Great Pacific Ocean Garbage Patch is constantly moving in response to shifts in winds and currents. Secondly, much of the garbage is in the form of minute particles suspended beneath the ocean's surface. To scoop it up would mean collecting marine life that inhabits these waters as well.

FIGURE 1 What message is this advertisement sending?

WHAT GOES
IN THE OCEAN
GOES IN YOU.

What can you do?

The Surfrider Foundation in Australia and the United States is responsible for the 'Rise Above Plastics' campaign. The aim of the campaign is to get people to think about how they can make a difference and prevent marine debris. They suggest 10 ways to reduce your personal plastic footprint. There are also many innovative ways to recycle plastic products that can be found on YouTube, such as converting plastic bags to rope or handbags (use the **Plastic to rope** and **Plastic to handbags** weblinks in the Resources tab to view these videos).

learn on
RESOURCES — ONLINE ONLY

Explore more with these weblinks: Plastic to rope, Plastic to handbags

TEN WAYS TO REDUCE YOUR PERSONAL PLASTIC FOOTPRINT

1. Choose to reuse when it comes to shopping bags and bottled water. Use cloth bags and metal or glass reusable bottles if possible.
2. Refuse single-serving packaging, excess packaging, straws and other 'disposable' plastics. Carry reusable utensils in your bag, backpack or car.
3. Reduce everyday plastics such as sandwich bags and juice cartons by replacing them with a reusable lunch bag or box that includes a thermos.
4. Bring your to-go mug with you to the coffee shop, smoothie shop or restaurants that let you use them. This is a great way to reduce lids, plastic cups and/or plastic-lined cups.
5. Go digital! No need for plastic CDs, DVDs and jewel cases when you can buy your music and videos online.
6. Seek alternatives to the plastic items you use.
7. Recycle. If you must use plastic, try to choose #1 (PETE) or #2 (HDPE), which are the most commonly recycled plastics. Avoid plastic bags and polystyrene foam as both typically have very low recycling rates.
8. Volunteer at a beach clean-up. Surfrider Foundation Chapters often hold clean-ups monthly or more frequently.
9. Support plastic bag bans, polystyrene foam bans and bottle recycling bills.
10. Spread the word. Talk to your family and friends about why it is important to 'Rise Above Plastics'!

NO PLASTICS!

What can communities do?

Over 100 billion plastic bags are used each year in the United States, with less than 12 per cent recycled. Many governments and communities around the world now actively discourage the use of plastic bags. When Ireland introduced a bag levy in 2002, plastic bag usage dropped by 90 per cent. On one day of each year, volunteers from over 152 countries clean up the shores of beaches, lakes and streams, by classifying, counting and collecting garbage, as part of the International Coastal Cleanup Campaign. Over the past 25 years, this campaign has led to the removal of more than 66 million kilograms of litter, the equivalent of 330 kilometres of cars nose to tail, or 66 000 average-sized cars! The data collected via the campaign have contributed to new littering laws (see figure 2).

FIGURE 2 Marine pollution restrictions in the United States

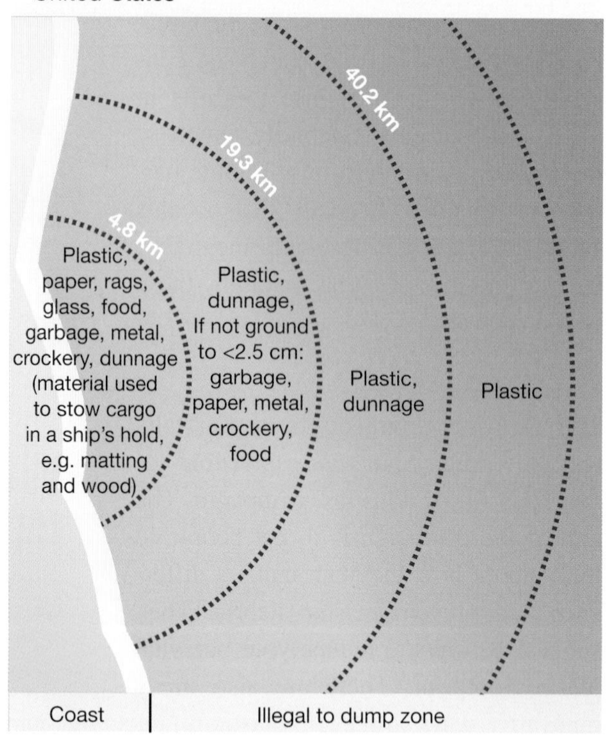

What can fishermen do?

A 'Fishing for Litter' scheme has been set up in Scotland where fishermen and port authorities have collaborated to collect all litter caught in nets. Instead of throwing this litter overboard, the debris is collected and brought back to port for managing. From 2005 to 2015, Scottish fishermen collected more than 900 tonnes of marine litter. Recreational fishermen in the United States can recycle fishing lines back to the manufacturer via collection points. Since the US scheme started in 1990, it has prevented over 15 million kilometres of fishing line potentially entangling wildlife.

What can manufacturers do?

In recent years, manufacturers have become much more environmentally aware. New biodegradable packaging materials and improved recycling methods have been developed. The United States has passed laws to phase out the use of microbeads in cosmetic and personal products from 2017 onwards. Microbeads are tiny plastic fabrics found in such things as toothpastes, body washes and other cosmetic products. On packaging they are labelled as polyethylene (PE). The beads pass through water filtration plants and are swept out to sea where they are ingested by sea creatures. These microbeads can contain toxic substances which can be passed up the food chain to people.

In Australia, major supermarket chains and beauty product manufacturers are also starting to phase out microbeads, although imported goods might still contain them.

What can the international community do?

The United Nations Environment Programme (UNEP) has launched an intensive publicity campaign to help raise awareness of marine debris. They are working at a regional level to promote schemes such as recycling, waste separations and other land-based actions. International agreements such as the International Convention for the Prevention of Pollution from Ships (known as MARPOL) prohibit the disposal of all plastic into the sea, and ships cannot dispose of food waste within 12 **nautical miles** of land. Such regulations are extremely difficult to police and have no impact on the amount of waste entering the ocean from land-based sources.

6.4 Activities

To answer questions online and to receive **immediate feedback** and **sample responses** for every question, go to your learnON title at www.jacplus.com.au. *Note*: Question numbers may vary slightly.

Think

1. Is the saying, 'Think global, act local' applicable to marine pollution? Justify your answer.
2. How successful would an international agreement where all countries decide to reduce land-based marine pollution be? What would be the advantages and disadvantages?
3. (a) Construct a table, similar to the one below, to evaluate each of the proposals to help reduce ocean debris.

Response	Economic criteria (Cost)	Social criteria (Time and effort required)	Environmental criteria (Effectiveness)
Individual actions			
Manufacturers			
International community			

   (b) What conclusions can you draw from your table?
4. How can you reduce your school's plastic footprint? You may like to use the **School** weblink in the Resources tab for ideas. Brainstorm ideas as a class and then develop one idea in detail. How can you promote your idea? You may like to create a slogan and poster, address a school group or assembly or write a proposal to the school administration.
5. Undertake a plastic bottle survey at home. Check your kitchen, laundry and bathroom and count the number of plastic bottles, jars and other containers you find (only count containers). Collate your results, in graph form, with other students in your class and then write a summary of your findings.

learnon RESOURCES — ONLINE ONLY
Explore more with this weblink: School

6.5 Where else is marine debris a problem?

6.5.1 What are ghost nets?

The humble fishing net, once a simple handmade rope construction, has largely been replaced by thousands of metres of nylon webbing. If accidentally lost or purposely discarded, these massive rafts of netting drift around the oceans as **ghost nets**, waiting to trap any unwary sea creature or bird.

6.5.2 Where are ghost nets a problem in Australia?

Marine debris occurs around the coast of Australia, especially in places close to major population centres. It is also a major problem in northern Australia, particularly in the Gulf of Carpentaria. Here densities of nets can reach up to three tonnes per kilometre, among the highest in the world. The coastlines in this region are pristine environments and support six of the world's seven marine turtles. Turtles make up 80 per cent of marine creatures captured in the nets. Over 90 per cent of the debris that collects is derived from the fishing industry, most of it originating from South-East Asia, with the remaining 8.6 per cent being Australian in origin. Most of the nets come from the Arafura Sea, an important fishing ground, especially for the Indonesian fishing industry. More than 62 per cent of the nets are trawling nets – the Arafura Sea being the only region of Indonesian waters where trawling is not banned. Under the influence of the south-east trade winds and north-west monsoon winds, a circular gyre pattern develops, which allows the build-up of ghost nets to develop, similar to the Great Pacific Ocean Garbage Patch (see figure 1).

FIGURE 1 Distribution of ghost net hot spots around northern Australia

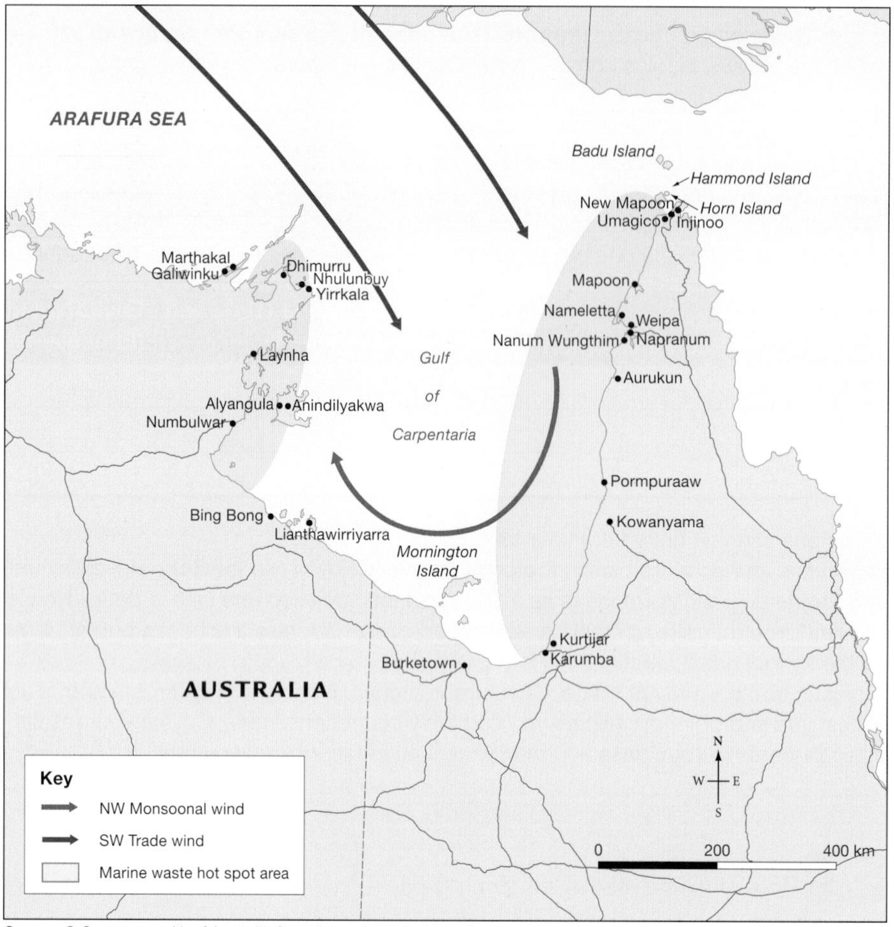

Source: © Commonwealth of Australia Geoscience Australia 2013. Ghost Nets Australia, www.ghostnets.com.au/index-.html

6.5.3 What is being done?

GhostNets Australia is an alliance of over 22 Indigenous communities in remote coastal places of Western Australia, Queensland and the Northern Territory, funded by the Federal Government. Since its establishment in 2004, over 13 000 ghost nets have been captured by locally trained rangers (see figure 2).

FIGURE 2 Captured trawler nets being collected by rangers

Often, helicopters are used to spot the ghost nets washed ashore, which are then checked for trapped wildlife. Live turtles are tagged and data recorded before they are returned to the sea. Nets are dragged up above the **high tide line** to be identified, collected and disposed of later. The project works on a '6R' principle:

1. **R**emove ghost nets from waters and coastline of the Gulf of Carpentaria.
2. **R**ecord the number, size, type and location of nets.
3. **R**escue animals trapped in nets.

4. **R**eport the activities that the community has done to increase awareness.
5. **R**educe the number of nets in the Gulf by working together.
6. **R**esearch factors that influence the distribution, movement and impact of ghost nets.

This program is part of a Caring for our Country initiative in the region, which promotes stewardship of Indigenous customary lands and seas.

What can be done with the debris?

Traditionally, fishing nets were made of more eco-friendly materials, such as flax or hemp, but they are now increasingly made of nylon, which makes them stronger, cheaper and more buoyant. However, they are also harder to dispose of as they take a very long time to break down. Nets can also range in size from 30 cm to 6 km in length. There are three options for disposal of the waste: burning, placing in landfill or recycling. Each, however, has disadvantages, and all methods require the waste to be collected over long distances and difficult terrain.

Disadvantages of burning fishing nets include:
- burning plastic is illegal in most countries
- after burning, the residue is a huge, heavy, immovable mass of melted plastic, which is a visual eyesore
- health risks associated with burning plastic.

Disadvantages of disposing of fishing nets in landfill include:
- expense of transporting the waste over large distances to a landfill site
- often waste is burned in tips, and these tips are close to settlements.

Disadvantages of recycling or reusing fishing nets include:
- remoteness of and distances to recycling plants (South Australia and Taiwan have plants big enough to cope with fishing nets)
- expense of transporting the waste over large distances
- the need for large machinery to chop plastic into manageable pieces
- the need to find a local use for the recycled waste material.

What is GhostNets Australia's solution?

While only a partial solution to the large quantity of nets accumulating, GhostNets Australia promotes the reuse of nets by providing local artists with netting material. The artists use traditional weaving techniques to create artworks (see figure 3). This type of **cottage industry** brings economic and social benefits as well as raising awareness of the problem of marine debris.

FIGURE 3 Woven basket made of recycled fishing net

6.5 Activities

To answer questions online and to receive **immediate feedback** and **sample responses** for every question, go to your learnON title at www.jacplus.com.au. *Note*: Question numbers may vary slightly.

Remember

1. Why are ghost nets a problem in northern Australia?
2. Why are fishing nets an ***environmental*** problem?

Explain

3. Refer to figure 1. On which side of the Gulf would you expect ghost nets to build up:
   (a) during the north-west monsoon season
   (b) during the south-east trade wind season?
4. Why is an understanding of local wind patterns useful to rangers?
5. Why is transporting nets to South Australia for recycling not a viable option?
6. Evaluate the ***environmental***, economic and social aspects of the GhostNets program.

Discover

7. If you have access to a beach, walk along the high tide line and see if you can collect and identify different forms of marine litter. Collate and record your findings. What were the most common forms of litter than you identified? Where have they come from?
8. Research information on the different types of fishing nets used: gill, purse, seine and trawl nets.
   (a) Construct a table to list the advantages and disadvantages of each from a fishing and an ***environmental*** perspective.
   (b) Which net design might prove to be the most damaging to the ***environment*** if lost or discarded?

6.6 SkillBuilder: Using geographic information systems (GIS)

WHAT IS GIS?

GIS is a computer-based system of layers of geographic data. Just as an overlay map allows you to interchange layers of information, GIS allows you to turn layers on and off to make comparisons between pieces of data.

Go online to access:

- a clear step-by-step explanation to help you master the skill
- a model of what you are aiming for
- a checklist of key aspects of the skill
- a series of questions to help you apply the skill and to check your understanding.

FIGURE 1 Studying marine reefs using GIS on the Red Sea

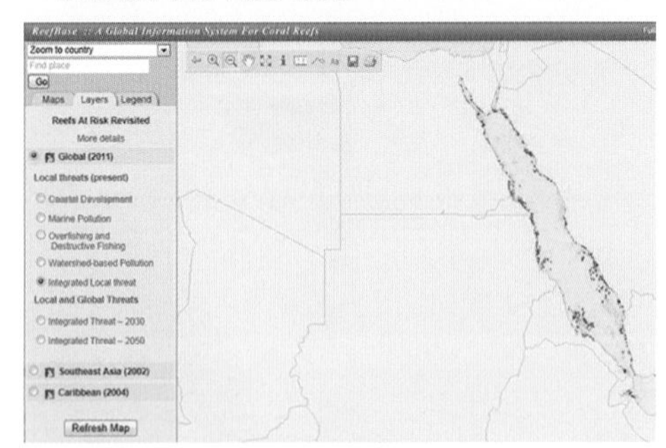

Source: © Reefbase/Worldfish

RESOURCES — ONLINE ONLY

Watch this eLesson: Using geographic information systems (GIS) (eles-1752)

Try out this interactivity: Using geographic information systems (GIS) (int-3370)

6.7 Where does oil in the sea come from?

6.7.1 How does oil get into the ocean?

You have probably seen images of birds covered in sticky oil, usually as a result of the most dramatic type of marine pollution: oil spills and shipping accidents. The impact of oil on ocean and coastal ecosystems is often localised over a relatively small area, but may last for many years.

Almost all of the Earth's supply of oil and natural gas is found in deep underground reservoirs. Reservoirs can be under a landmass, under the seabed and under **continental shelves**. Extracting oil from the seabed accounts for nearly 30 per cent of the world's production. Offshore drilling takes place on huge floating platforms, in waters up to 2 kilometres deep and as far as 300 kilometres from the coast. More than 50 per cent of countries around the world drill for offshore oil and gas.

The most obvious and visible kinds of marine oil spills usually involve tanker accidents, or leaks from offshore oil rigs. However, oils enter the ocean from a variety of sources, with both natural and land-based sources accounting for a much larger proportion than disasters (see figure 1). There has been a decrease in the number of tanker accidents in recent years, mostly due to improved ship design and greater safety methods. However, with more ships and supertankers being built, the potential risk of an accident is still high.

FIGURE 1 Main sources of marine oil pollution

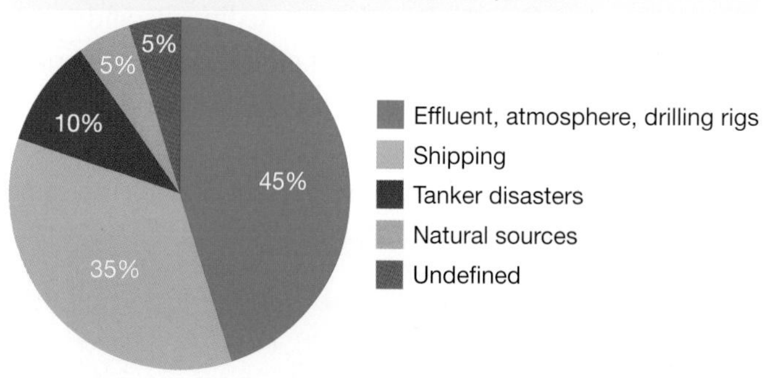

FIGURE 2 What happens to oil in the ocean?

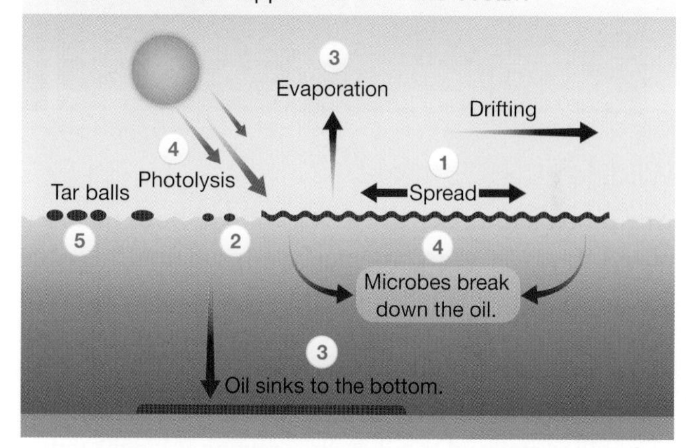

1. When oil is released into the ocean it immediately forms large slicks which float on the surface. It can take only 10 minutes for one ton of oil to disperse over a radius of 50 m and be 10 mm thick.
2. After a few hours, weathering by wind and waves breaks down the slick into narrow bands, or windrows, that float parallel to the wind. The oil may be less than 1 mm thick but can now cover 12 km^2. After the slick thins down it breaks up into fragments and fine droplets that can be transported over larger distances.
3. Some of the oil evaporates or sinks.
4. Some of the oil can be chemically broken down by sunlight or bacteria.
5. Finally the oil solidifies into tar balls (clumps), which are more resistant to bacterial decomposition.

6.7.2 What happens to oil in the ocean?

Each oil spill is different and there are various physical, chemical and biological factors that will influence the behaviour of spilt oil. The type of oil, temperature of the water, wave and current action, and the nutrient content of the water are all critical influences. The stages in the breakdown of oil can be seen in figure 2.

6.7.3 What does oil do to the environment?

Oil spills can result in both short- and long-term environmental change, with some damage lasting for decades. A spill in open waters is usually less destructive than a spill near coastal waters, where most fish and bird breeding takes place. Oil pollution is less visible in the open ocean, especially once it disappears from the surface, but it is still capable of being moved via ocean currents.

Coastlines

The geography of the coastline can influence the degree of impacts from an oil spill. Impacts are less on exposed coasts due to strong wave action. A long, sheltered, sandy coastline is vulnerable as the oil can soak into the sand, which is extremely difficult to clean. Mangroves, salt marshes and extensive sandbanks are also sensitive as the oil soaks into the fine sediments and can be taken up by plants. This affects wildlife that live in this habitat, and the loss of vegetation increases the risk of coastal erosion, as shown in figure 3 (a) and (b). Coral reefs are possibly the most vulnerable to oil spills, and they are extremely slow to recover.

FIGURE 3 (a) Oil damage to wetland habitat (b) The same area of wetlands one year after the oil spill

Wildlife

Any oil on the surface of the sea will kill birds that swim and dive for their food there. Feathers covered in oil rob birds of waterproofing and insulation. Ingesting the oil can poison them. Oil spills also damage coastal nesting and breeding grounds. Oil can block the blow holes of marine mammals such as whales, dolphins and seals, making breathing difficult. If oil coats their fur, they become vulnerable to hypothermia. Animals' food supply is also poisoned by floating oil. Fish, especially shellfish, suffer immediate effects of an oil accident. Reduced reproduction, birth defects and other abnormalities develop in the next generation of wildlife exposed to oil spills, creating a longer term impact.

6.7 Activities

To answer questions online and to receive **immediate feedback** and **sample responses** for every question, go to your learnON title at www.jacplus.com.au. *Note*: Question numbers may vary slightly.

Remember

1. What percentage of the world's oil comes from the seabed?
2. Examine figure 2. Why is it important that oil spills are treated as quickly as possible?
3. Examine figures 3 (a) and (b) and describe the ***changes*** that you can see in the two ***environments***.
4. Use the **Oil spill** weblink in the Resources tab to examine a sequence of maps that track the distribution of the oil spill from the Deepwater Horizon oil rig explosion on 20 April 2010 until 3 August 2010. Select the 'Loop current' button and the 'Oil on shoreline' button to view the additional features these show. You may also like to view the satellite images.
   (a) What were the main directions that the spill travelled in? What factors would have influenced the directions?
   (b) What other ***places*** may have been affected had the oil spill moved in the Loop Current?

Explain

5. (a) Examine figure 2. Select the conditions from those listed below that would be most likely to encourage the rapid breakdown of an oil spill.
   - Cold ocean water/warm ocean water
   - Calm seas/choppy seas
   - Ready supply of bacteria/limited supply of bacteria
   - High level of oxygen in the water/low level of oxygen in the water
   - High number of bacteria-eating organisms/low number of bacteria-eating organisms

(b) Justify each of your choices in part (a).
6. Suggest one ***environmental***, economic and technological factor that can contribute to marine oil pollution.
7. List the ways in which oil creates ***environmental change*** in the ocean.
8. Compare some of the advantages and disadvantages of drilling for oil in the ocean compared to drilling for oil on land.

Discover

9. Research the potential impact of oil spills on the Great Barrier Reef. How does the Great Barrier Reef Marine Park Authority manage the park ***sustainably*** to prevent oil spills?

Try out this interactivity: Oil slick (int-3300)

Explore more with this weblink: Oil spill

6.8 SkillBuilder: Describing change over time

online only

WHAT IS A DESCRIPTION OF CHANGE OVER TIME?

A description of change over time is a verbal or written description of how far a feature moves, or how much it is altered, over an extended time period, and can alert us to the possible impacts of a change or changes over a wider region.

Go online to access:

- a clear step-by-step explanation to help you master the skill
- a model of what you are aiming for
- a checklist of key aspects of the skill
- a series of questions to help you apply the skill and to check your understanding.

FIGURE 1 Tsunami mapping from Peru, 2007. A magnitude 8.0 earthquake occurred on 15 August 2007 near the coast of Peru. A tsunami was detected by Seaframe (sea-level fine resolution acoustic measuring equipment) stations located on Pacific islands.

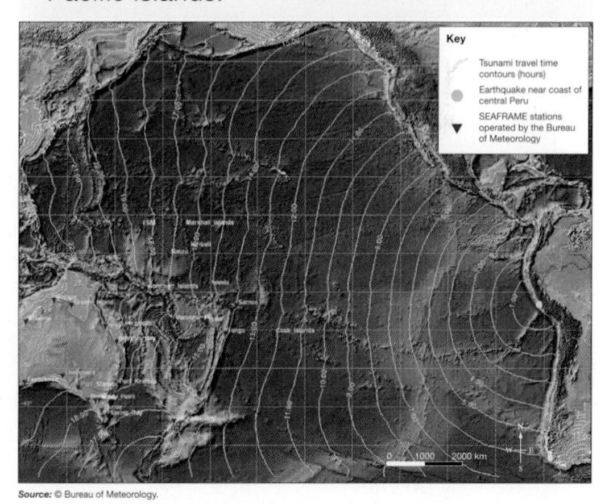

Source: © Bureau of Meteorology.

learn on RESOURCES — ONLINE ONLY

Watch this eLesson: Describing change over time (eles-1753)

Try out this interactivity: Describing change over time (int-3371)

6.9 What is the solution to marine pollution?

6.9.1 How can we clean up oil spills?

The solution to pollution is not dilution. For many years it was thought that any pollutants that ended up in the ocean would just disappear — the oceans were so vast and so deep. We now know better. Despite wonderful advances in technology, we still have marine accidents, specifically related to the oil industry.

There are two ways to deal with oil spills: **remediation** and prevention. A combination is used, depending on location, weather and the type of spill.

Remediation

It is extremely difficult to contain and clean up any size oil spill, and many of the earlier methods used often caused more environmental damage than the oil itself. For instance, consider the impact of a high pressure hose on a fragile ecosystem, or the spraying of toxic chemicals to absorb the oil. There is a range of remediation methods for cleaning up an oil spill (see figure 1).

FIGURE 1 Different methods of cleaning up oil spills

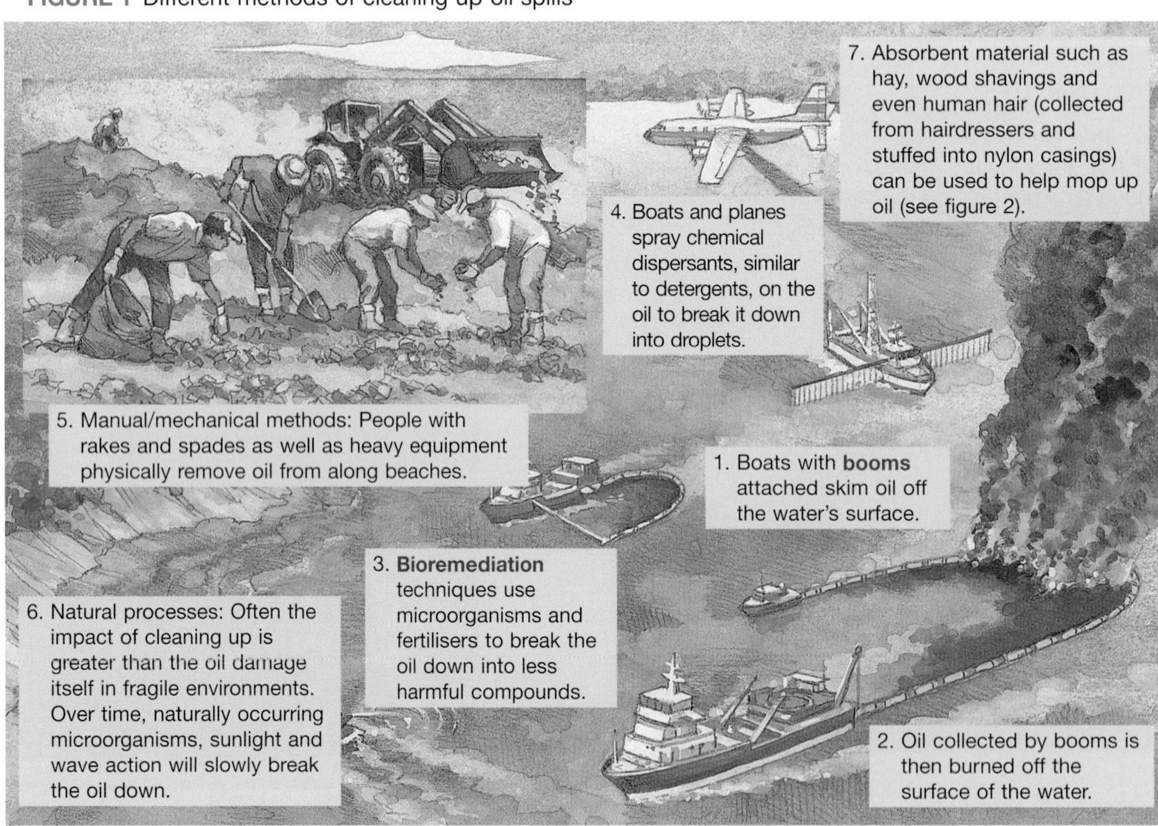

1. Boats with **booms** attached skim oil off the water's surface.
2. Oil collected by booms is then burned off the surface of the water.
3. **Bioremediation** techniques use microorganisms and fertilisers to break the oil down into less harmful compounds.
4. Boats and planes spray chemical dispersants, similar to detergents, on the oil to break it down into droplets.
5. Manual/mechanical methods: People with rakes and spades as well as heavy equipment physically remove oil from along beaches.
6. Natural processes: Often the impact of cleaning up is greater than the oil damage itself in fragile environments. Over time, naturally occurring microorganisms, sunlight and wave action will slowly break the oil down.
7. Absorbent material such as hay, wood shavings and even human hair (collected from hairdressers and stuffed into nylon casings) can be used to help mop up oil (see figure 2).

Prevention

The most important way to deal with oil spills is to prevent them from happening. International cooperation has seen the United Nations treaty MARPOL (MARine POLlution) established in 1983 to deal with the growing problem of marine pollution. Individual countries have also established new rules and regulations. For example, by 2015, all tankers operating in United States waters must be double hulled, so that if the outer hull is damaged the inner hull can still hold the fuel. The oil industry must now have detailed response plans for cleaning up any spills.

FIGURE 2 Recycled human hair turned into oil-absorbent logs for soaking up oil spills

6.9.2 How was the Gulf disaster cleaned up?

In April 2010, an explosion aboard the Deepwater Horizon oil rig in the Gulf of Mexico created a large-scale environmental disaster (see subtopic 6.10). Figure 3 shows the results of the clean-up following the oil spill, 103 days after the accident. Favourable weather conditions at the time enabled authorities to put some defensive measures in place, including more than 4000 kilometres of booms, to protect coastal land.

In all, an estimated 6.4 million litres of dispersants were used on the spill. Scientists believe that nearly 50 per cent of the oil spilt and nearly 100 per cent of the methane gas released has stayed deep in the ocean. As much as 3200 square kilometres of ocean floor is thought to be polluted. Patches of oil are still emerging in different locations, years after the accident. Some is being swept up the sea bed by currents and moved by storm waves. On the sea bed, coral reefs are still showing signs of damage, and on land, some marshes are still giving off toxic fumes.

Due to the high number and distribution of oil wells in the Gulf of Mexico (see figure 4), the threat of future accidents remains. In Florida, it has been estimated that the annual value of tourism and fishing along the state's eastern Gulf coast is three times higher, and considerably more sustainable, than the value of any oil or gas that might be found there. The US Federal Government has a ban on offshore oil and gas drilling in any new areas.

FIGURE 3 How the Gulf of Mexico oil spill was cleaned up

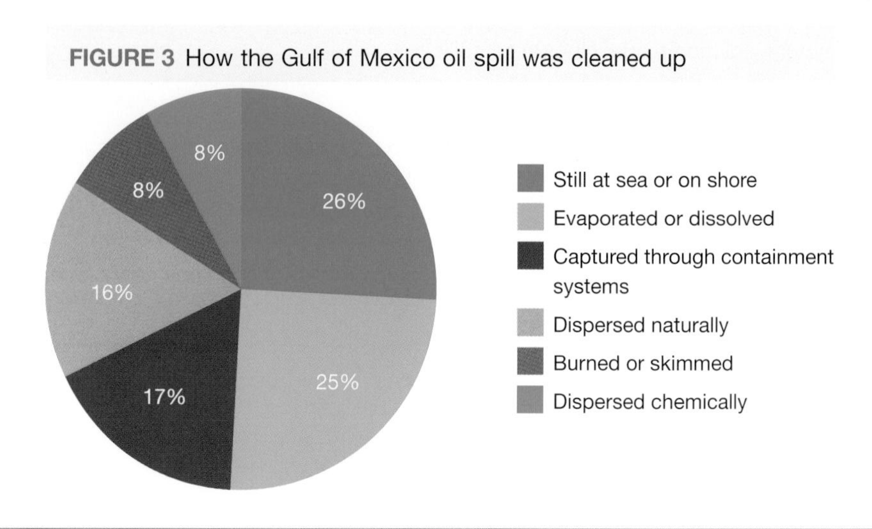

FIGURE 4 Distribution of oil wells in the northern Gulf of Mexico

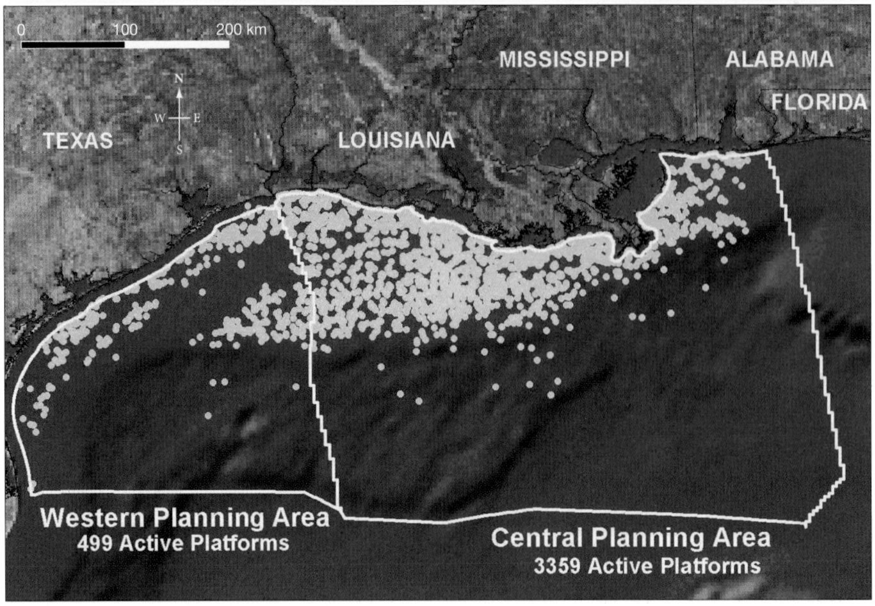

Source: National Oceanic and Atmospheric Administration, Office of Ocean Exploration and Research, U.S. Department of Commerce. Adapted by Spatial Vision

6.9 Activities

To answer questions online and to receive **immediate feedback** and **sample responses** for every question, go to your learnON title at www.jacplus.com.au. *Note*: Question numbers may vary slightly.

Remember

1. Why is it important to treat an oil spill quickly?
2. (a) Study figure 3. What percentage of the oil spill had been treated at this time? What percentage of the oil had dispersed naturally or evaporated?
   (b) Why do you think only a small percentage was chemically dispersed?
3. Imagine that you spill a whole bottle of cooking oil on your kitchen floor. Describe three different remediation methods that you could use to clean up the spill. Select from booms, skimmers, bioremediation, manual/mechanical, dispersants and absorbers. Would one method be more effective than another? Give reasons.

Explain

4. Construct a table to suggest the possible advantages and disadvantages of the seven methods of remediation shown in figure 1. Consider the influence of the following factors: weather conditions, timing, location of treatment area (at sea or on coast), size of area to be treated, ***environmental*** impacts, practicality, economic viability and social justice.
5. Why is there is no one solution to cleaning up oil spills?

Discover

6. Investigate the cause and impacts of the Montara oil rig explosion off the north-west coast of Western Australia in 2009. Why was this not as severe as the Gulf accident?

Think

7. Has the US Government made the right decision in banning drilling for oil in any new ***places***? What did it factor into its decision? Discuss.

6.10 The world's worst oil spill?

Access this subtopic at **www.jacplus.com.au**

6.11 Review

6.11.1 Review

The Review section contains a range of different questions and activities to help you revise and recall what you have learned, especially prior to a topic test.

6.11.2 Reflect

The Reflect section provides you with an opportunity to apply and extend your learning.

Access this subtopic at **www.jacplus.com.au**

TOPIC 7 Sustaining urban environments

7.1 Overview

Numerous **videos** and **interactivities** are embedded just where you need them, at the point of learning, in your learnON title at www.jacplus.com.au. They will help you to learn the content and concepts covered in this topic.

7.1.1 Introduction

Urban environments provide homes, places of work and all the conveniences of modern-day life for their citizens. They are often a magnet for people living in small rural townships, as goods and services abound and social and economic opportunities for a better life are seen as more possible in the big cities.

The complexity of urban environments can be seen in a modern city such as Shanghai, with all its multi-layered buildings, bridges, roadways, electricity, water supplies and services. The need to deal with the huge amounts of waste generated by the population of a city of this size is a concern for its urban planners and managers. To ensure viability into the future, sustainable solutions to the wide range of problems that exist in big cities must be found.

Shanghai urban environment

Starter questions

1. What ***changes*** would have been made to the natural ***environment*** to build the urban ***environment*** shown in section 7.1.1?
2. How might this urban ***environment*** affect wind movement and run-off of rainfall?
3. Would you like to live in this urban ***environment***? Why?

INQUIRY SEQUENCE

7.2 How do urban environments develop?

7.2.1 Early settlements

The earliest forms of the **urban environment** consisted of shelters to protect people from the elements and provide security from the attacks of predators. From these simplest forms, the highly complex modern urban environment has developed.

The United Nations predicts that by the year 2050, 70 per cent of the population in developed nations and 40 per cent in **developing nations** will live in large urban complexes.

7.2.2 Impacts of the growth of urban environments

Rapid growth in city populations has led to problems such as urban sprawl, traffic congestion and air and water pollution, with significant impacts on the natural environment. Social problems such as unemployment; inadequate housing, **infrastructure**, water, sewerage and electricity supplies; pollution; and the spread of slums and crime are further problems. In addition, with prospects of climate change through global warming, many of the world's coastal cities are under threat of rising sea levels. The application of **human–environment systems thinking** will be the key to evaluating and solving these economic and social issues.

The United Nations estimates that a staggering 90 per cent of the worlds' population growth is taking place in the cities of the developing nations. For many of the people in these countries, pressures such as extreme poverty, famine and civil unrest often draw them away from rural areas towards cities, 'pulled' by the promise of jobs, shelter and protection.

FIGURE 1 Contrasts in urban development in Manila, Philippines

7.2.3 Development of cities

The earliest cities emerged five to six thousand years ago in Mesopotamia (present-day Iraq), Egypt, India and China. These cities became centres for merchants, craftspeople, traders and government officials and were dependent on agriculture and domesticated animals from surrounding rural areas.

Before the year 1800, over 90 per cent of the world's population lived in rural agriculture-based societies. With the **Industrial Revolution**, people started moving from rural areas to cities or migrated to other countries to take advantage of jobs in factories and to improve their standard of living. In 1850, only two cities in the world — London and Paris — had a population above one million. By 1900 there were 12, by 1950 there were 83, and by 1990 there were 286. 2015 figures indicate that there are now in excess of 500 cities that are home to a million or more people.

Megacities

The term 'megacity' commonly refers to urban settlements of 10 million inhabitants or more. Currently, about 500 million people live in at least 35 megacities across the world. This population figure is expected to rise to 5 billion by 2030.

FIGURE 2 Distribution of the world's population, 2014

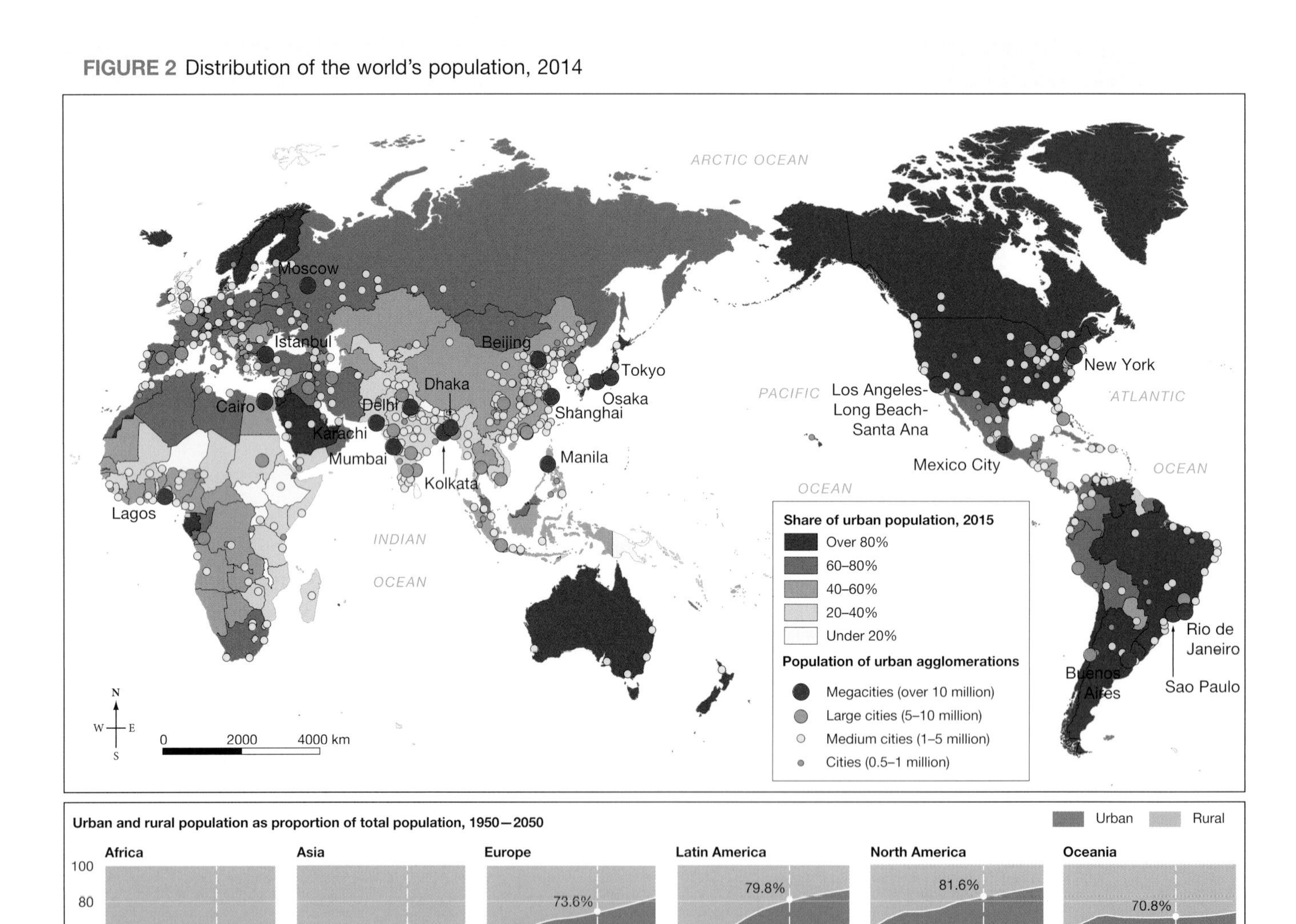

Source: United Nations Department of Economic and Social Affairs

Europe and Asia have many large cities, and Africa and Oceania have proportionally fewer (see figure 2). In Papua New Guinea and Burundi, only 10 per cent of the population is urbanised, whereas in Singapore this figure is 100 per cent.

Of the 15 largest megacities, only four are located in highly industrialised countries: Tokyo, New York, Los Angeles and Osaka–Kobe–Kyoto. Three-quarters of these cities are in developing countries; they include gigantic **conurbations** such as Jakarta, Manila and Karachi (see table 1).

TABLE 1 Populations of the 10 largest cities, 1950 to 2025

Rank	1950 population	2015 population	2025 population (projected)
1	New York (12.3)	Tokyo (37.8)	Tokyo (35.2)
2	London (8.7)	Jakarta (30.5)	Mumbai (26.4)
3	Tokyo (6.9)	Delhi (24.9)	Delhi (22.5)
4	Paris (5.4)	Manila (24.0)	Dhaka (22.0)
5	Moscow (5.4)	Seoul (23.4)	São Paulo (21.4)
6	Shanghai (5.3)	Shanghai (23.4)	Mexico City (21.0)
7	Essen (5.3)	Karachi (22.1)	New York (20.6)
8	Buenos Aires (5.0)	Beijing (21.0)	Kolkata (20.6)
9	Chicago (4.9)	New York (20.6)	Shanghai (19.4)
10	Kolkata (4.4)	Guangzhou (20.6)	Karachi (19.1)

Sources: United Nations (1950 & 2025); Demographia (2015)
Note: All population figures are in millions.

7.2 Activities

To answer questions online and to receive **immediate feedback** and **sample responses** for every question, go to your learnON title at www.jacplus.com.au. *Note*: Question numbers may vary slightly.

Remember

1. When and where did the first cities develop?
2. Why did cities experience rapid growth and development after the Industrial Revolution?

Explain

3. (a) What factors are driving the process of urbanisation in the world?
   (b) In which regions of the world is this occurring most quickly? Why?
4. What are some of the major economic and social issues facing rapidly developing cities in the world?
5. Construct a bar graph to represent the expected populations for the projected 10 largest cities in the world in 2025. Also research and represent these cities' current populations on the same graph. Locate each of these cities on a map and annotate this map to show which cities will grow the most and which will grow the least. Suggest why.

Predict

6. What impact do you think global warming and rising sea levels will have on coastal cities around the world by 2025 and beyond?

Think

7. What are some other urban problems, besides those mentioned in this section, that arise as cities develop?
8. What do you think are some of the advantages of living (a) in a large city, (b) in a small town and (c) on a farm? Which would you prefer? Why?

Deepen your understanding of this topic with related case studies and questions.
- World urbanisation
- Urbanisation in Australia

7.3 How do cities change the environment?

7.3.1 What are the limits and changes?

All forms of urban environments are interconnected with the **biophysical environment**. The 'bio' elements are all forms of plant and animal life including people and all their activity and industry. The 'physical' elements are the atmosphere, hydrosphere and lithosphere or Earth's surface.

The biophysical elements impose limits on the development and sustainability of all forms of urban environment. On the other hand, the urban environment imposes significant human-induced change on the biophysical world. The understanding of this interconnection is particularly important in a world of increasing human numbers and pressure for resources on the biophysical environment (see figure 1).

FIGURE 1 Interaction between the urban environment and the biophysical environment

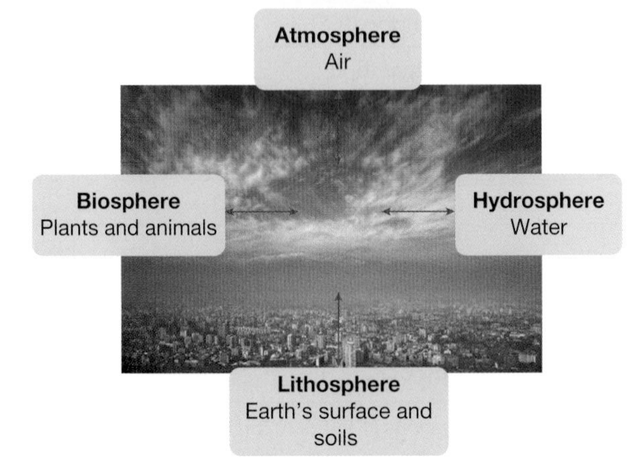

7.3.2 How does the urban environment affect the atmosphere?

Where sources of potentially dangerous gaseous emissions are high, such as from buildings, transport systems and industry, atmospheric pollution can be a problem. Examples such as hazy conditions, photochemical smogs, light and noise pollution, and acid rain are significant problems that need to be addressed. Hence, the development of clean air policies controlling emissions of gases into the atmosphere is important. Some examples are the introduction of lead-free petrol, banning the burning of household waste and emission control systems on factory furnaces.

Cities and industries have huge demands for energy, and the by-product of this is heat. What is known as the 'heat island effect', whereby urban environment structures such as buildings and roads absorb heat from the sun, raises the temperature of the city environment compared to rural surrounds (see figure 2).

The production of greenhouse gases such as carbon dioxide and methane by urban environments is recognised as probably the greatest contemporary climate issue. Global warming leading to climate change is the result of emissions of these gases into the atmosphere, particularly in large urban centres.

FIGURE 2 The 'heat island effect' of cities

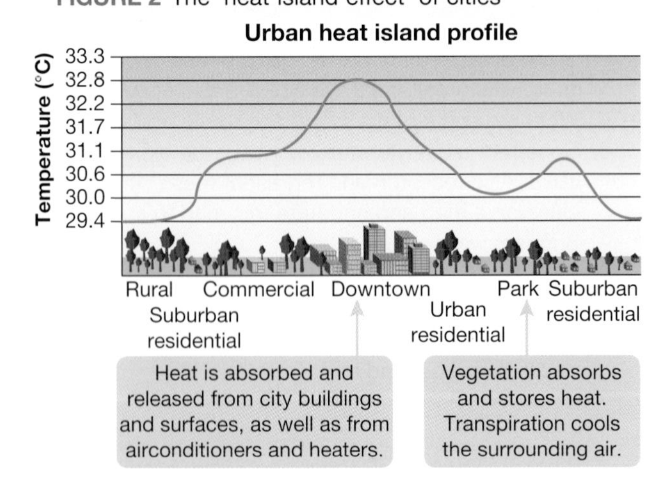

7.3.3 How does the urban environment affect the hydrosphere?

As the urban environment is closely dependent on the hydrosphere, it is not surprising that the provision of clean water and/or what is known as water security and **water rights** are important management objectives for a sustainable future. One of the most important aims for urban planners in cities is to ensure supplies of clean water and to manage the waste water from cities by schemes such as urban wetlands.

In general, all urban centres are trying to find increasing supplies of water for domestic and industrial consumption from rivers, groundwater and, more recently, desalinisation sources.

Infrastructure in the form of dams, pipelines, and artesian waters at the local level and major water management schemes such as the Snowy Mountains Scheme in Australia are ways that water is gathered. Water pollution caused by urban environments is also important as polluted waters are a risk to all life forms in any environment. Considerations for biomes and ecosystems of rivers, wetlands and swamps in terms of protecting habitats and maintaining biodiversity is also a major management aim.

FIGURE 3 Like those of many rivers in the world, the banks and channel of the Seine River have been heavily modified by people.

7.3.4 How does the urban environment affect the lithosphere?

Apart from the impact on the land of the 'tar and cement' structures of cities, there is the sheer size of today's megacities. These urban structures can cover areas of up to 100 square kilometres and this can lead to problems of urban sprawl, such as extended commuting times, congestion, air pollution and slums. Other problems include the disposal of the enormous amount of garbage and waste that cities produce, and the general impacts on agriculture, plants and animal life in adjacent habitats and ecosystems.

Urban environment surfaces, such as car parks, generate two to six times more run-off than a natural surface. Rain that falls on car parks can be contaminated with petroleum residues, fertilisers, pesticides and other pollutants.

FIGURE 4 One of the world's largest car parks will be located in the Dubai World Central International Airport (shown in the model below). It will have over 100 000 parking spaces for its employees, Dubai residents, tourists and other users.

7.3 Activities

To answer questions online and to receive **immediate feedback** and **sample responses** for every question, go to your learnON title at www.jacplus.com.au. *Note*: Question numbers may vary slightly.

Remember

1. What is the 'bio' part of the biophysical ***environment***?

Explain

2. Give reasons why urban ***environments*** can have such a major impact on the Earth's atmosphere.
3. Why do most of the large urban centres of the world have high-rise buildings?

Discover

4. How do wetland systems operate and why are they successful in 'cleaning up' storm waters from cities?

Predict

5. How might rising sea levels, predicted to be a result of global warming, affect the ***place*** and ***space*** of a city such as New York?
6. How will the supply of fresh water affect the development of cities in the future?

Think

7. Identify any transport infrastructure problems that exist in the capital city of your region and comment on how they are being overcome and how this will lead to a more ***sustainable*** urban ***environment***.

Deepen your understanding of this topic with related case studies and questions.

- **Polluted cities**
- **Mexico City**

Try out this interactivity: Urban impacts on the environment (int-3301)

7.4 Why do urban areas decline?

7.4.1 Reasons for the decline of urban environments

Over the ages, and even in recent times, there have been numerous cities, towns and villages that have declined or been abandoned. The factors that lead to such decline may be environmental or human.

Environmental factors

Over time, all forms of urban environments will deteriorate with age and require renovation or renewal. Extreme atmospheric events such as cyclones, hurricanes and tornadoes, which exhibit strong winds and

FIGURE 1 Ruins of the Roman city of Pompeii with Mt Vesuvius in the background

flooding rains, can have devastating short-term impacts on urban environments. Longer term events such as **desertification** and climate change can also have negative impacts.

Movements of the earth such as those due to earthquakes, volcanoes and tsunamis can also destroy urban environments. Two well documented examples of such events are the eruption of Mt Vesuvius in Italy in 79 CE, which completely buried the cities of Pompeii and Herculaneum under volcanic ash, and the destruction of coastal communities in Japan by the earthquake and tsunami of 2011. Towns such as Otsuchi are now just starting to be rebuilt, with buildings being constructed on 2-metre high mounds behind a 15-metre high seawall, as protection against future tsunamis. Thousands of people are still living in temporary accommodation or have simply left the region — the lack of employment opportunities and the risk associated with living in a disaster-prone region combining to drive people to move elsewhere.

Human factors

Human factors, which include changes in the social, economic and political elements of a region, can be a cause of the decline of cities and their urban environment. The destructive elements of war on the social fabric and economy of a nation, which have significant impacts on urban environments, is one example.

FIGURE 2 Main temple complex, Angkor Wat, Cambodia

Angkor, the capital city of the Khmer Empire in Cambodia, and thought to be the largest city in the world at the time, was abandoned in the fifteenth century due to a combination of wars and a series of droughts (see figure 2). The destruction of its economy, which was based on management of water and rice production, meant the city was no longer viable.

In modern times, there are many examples of towns and cities with extensive urban environments that have declined. Some reasons for change include depletion of mineral supplies and mining operations, changes in demand for industrial production and manufactured goods (**manufacturing and industrial base**), and **downturn of the global economy**. Paradoxically, the city of Ordos in China has been built ahead of its time and remains unoccupied due to insufficient population (see figure 3).

FIGURE 3 Ordos, inner Mongolia, China. Deserted roads and unoccupied high-rise apartments in 2011. A larger population needs to move to this area to take up ownership and use the facilities.

Other examples of urban environments that have declined due to human-induced factors can be seen in figure 4.

FIGURE 4 Cities abandoned due to changing human and physical factors

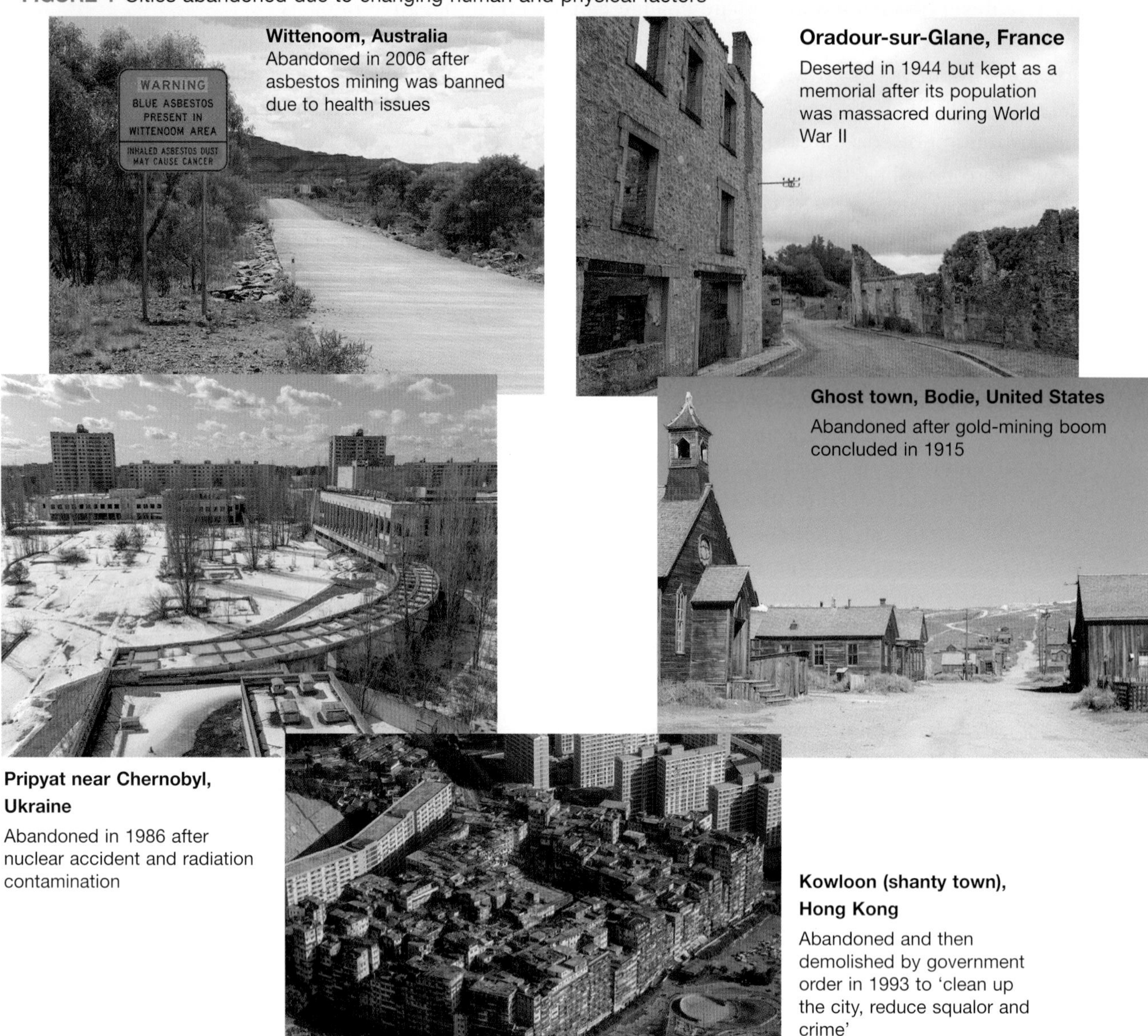
WARNING
BLUE ASBESTOS PRESENT IN WITTENOOM AREA
INHALED ASBESTOS DUST MAY CAUSE CANCER

Wittenoom, Australia
Abandoned in 2006 after asbestos mining was banned due to health issues

Oradour-sur-Glane, France
Deserted in 1944 but kept as a memorial after its population was massacred during World War II

Ghost town, Bodie, United States
Abandoned after gold-mining boom concluded in 1915

Pripyat near Chernobyl, Ukraine
Abandoned in 1986 after nuclear accident and radiation contamination

Kowloon (shanty town), Hong Kong
Abandoned and then demolished by government order in 1993 to 'clean up the city, reduce squalor and crime'

Note: This photograph was taken prior to the demolition of the shanty town.

7.4 Activities

To answer questions online and to receive **immediate feedback** and **sample responses** for every question, go to your learnON title at www.jacplus.com.au. *Note*: Question numbers may vary slightly.

Remember

1. What ***environmental*** hazards can lead to destruction or damage to urban ***environments***?
2. Why was the town of Pripyat near Chernobyl in the Ukraine abandoned?

Explain

3. The Miss World Competition was held in 2012 in the town of Ordos, Inner Mongolia, China. Suggest reasons why China would be keen to have the event held in that ***place***.
4. Why was water essential to the survival and decline of the city of Angkor?

Predict

5. What impact might a rise in sea level have on a city such as Venice?
6. If ore bodies are depleted in the mining town of Broken Hill, how might the town ***sustain*** its existence into the future?

Think

7. Is there a future for ghost towns?
8. Use the internet to find out more about the decline of the cities of Gary and Detroit in the United States, and make a list of reasons why these cities have declined and how they might be reinvigorated.

7.5 What are the challenges for fast-growing cities?

7.5.1 The growth of Mumbai

Located in the western Indian state of Maharashtra, Mumbai, according to United Nations data, is the second most populous city in India, and the equal-fourth most populous city in the world, with a total metropolitan area population of approximately 22 million. In 2009, Mumbai was named an **alpha world city**. Although the richest city in India, with the highest **GDP** of any city in south, west or central Asia, it also has much substandard housing and many of its residents live in squalor.

The large numbers of people and rapid population growth have contributed to serious social, economic and environmental problems for Mumbai. Mumbai's business opportunities, as well as its potential to offer a higher standard of living, attracted migrants from all over India seeking employment and a better way of life. In turn, this made the city a melting pot of many communities and cultures. The population density is estimated to be about 21 000 persons per square kilometre, and the living space 4.5 square metres per person.

Despite government attempts to discourage the influx of people, the city's population grew at an annual rate of more than 4 per cent per year (see table 1). The number of migrants to Mumbai from outside Maharashtra during the ten year period from 2001 to 2011 was over one million, which amounted to 54.8 per cent of the net addition to the population of Mumbai.

Many newcomers end up in abject poverty, often living in slums or sleeping in the streets. An estimated 42 per cent of the city's inhabitants live in slum conditions. Some areas of Mumbai city have population densities of around 46 000 per square kilometre — among the highest in the world.

TABLE 1 Population growth in Mumbai

Census	Population	% change
1971	5 970 575	—
1981	8 243 405	38.1
1991	9 925 891	20.4
2001	11 914 398	20.0
2011	12 478 447	4.7

Source: Mumbai Metropolitan Region Development Authority. Data is based on Government of India Census.

7.5.2 Challenges

Mumbai suffers from the same major urbanisation problems that are seen in many fast-growing cities in developing countries: widespread poverty and unemployment, urban sprawl, traffic congestion, inadequate sanitation, poor public health, poor civic and educational standards, and pollution. These pose serious threats to the quality of life in the city for a large section of the population. Automobile exhausts and industrial emissions, for example, contribute to serious air pollution, which is reflected in a high incidence of chronic respiratory problems. With available land at a premium, Mumbai residents often reside in cramped, relatively expensive housing, usually far from workplaces and therefore requiring long commutes on crowded public transport or clogged roadways (see figure 1). Although many live in close proximity to bus or train stations, suburban residents spend a significant amount of time travelling southwards to the main commercial district. Dharavi, Asia's second-largest **slum**, is located in central Mumbai and houses more than one million people (see figure 2).

FIGURE 1 Mumbai train rush hour

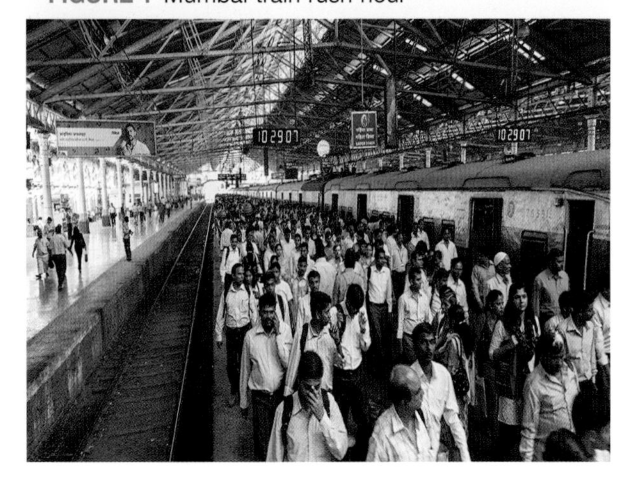

FIGURE 2 Dharavi

7.5.3 Dharavi's recycling entrepreneurs

Dharavi sprawls across 222 hectares of land in the centre of Mumbai. Hidden amid the labyrinth of ramshackle huts and squalid open sewers are an estimated 15 000 single-room factories, employing around a quarter of a million people and turning over a staggering US$1 billion each year through recycling, and other trades such as the production of pottery, textiles and leather goods.

In developed countries, communities recycle because there is the understanding that it contributes to sustaining the planet's resources. However, for some of the poorest people in the developing world, recycling often isn't a choice, but rather a necessity of life.

In India, almost 300 000 people make their living by recycling waste. These people are known as 'ragpickers' and are made up of India's poorest and most marginalised groups (see figure 3). The ragpickers wade through piles of unwanted goods to salvage easily recyclable materials such as glass, metal and plastic, which are then sold to scrap dealers who process the waste and sell it on either to be recycled or to be used directly by the industry.

FIGURE 3 Ragpickers

Due to the lack of formal systems of waste collection, it falls to Mumbai's ragpickers to provide this basic service for fellow citizens. Without them, solid waste and domestic garbage would not be collected or recycled, let alone sorted. Despite many of the social and ethical controversies surrounding the recycling industry in India, Dharavi is seen as the 'ecological heart of Mumbai', recycling up to 85 per cent of all waste material produced by the city, an excellent example of human–environment systems thinking in action.

If we consider the United Kingdom's recycling figures, it is pleasing to see that in recent years, the recycling rate has increased to 44.9 per cent of the waste produced, and the total amount of waste has decreased to 26.7 million tonnes per year. After recycling, however, this still equates to around 14.7 million tonnes of waste that ends up in landfills. If the United Kingdom could sustain the recycling rate of Mumbai (i.e. up to 85 per cent), it would result in only a little more than a quarter of the existing waste per year entering landfills (around 4 million tonnes), thus significantly reducing costs in sourcing materials.

In Dharavi, there is only one toilet for every 1440 people. This results in floods of human excrement during the monsoon season. Much of the water becomes contaminated because of this, and death rates tend to be 50 per cent higher in Mumbai's slums than in upper- and middle-class areas.

There are currently plans to demolish and redevelop Dharavi, as Mumbai is working on a facelift in order to become a world city. This redevelopment would transform the slum into a series of high-rise housing facilities, and each of Dharavi's 57 000 registered families would get 21 square metres of living space.

However, many Dharavi residents do not support this plan, as they are content with their current lifestyle. Most residents of the slum do not mind squatting near Mahim Creek, and prefer not to have their own flush toilets. Most are working and making a living, and many have lived their entire lives in Dharavi and do not want to trade their culture for a redeveloped life.

7.5 Activities

To answer questions online and to receive **immediate feedback** and **sample responses** for every question, go to your learnON title at www.jacplus.com.au. *Note*: Question numbers may vary slightly.

Remember

1. List Mumbai's challenges.
2. Why are migrants attracted to Mumbai?

Explain

3. What human influences caused Mumbai to develop over time?

Predict

4. What would be the economic, social and ***environmental*** benefits of ragpickers?

Think

5. List the advantages and disadvantages of replacing slums with high-rise low-income housing.
6. Use the **Mumbai slums** weblink in the Resources tab to watch the video *Slums in Mumbai*.
   (a) Describe the everyday living conditions.
   (b) What percentage of Mumbai's population lives in Dharavi? What percentage is this of Mumbai's area?
   (c) What are the challenges facing the residents?
7. Find out more about the natural and human influences on the development of cities. Examples for research could include Canberra, Australia (a planned city); Cape Town, South Africa (a port city); Rotenburg, Germany (a walled city); Geneva, Switzerland (where a river meets a lake); Johannesburg, South Africa (near a mining site); Chicago, United States (where north–south and east–west railway routes cross); Jerusalem, Israel (an ancient religious city); Bath, England (located at the site of a natural supply of mineral waters).

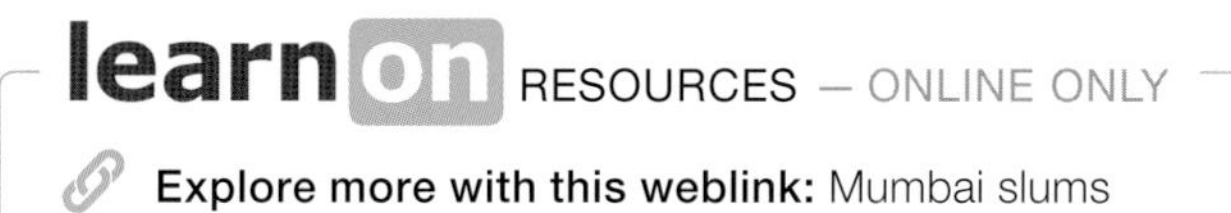

learnON RESOURCES — ONLINE ONLY

Explore more with this weblink: Mumbai slums

7.6 Has Melbourne sprawled too far?

7.6.1 Where do you put all the people?

What does a city do when it runs out of room? Across the world cities are expanding at a rapid rate, bringing unprecedented change to the built and natural environments. To accommodate growing populations there is a need for more housing and the infrastructure to support so many people. How can this be done?

Melbourne today sprawls over a huge area of 9000 square kilometres and has a population of more than 4.4 million. It is the fastest growing Australian capital city, increasing at a rate of 1800 people per week. By the year 2050, this population will increase to 7.7 million, with a need for over 500 000 additional households. To house this number, urban planners have basically two choices. One option is to infill (**urban infilling**), or 'create' land in the inner and middle suburbs. This can be done by dividing older larger blocks into smaller new blocks, by the **urban renewal** of old industrial sites, or by building up and increasing population density with high-rise apartments as seen in figure 1.

The second option is to expand space by extending the city outwards into a zone known as the rural–urban fringe. Figure 2 shows the predicted population growth for Melbourne. Note the location of those suburbs expected to have the greatest population increases.

FIGURE 1 These apartments are an example of high-rise housing in a large-scale urban renewal project at Docklands in Melbourne.

FIGURE 2 Future population growth for Melbourne

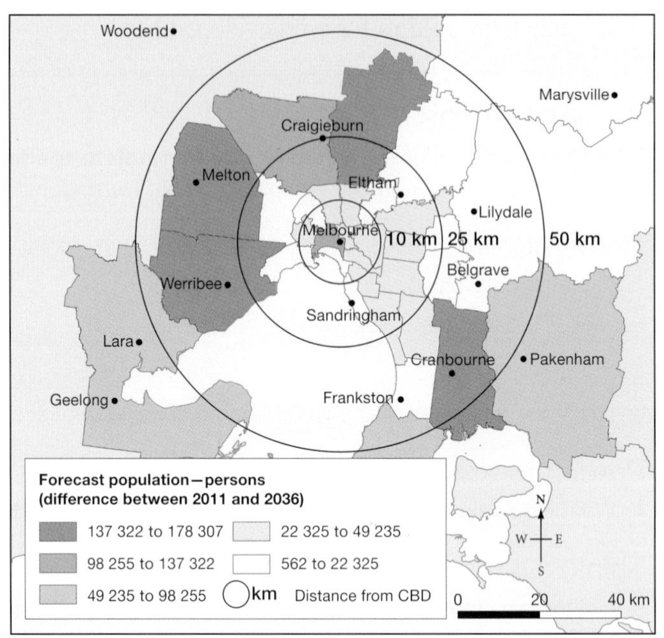

Source: © The State of Victoria, Department of Environment and Primary Industries 2013 © Commonwealth of Australia Geoscience Australia 2013

7.6.2 How can we increase density in established suburbs?

To increase the density of housing in older suburbs, one concept is to establish activity centres. These consist of higher-density housing in specific locations, where people shop, work, meet, relax and live in the local environment. These centres are focused on existing infrastructure, transport networks, popular shopping centres, employment opportunities and community facilities. New housing tends to be medium-rise (three to five storeys) apartments built along main transport routes.

7.6.3 What changes are taking place on the rural–urban fringe?

The rural–urban fringe is typically the urban zone that is undergoing the most rapid change. Former farmland, often market gardens and orchards, are sold off and new housing and industrial estates are built. These are usually low-density, planned estates sometimes built around a theme or geographical feature such as a built lake or wetland.

Urban expansion into the rural–urban fringe brings environmental, economic and social impacts.

What are the effects of change on the rural–urban fringe?

Cost of infrastructure

A major problem of **urban sprawl** is the cost to provide infrastructure (for example, roads and other transport systems and services such as water, gas and electricity) to new areas on the rural–urban fringe. New suburbs such as Narre Warren, in the City of Casey, are examples of planned developments designed to meet the needs of the families who will live there. Being on the urban fringe, the blocks vary in size, and are more affordable for first home buyers and young families. These suburbs will develop activity centres that will include shopping complexes, medical centres, open spaces, schools and recreational facilities such as green zones, although there are often time delays in construction until a population is established.

Loss of fertile farm lands

Possibly the largest issue with urban sprawl is the loss of fertile farmlands. The Casey Council, 48 kilometres south-east of Melbourne's city centre, has resisted moves for subdivision of farmlands at Clyde, arguing that the sandy loam soils that produce most of Melbourne's fruit and vegetables should be set aside for growing food, not houses (see figure 3).

FIGURE 3 Growing over our food

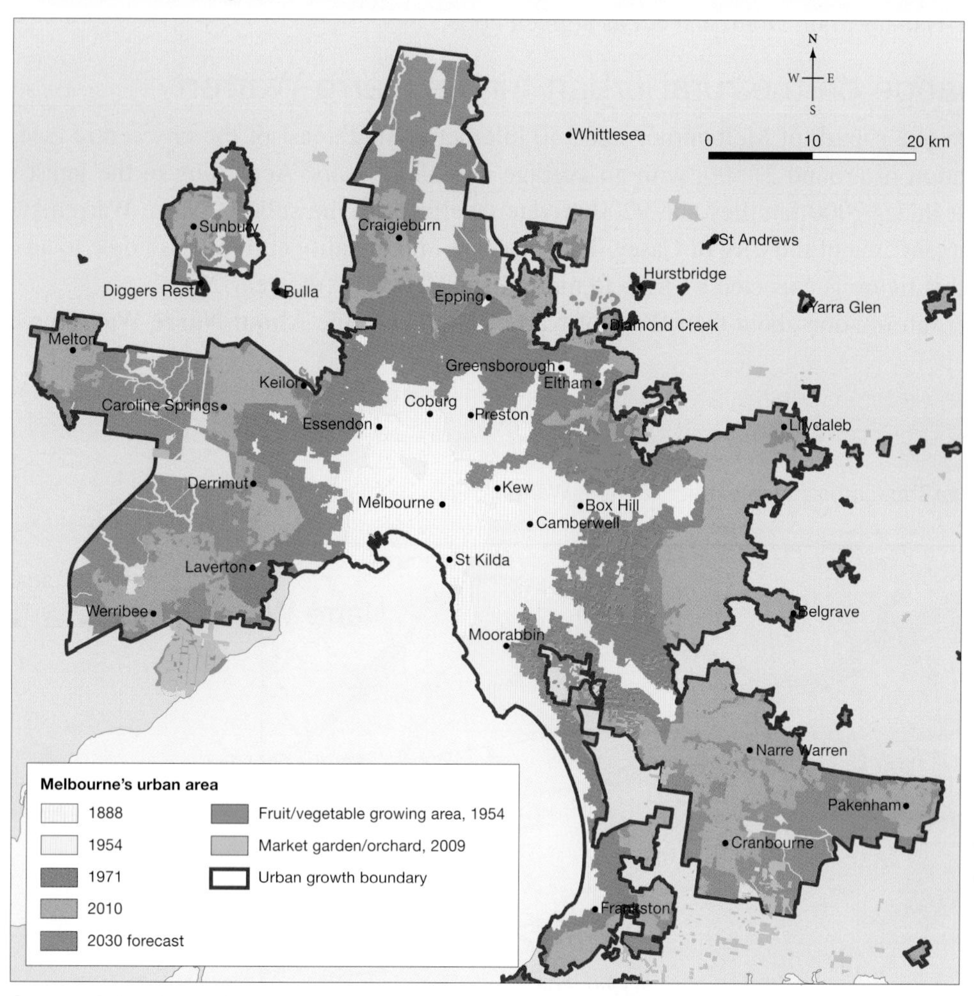

Source: © The State of Victoria, Department of Environment and Primary Industries 2013

As a newspaper article has put it: 'We've already built over the best soils in this state — the soils around Melbourne. Why would you keep building over it and subdividing it when in the next 50 years we're facing an era of incredible uncertainty and major changes to climate, to fuel supplies and to energy markets?'

Some local farmers have different viewpoints on this matter as rezoning their properties into the urban boundary immediately boosts the value of their land. These farmers prefer to relocate further out and reap the financial gains from the sale of the land for housing.

Loss of green spaces

The expansion of urban areas can significantly alter the natural environment. Clearing of vegetation can reduce habitat and biodiversity. Natural drainage and topography can be altered, with streams redirected or even converted to pipes. Today there is a growing awareness of the need to preserve environments for the important functions they provide for wildlife and people. There is a trade-off, however, as the push for housing into rural areas can increase bushfire risks. Planners these days try to incorporate and retain as much of the natural environment as possible when developing housing estates.

'Green zones' are open landscapes set aside to conserve and protect significant natural features as well as resources such as farms, bushland and parks. They ensure habitats for native flora and fauna are preserved. The construction or expansion of wetlands in the rural–urban fringe can have many benefits, including:

- acting as flood retention basins and receiving and purifying stormwater run-off from residential areas
- providing habitats that can increase plant and animal biodiversity
- providing recreational opportunities.

Use the **Narre Warren wetlands** weblink in the Resources tab to view how the wetlands help control waters in the Hallam Main Drain and act as a green zone.

7.6.4 Change on the rural-urban fringe: Narre Warren

Narre Warren is a suburb of Melbourne, some 40 kilometres south-east of the city centre (see figure 3). It has a population of around 27 500, with an average age of 34 years. According to the latest census data, there are just under 7000 families and 9200 private dwellings in the suburb. Narre Warren sits within the local government area of the City of Casey. This growing municipality currently is home to around 297 000 people, but this figure is expected to grow to more than 492 000 by the year 2041.

For further information about the suburb of Narre Warren, use the **About Narre Warren** weblink in the Resources tab.

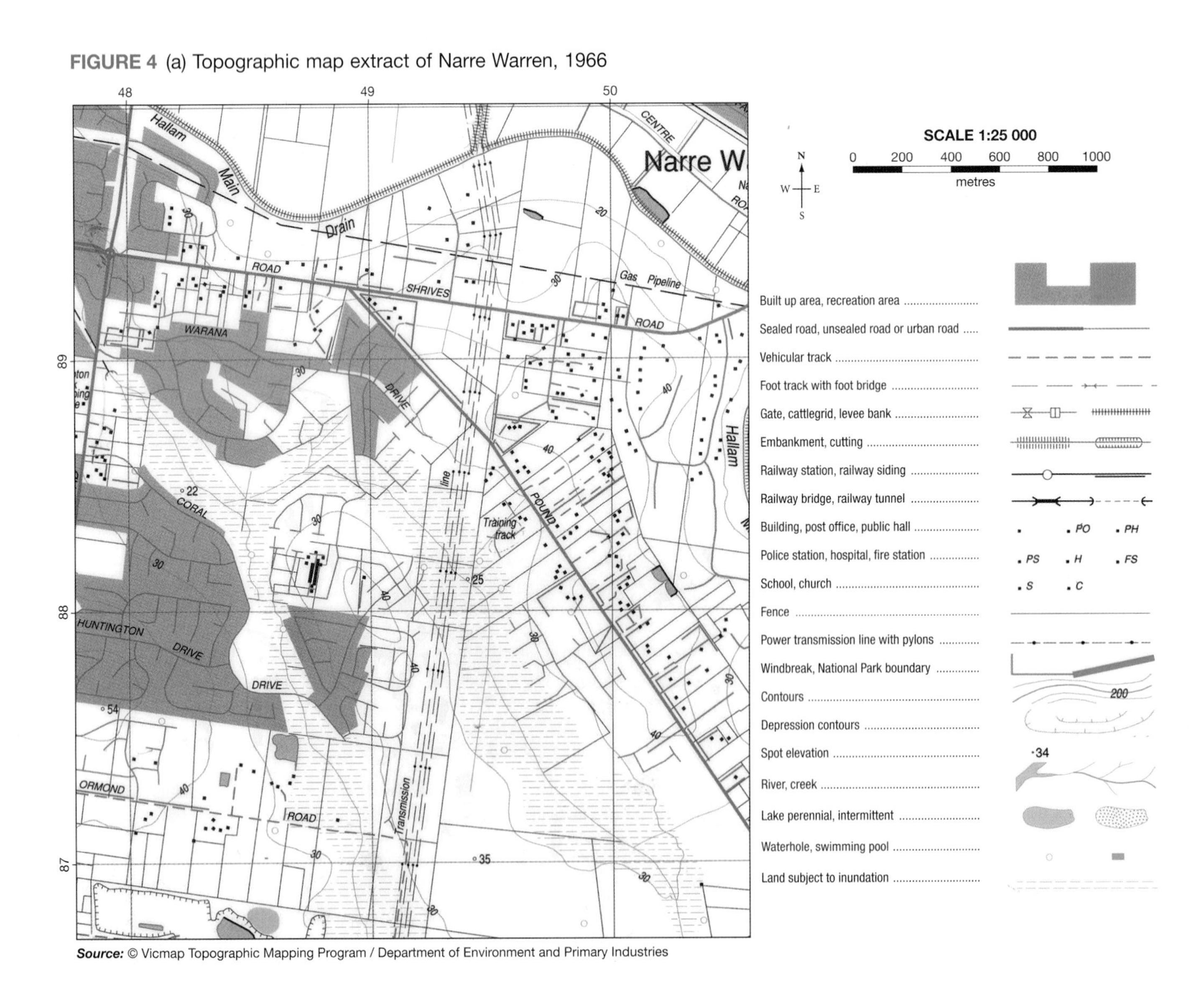

FIGURE 4 (a) Topographic map extract of Narre Warren, 1966

Source: © Vicmap Topographic Mapping Program / Department of Environment and Primary Industries

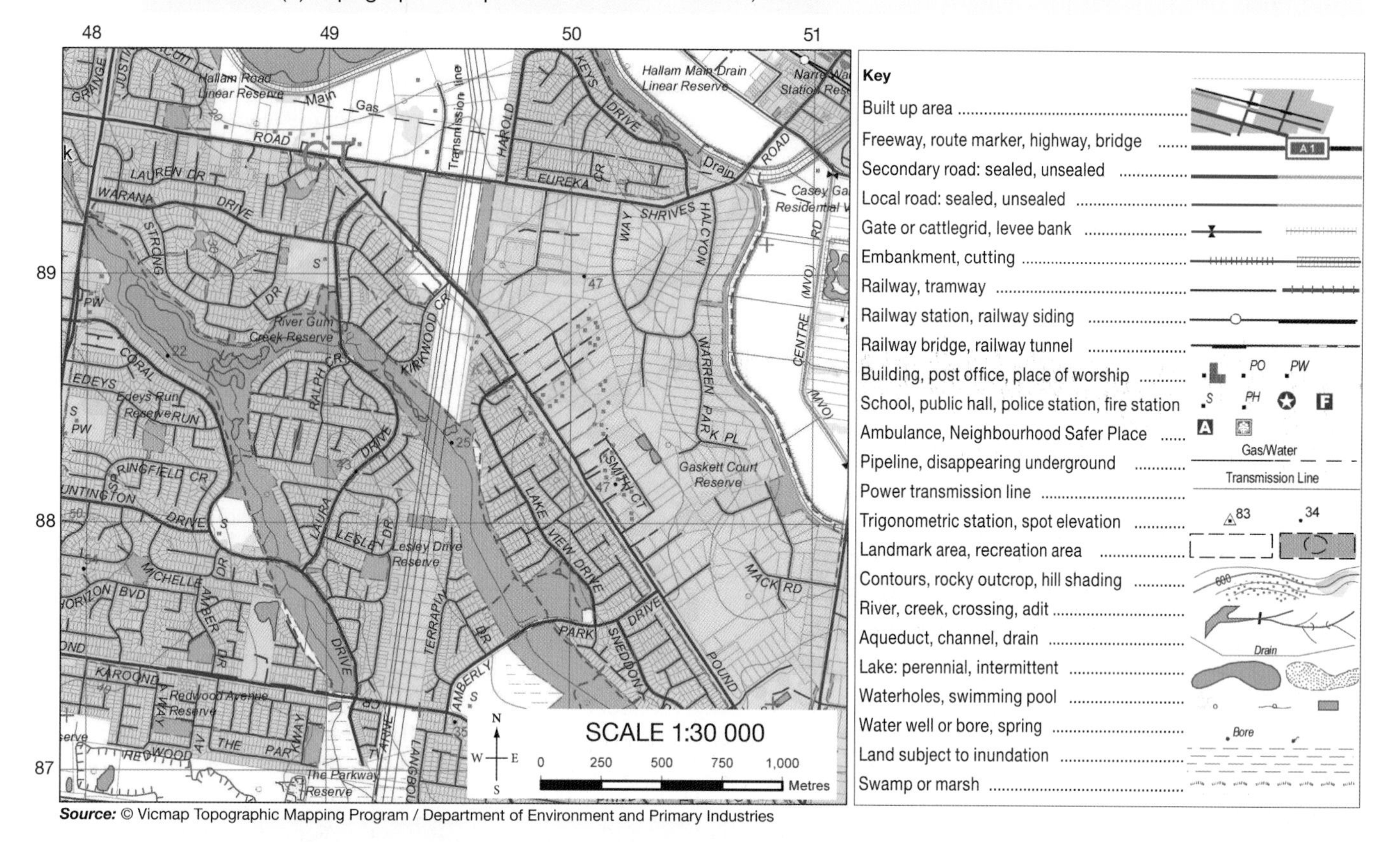

FIGURE 4 (b) Topographic map extract of Narre Warren, 2013

Source: © Vicmap Topographic Mapping Program / Department of Environment and Primary Industries

7.6 Activities

To answer questions online and to receive **immediate feedback** and **sample responses** for every question, go to your learnON title at www.jacplus.com.au. *Note*: Question numbers may vary slightly.

Remember

1. What are the two main ways that additional housing can be established in an expanding city?
2. Study figure 1 and the image available via the **Narre Warren wetlands** weblink in the Resources tab. Construct a table to compare the advantages and disadvantages of living in an inner-city high-rise apartment and a housing estate on the rural–urban fringe.
3. Refer to the image available via the **Narre Warren wetlands** weblink. How do you think the natural ***environment*** has been ***changed*** for this housing estate?

Explain

4. Study figure 2. Describe the location of those suburbs of Melbourne that are expected to show the greatest increase. What is the average distance of those suburbs from Melbourne's CBD?
5. (a) Refer to figure 3. What has happened to the areas of market gardening (fruit and vegetable farming) between 1954 and 2009? Use distances and directions in your answer.
   (b) What would be the benefits of market gardening being located close to urban areas?
   (c) Predict the future location of this land use in Melbourne in 2030.
6. Suggest how urban planners can reduce some of the ***environmental***, social and economic impacts of expansion into the rural–urban fringe.

Discover

7. Study figure 4(a). What evidence is there on the map to suggest that this area is part of the rural–urban fringe?
8. (a) Create a map to show the main land uses in Narre Warren in 1966. Include other key features such as main roads and railway lines. Refer to the 'Constructing a land use map' SkillBuilder in subtopic 7.7 in the Resources tab.
   (b) Using tracing paper, make an overlay map of built-up areas from the 2013 map, and attach your overlay to your base map. Complete your map with full BOLTSS.

9. Study your complete map and overlay.
   (a) What was the main land use in this area in 1966?
   (b) What is the main land use for the area in 2013?
   (c) Study both maps and describe three other ***changes*** in land use in Narre Warren from 1966 to 2013.
10. (a) The area at GR485880 on figure 4(b) is subject to flooding (inundation). Use evidence from the maps to suggest two reasons why it is flood prone.
    (b) How have planners used this flood prone land when designing this housing estate? Refer also to the image available via the **Narre Warren wetlands** weblink the Resources tab.
11. Study figure 4(b). List and give grid references for any new forms of infrastructure established. Consider schools, shopping centres, parks and transport.
12. Suggest one human and one ***environmental*** factor that make this ***place*** suitable for a housing estate.
13. Working with a partner or in small groups, undertake a fieldwork investigation of your local area in terms of:
    - the types of dwellings
    - transport facilities and issues
    - shopping and other community services available
    - amount of open spaces and parkland, and associated recreation facilities
    - how 'liveable' it is. This might mean checking with your local council or conducting surveys of local residents.

    Document your findings in a report, including maps, photographs and data (e.g. tables, pie charts), and listing any references.

RESOURCES — ONLINE ONLY

Try out this interactivity: Changes on the rural–urban fringe (int-3302)

Explore more with these weblinks: Narre Warren wetlands, About Narre Warren

myWorldAtlas Deepen your understanding of this topic with related case studies and questions.
Urbanisation in Australia

7.7 SkillBuilder: Constructing a land use map

WHAT IS A LAND USE MAP?

A land use map may be drawn from a topographic map, an aerial photograph, a plan or during fieldwork. A land use map shows simplified information about the uses made of an area of land.

Go online to access:

- a clear step-by-step explanation to help you master the skill
- a model of what you are aiming for
- a checklist of key aspects of the skill
- a series of questions to help you apply the skill and to check your understanding.

FIGURE 1 Land use map of Blue Lake Shopping Centre

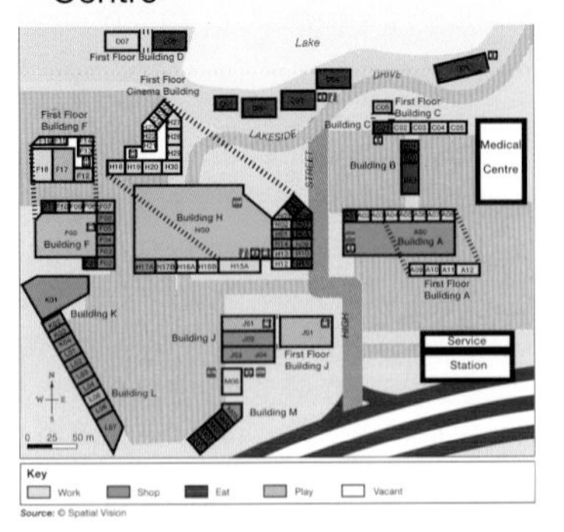

RESOURCES — ONLINE ONLY

Watch this eLesson: Constructing a land use map (eles-1755)

Try out this interactivity: Constructing a land use map (int-3373)

7.8 SkillBuilder: Building a map with geographic information systems (GIS)

online only

WHAT IS GIS?

GIS is a computer-based system that consists of layers of geographic data. Just as an overlay map allows you to interchange layers of information, GIS allows you to turn layers on and off to make comparisons between data.

Go online to access:

- a clear step-by-step explanation to help you master the skill
- a model of what you are aiming for
- a checklist of key aspects of the skill
- a series of questions to help you apply the skill and to check your understanding.

FIGURE 1 St Arnaud and its environs

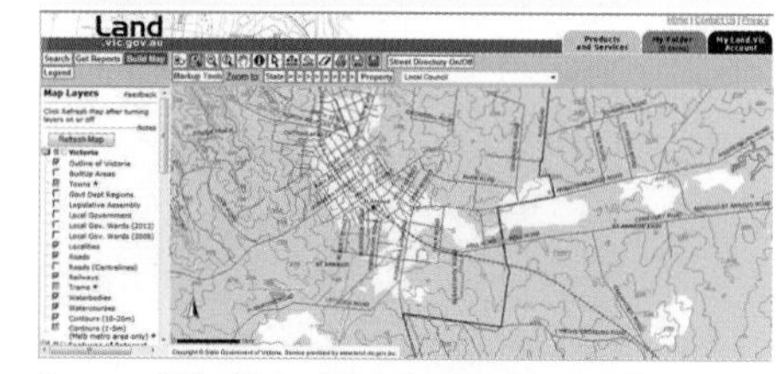

Source: © The State of Victoria, Department of Environment and Primary Industries, 2013. Reproduced by permission of the Department of Environment and Primary Industries.

learnon RESOURCES — ONLINE ONLY

Watch this eLesson: Building a map with geographic information systems (GIS) (eles-1754)

Try out this interactivity: Building a map with geographic information systems (GIS) (int-3372)

7.9 Can we stop Venice from sinking?

7.9.1 Venice

Urban centres that are built on low-lying coastal plains, where features such as **river deltas**, wetlands, lagoons, sand dunes, bars and barriers are found, are susceptible to the environmental impacts of the sea.

Storms and high tides, when added together, can lead to destructive surges causing erosion and damage to cities. The prospect of rising sea levels as a result of the melting of ice caps due to global warming will require specialised management techniques such as the construction of coastal defence works to protect property and life.

Venice, Italy, is a city built on mud islands in a coastal **lagoon** at the head of the Adriatic Sea (see figures 1 and 2).

FIGURE 1 Venice and surrounding areas

Source: © OpenStreetMap contributors

FIGURE 2 Aerial view of Venice showing the built area in the lagoons and the canals

Although Venice has a population of only 60 000, its **historical architecture**, life on the canals and cultural events such as Carnivale attract over two million tourists per year. Not surprisingly, the Venetians are keen to protect their heritage and manage the impacts of erosion and rising sea levels into the future.

7.9.2 Why is Venice sinking?

When Venice was established almost 2000 years ago, the sea level was two metres lower than current levels and buildings seemed secure from the impacts of the sea. Over time, the sea level has risen, and in more recent times this rate of increase has accelerated due to global warming. Also affecting the stability of buildings was the removal of fresh water from artisan wells near Venice in the 1950s. This practice, which fortunately has stopped, led to building subsidence. Another problem has been the erosive force of waves generated by powerful motor boats splashing corrosive sea salt onto the buildings.

7.9.3 Floods or 'aqua alta'

Venetians refer to floods as 'aqua alta' or high water. In December 2008, high sea waters with a depth of 1.56 metres above average sea levels caused the fourth highest flood level in Venice for the past 22 years. A combination of high tides and winds forced waters over the canal banks and into buildings and public areas (see figure 3). This type of flooding, which has increased dramatically in the recent past, has seen up to 40 flooding events a year. This has led to reduced tourist numbers as buildings and public areas are flooded and transport is curtailed because boats cannot fit under bridges.

FIGURE 3 Flooding in Venice

7.9.4 How will Venice reduce the impact of floods?

The MOSE (MOdulo Sperimentale Elettromeccanico) or Experimental Electromechanical Module Project has been proposed to reduce the impact of floods on Venice. It consists of rows of mobile gates that are able to isolate the lagoon and canals from high tides that are above a level of 110 centimetres to a maximum of three metres. The project is on schedule to be finished in 2020 with most foundations and some flood gates ready to be set. Use the **MOSE** weblink in the Resources tab to read about this project in more detail. The project has been criticised by some who say that flushing of the canals would be reduced and the huge cost of the project could not be justified as it may be effective for only a few years if sea levels continue to rise. If this project is not to go ahead, then clearly other management strategies need to be found or Venice will one day sink into the sea.

7.9 Activities

To answer questions online and to receive **immediate feedback** and **sample responses** for every question, go to your learnON title at www.jacplus.com.au. *Note*: Question numbers may vary slightly.

Remember

1. Where is Venice located?

Explain

2. What aspects of its landscape make the city of Venice vulnerable to flooding?
3. How would you employ human–environment systems thinking to solve the flooding of Venice? (*Hint:* Make a list of ***environmental*** impacts and human management responses.)

Discover

4. (a) Give details of the MOSE Project in Venice and explain how it works in holding back the sea.
   (b) Evaluate the MOSE Project in terms of its:
      (i) ***environmental*** impact
      (ii) social impact (i.e. its value in preserving a unique city with a long history.
5. Select a capital city in Australia and find out more about the impact of rising sea levels on suburbs close to the coast. How might the social and economic impacts of rising seas be managed?

RESOURCES — ONLINE ONLY

Explore more with this weblink: MOSE

7.10 What is the future of our urban environment?

7.10.1 The influence of technology on urban environments

Cities throughout human history have changed as new forms of technology have developed. For instance, high-rise buildings such as skyscrapers could not exist without modern cement-and-steel methods of construction and the development of high-speed lifts. What will be the nature of cities as technology progresses, and how can the social, economic and environmental elements of cities develop and be managed in a fair and sustainable manner?

7.10.2 How is urbanisation being managed?

The United Nations (UN) established the Millennium Development Goals (MDGs) in the year 2000 with a date for review of 2015. Now a new set of goals, called the Sustainable Development Goals (SDGs), has been developed and these will inform other bodies that have an interest in urban development. These bodies include UN-Habitat, ComHabitat (Commonwealth Habitat), the Cities Alliance and the World Bank, and all are aiming to address the urban challenges of the twenty-first century with a focus on social and economic management criteria. The new SDGs are shown in figure 1.

It has been estimated by UN studies that the global urban population, which is currently 50 per cent, will increase to about 4.9 billion in 2030, which means 60 per cent of the total world's population will be living in cities. This increase means that approximately 2 billion people will need new housing, basic urban infrastructure and services. To achieve this, the equivalent of seven new megacities will need to be created annually (see figure 2).

FIGURE 1 The 17 Sustainable Development Goals

FIGURE 2 Urban population projections (a) 2011 and (b) 2025

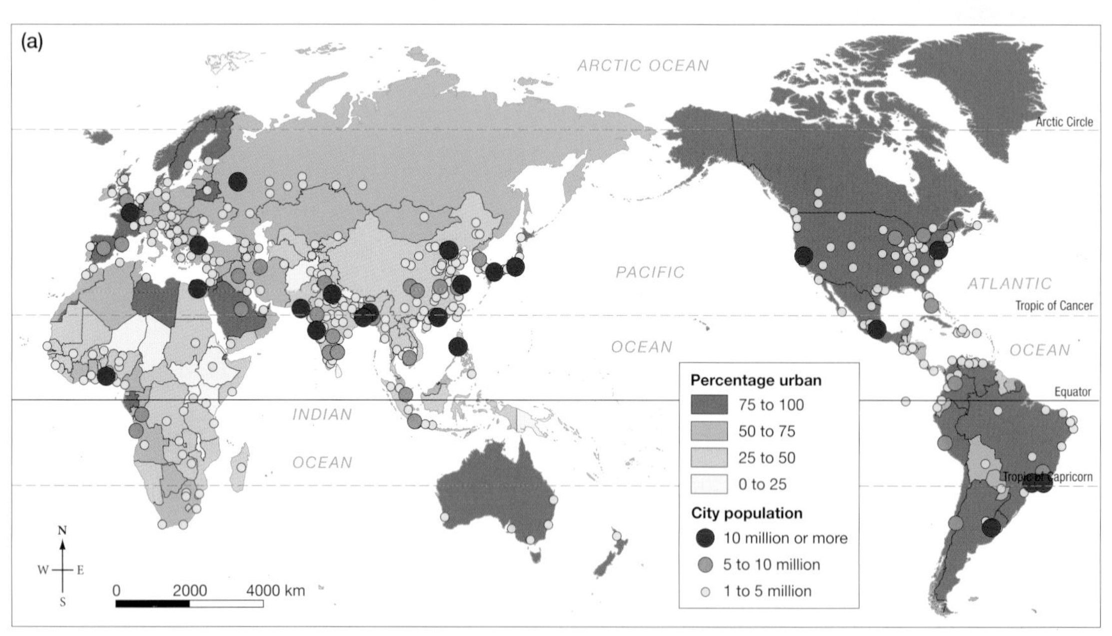

Source: United Nations, Department of Economic and Social Affairs, Population Division 2012. World Urbanization Prospects: The 2011 Revision

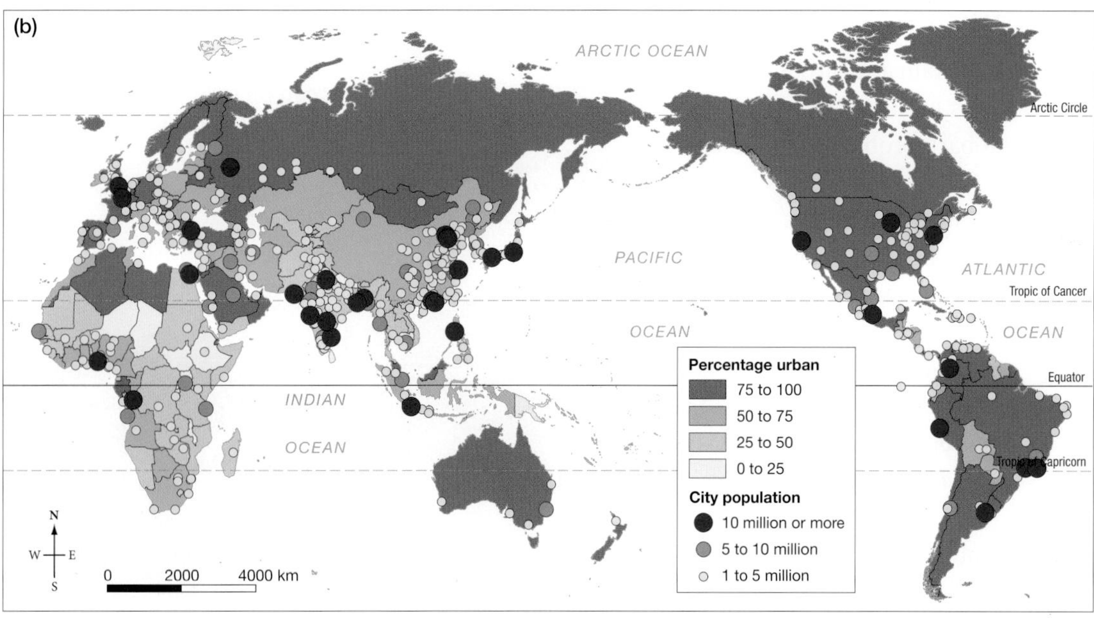

Source: United Nations, Department of Economic and Social Affairs, Population Division 2012. World Urbanization Prospects: The 2011 Revision

7.10.3 What are the challenges for cities?

Many cities throughout the world have extensive areas of slum housing and a general lack of infrastructure to support what may be called a socially just and economically fair lifestyle. Particularly in the poorest or least developed countries there are significant environmental management issues associated with large cities. Some of these issues are detailed in table 1. It should also be noted that there are issues even in cities that could be called wealthy or most developed.

TABLE 1 Urban challenges

Level	Challenges to be addressed to ensure a sustainable urban environment
Least developed countries	• Poverty and inequality • Rapid and chaotic development of slum housing • Increasing demand for housing, urban infrastructure, services and employment • Education and employment needs of the majority population of young people • Shortage of skills in the urban environment sector
Transition countries	• Slow (or even negative) population growth and ageing • Shrinking cities and deteriorating buildings and infrastructure • Urban sprawl and preservation of inner-city heritage buildings • Growing demand for housing and facilities by an emerging wealthy class • Severe environmental pollution from old industries • Rapid growth of vehicle ownership • Financing of local authorities to meet additional responsibilities
Developed countries	• Recent mortgage and housing markets crises • Unemployment and impoverishment due to changing availability of jobs • Large energy use of cities caused by car dependence, huge waste production and urban sprawl • Slow population growth, ageing and shrinking of some cities

There are a wide range of cities in the world that face environmental, social and economic challenges. A short list, as defined by the UN and based on cities with large-scale slum conditions, includes Rio de Janeiro (Brazil), Phnom Penh (Cambodia), Kolkata (India), Durban (South Africa), Lusaka (Zambia) and Mexico City (Mexico).

7.10.4 How can we plan for the future?

By promoting sustainable urban environments at all levels of scale (local, regional, national and global), problems can be overcome.

Some management strategies that will foster socially, economically and environmentally sustainable urban environments include:

- building energy-efficient houses based on materials and energy sources that reduce the **ecological footprint** of cities
- reducing waste by recycling and reusing materials
- improving public transport systems to reduce reliance on cars
- redeveloping to include **medium-density housing** to reduce urban sprawl
- exchanging ideas between governments about planning and building policies and best and successful practice in design.

FIGURE 3 Vertical gardens can be used to add green spaces to medium- and high-density housing developments.

7.10 Activities

To answer questions online and to receive **immediate feedback** and **sample responses** for every question, go to your learnON title at www.jacplus.com.au. *Note*: Question numbers may vary slightly.

Remember

1. How many people in the world in 2030 will need new housing based on current predictions of **urbanisation**?

Predict

2. How would migration help solve the problems of ageing populations in developed Western cities?

Think

3. Use the **Energy-efficient technology** weblink in the Resources tab to view a diagram of a modern house. Comment on how this house can save energy and solve waste management problems. What features of this house do you have in your own ***place*** of residence? Are there other features you would like to include in your home? Why?
4. It has been said that if all nations have the same ecological footprint as the developed countries (e.g. the United States, Australia and most European nations), we will need four new worlds the size of planet Earth to accommodate the growth in resource consumption. In what ways can we achieve energy, food and water security with an aim of ***sustainability*** into the future?
5. Refer to the 17 SDGs.
   (a) Which one(s) will directly improve social conditions in urban ***environments***?
   (b) Which one(s) will directly improve ***environmental*** conditions in urban ***environments***?
6. Give reasons for each of your answers to 5 (a) and (b).

RESOURCES — ONLINE ONLY

Try out this interactivity: Where am I? (int-3303)

Explore more with this weblink: Energy-efficient technology

Deepen your understanding of this topic with related case studies and questions.
- **Brisbane: an eco-city**

7.11 Review

7.11.1 Review

The Review section contains a range of different questions and activities to help you revise and recall what you have learned, especially prior to a topic test.

7.11.2 Reflect

The Reflect section provides you with an opportunity to apply and extend your learning.

Access this subtopic at **www.jacplus.com.au**

TOPIC 8 Geographical inquiry: Developing an environmental management plan

8.1 Overview

Numerous **videos** and **interactivities** are embedded just where you need them, at the point of learning, in your learnON title at www.jacplus.com.au. They will help you to learn the content and concepts covered in this topic.

8.1.1 Scenario and your task

There are many environmental changes that have an impact on different environments. Organisations or their specialist consultants often prepare environmental management plans (EMPs). EMPs recommend the steps to be undertaken to solve identified problems in managing the environment. They are also useful for predicting and minimising the effects of potential future changes. These strategies are designed to either remove or control the problem(s).

Your task

Each class team will research and prepare an EMP that deals with a specific environmental threat and then present it to the class. Decide on an environment and the threat it faces and then devise three key inquiry questions you would like to answer.

8.2 Process

8.2.1 Process

- Go to **www.jacplus.com.au** to access and watch the introductory video lesson for this geographical inquiry.
- **Planning:** In pairs or groups, decide on a particular environmental issue and devise a series of three key inquiry questions that will become a focus of your study and a means of dividing the workload. Download the EMP planning template in the Resources tab to help you think about and decide which environments your team will choose to research. The following steps will act as a guide for your report writing.

8.2.2 Collecting and recording data

- Find out about the issue and why an EMP is needed. Identify potential environmental threats or changes that may occur. Describe the issue, the scale of potential changes and their significance. Prepare a map, or series of maps, to show the location of the issue. This may be sourced from a street directory, atlas, Google maps or an online reference. Additional data can be researched and collected; for example, you may wish to survey people's opinions on the issue, use census data to determine the number of people affected in the region or find climatic data for the area. (Your teacher may guide you at this point.) Decide on the most suitable presentation method for your data; for example, graphs, maps and annotated photographs. You may wish to refer to relevant SkillBuilders to help you present your data.

8.2.3 Analysing your information and data

1. Review and discuss with your team members the information that you have collected. Has it come from reliable sources? What patterns, trends and interconnections can you identify from your data?
2. Come up with two or three possible options that will address the issue(s) you have collected information about. It would be beneficial to include diagrams and/or photographs of strategies currently operating in different places that could be used or adapted to your site.
3. Evaluate which option would be most effective based on the criteria:
   - economic viability (affordable)
   - social justice (fair to all people involved)
   - environmental benefit (minimal environmental impact and with future sustainability).
4. Make concluding recommendations based on your research and evaluation of options. This should be in the form of a suggested course of action to follow in managing the environment and reducing any negative changes.

8.2.4 Communicating your findings

- Present your report to the class and be prepared to answer questions from the audience. Use the EMP template in the Resources tab to help you structure your report. Use graphics such as maps, graphs, images and charts in your EMP.

8.3 Review

8.3.1 Reflecting on your work

- Review your participation in the production of your EMP by completing the reflection document in the Resources tab.
- Submit your reflection document along with your completed EMP.

UNIT 2 GEOGRAPHIES OF HUMAN WELLBEING

Not everyone has the same life, so human wellbeing varies from place to place across the world. How do you measure and compare wellbeing, and why are there such spatial variations? Organisations and governments devise programs that attempt to improve wellbeing for their own as well as other countries.

How accurate is this image of human wellbeing around the world?

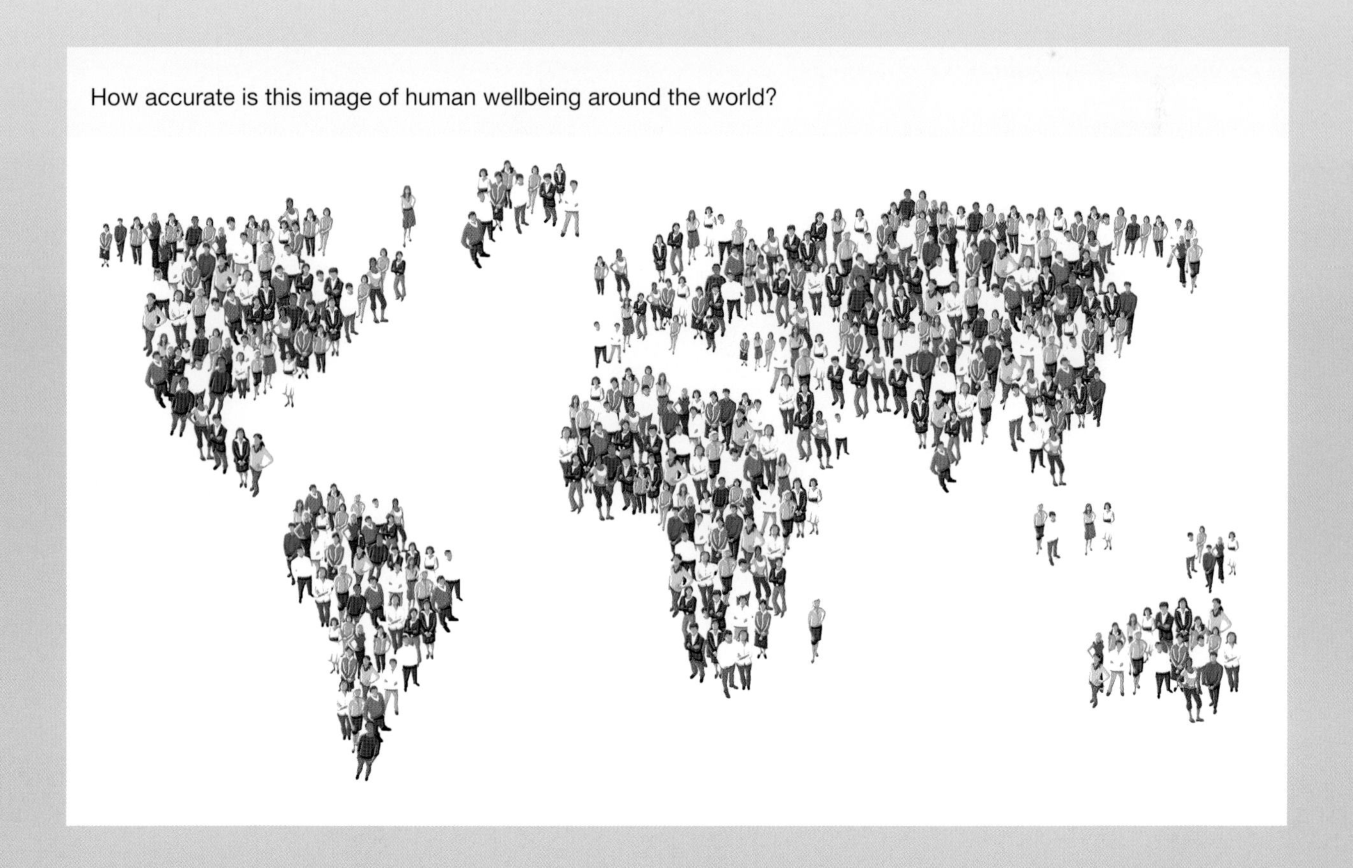

TOPIC 9 What makes a good life?

9.1 Overview

Numerous **videos** and **interactivities** are embedded just where you need them, at the point of learning, in your learnON title at www.jacplus.com.au. They will help you to learn the content and concepts covered in this topic.

9.1.1 Introduction

We all want a better life for ourselves, our families and our children, no matter where we live. We care about the progress of our communities, our state or territory, and our country. But how can we measure this progress? What does progress really mean? What do we count when we measure progress? How do we know if we are succeeding, and what is the concept of wellbeing?

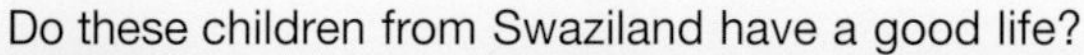

Do these children from Swaziland have a good life?

Starter questions

1. How would you define 'wellbeing'?
2. What does it mean to have a 'good life'?
3. Do all Australians have a good life?
4. What could be improved in your life or community?

INQUIRY SEQUENCE

9.2 Better off, worse off?

9.2.1 What is a good life?

In the past decade, a new global movement has emerged seeking to produce measures of progress that go beyond a country's income. Driven by citizens, policy-makers and statisticians around the world and endorsed by international organisations like the United Nations, the concept of **wellbeing** offers us a new perspective on what matters in our lives.

Wellbeing is experienced when people have what they need for life to be good. But how do we measure a good life? We can use **indicators** of wellbeing to help us. Indicators are important and useful tools for monitoring and evaluating progress, or lack of it. There are **quantitative indicators** and **qualitative indicators**.

Traditionally, **development** has been viewed as changing one's environment in order to enhance economic gain. Today, the concept of development is not only concerned with economic growth, but includes other aspects such as providing for people's basic needs, equity and social justice, sustainability, freedom and safety. We have built on this traditional concept for measuring progress by considering wellbeing, which emphasises what is positive and desirable rather than what is lacking. The most successful development programs address all areas of wellbeing, rather than simply focusing on economic, health or education statistics. There is a growing awareness that human beings and their happiness cannot simply be reduced to a number or percentage. We can measure development in a variety of ways, but the most common method remains to use economic indicators that measure economic progress using data such as **gross domestic product** (GDP).

How do we use indicators?

Indicators can be classified into a range of broad categories (see figure 1). Economic indicators measure aspects of the economy and allow us to analyse its performance. Social indicators include demographic, social and health measures. Environmental indicators assess resources that provide us with the means for social and economic development, and gauge the health of the environment in which we live. Political indicators look at how effective governments are in helping to improve people's standard of living by ensuring access to essential services. Wellbeing can also be influenced by technological indicators in such fields as transport, industry, agriculture, mining and communications.

FIGURE 1 Categories of wellbeing indicators

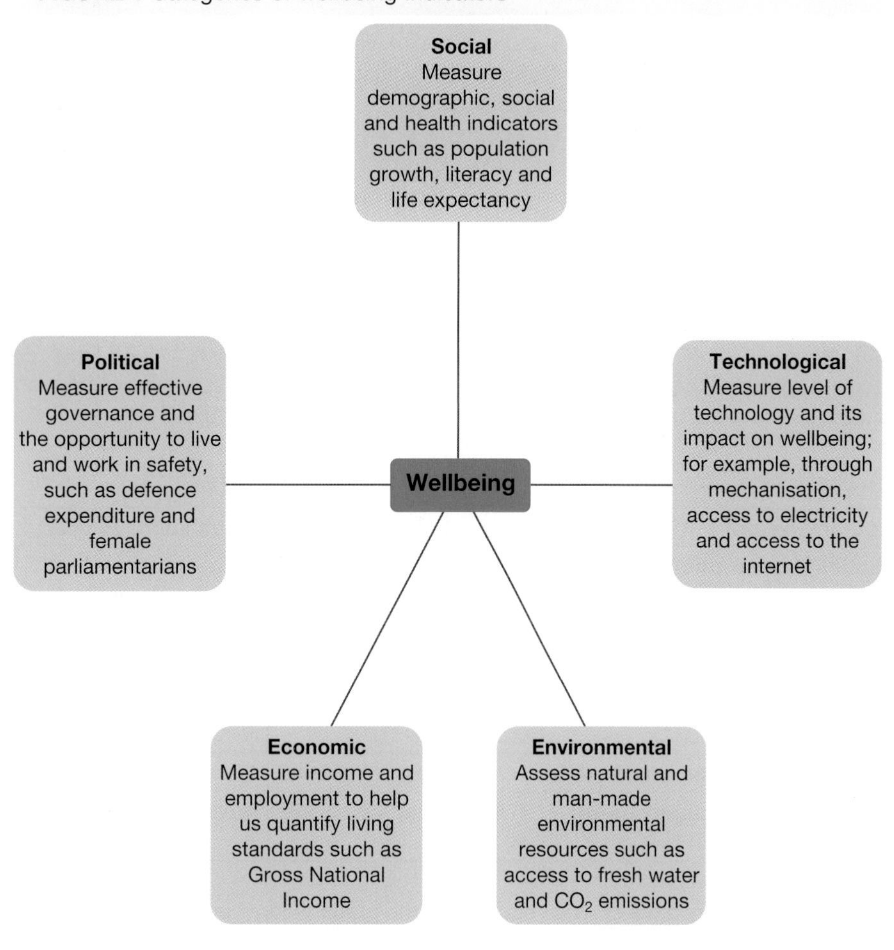

Social
Measure demographic, social and health indicators such as population growth, literacy and life expectancy
Political
Measure effective governance and the opportunity to live and work in safety, such as defence expenditure and female parliamentarians
Wellbeing
Technological
Measure level of technology and its impact on wellbeing; for example, through mechanisation, access to electricity and access to the internet
Economic
Measure income and employment to help us quantify living standards such as Gross National Income
Environmental
Assess natural and man-made environmental resources such as access to fresh water and CO_2 emissions

Where and why?

Geographers use the spatial dimension, which helps us to identify patterns of where things are located over Earth's space and attempt to explain why these patterns exist. Identifying patterns across the globe may help to explain why the world is so unequal. Factors that affect equality across areas in a positive way may include the availability of natural resources or an educated workforce, whereas susceptibility to natural disasters or corruption may create more inequality.

Inequalities may exist between individuals, but also within and between countries, regions and continents (often referred to as 'spatial inequality'). Just as each person has their own unique strengths and weaknesses, places are either endowed with, or lack, various resources.

9.2 Activities

To answer questions online and to receive **immediate feedback** and **sample responses** for every question, go to your learnON title at www.jacplus.com.au. *Note*: Question numbers may vary slightly.

Explain

1. (a) Classify the following as either quantitative or qualitative indicators: motor vehicles, proportion of seats held by women in national parliaments, unemployment, electric power consumption, forest area, obesity, quality of teaching at your school, freedom of speech, how safe you feel walking in the city at night, how much you trust your neighbours, access to public transport.
   (b) Can you categorise the indicators listed in part (a) using figure 1 as a guide? Which indicators were difficult to classify, and why do you think this is the case?
2. Does your pet dog or cat have a good life? What indicators would you use to measure the wellbeing of your pet? Write a selection of 10 quantitative and qualitative indicators to help determine their wellbeing.

Predict

3. Look back over the indicators in question 1. Indicators can also imply further information about a country's progress, rate of ***change*** or development. Could these indicators be clues to the factors affecting the development of a country? If so, what else do they tell you?
4. Select one of the indicator categories: social, economic or ***environmental***. In pairs or small groups, brainstorm the various indicators that you think might be used to measure it. Create a short list of at least five before checking the World Statistics section of your atlas to see which indicators are commonly used.

Think

5. Use the **Gauging interconnections** weblink in the Resources tab to discover some of the ***interconnections*** that exist between indicators. List two strong ***interconnections***.
6. Are you better off or worse off? As a teenager in Australia, you might think you have it tough. But, when we look at the indicators, is that really the case? Decide whether you are better off or worse off for each indicator in table 1 by evaluating the data. What reasons could account for these differences?

TABLE 1 Australia versus the world — a selection of quantitative indicators, 2013

Life expectancy (years)	Australia	82	Sierra Leone (Africa)	46
Mobile phones (subscriptions per 100 people)	Australia	131	Eritrea (Africa)	6
Adolescent fertility rate (births per 1000 women 15–19 years of age)	Australia	14	Denmark (Europe)	4
Proportion of seats held by women in national parliament (%)	Australia	27	Rwanda (Africa)	64
Gross National Income per capita ($US)	Australia	64 680	Qatar (West Asia)	94 410
Literacy rate (% of youth aged 15–24)	Australia	99	Mozambique (Africa)	Males 80 Females 57

7. The concept of wellbeing is relative to who you are and the ***place*** where you live. Consider the following statements. Does the term 'wellbeing' have any relevance to these people? Does wellbeing hold any relevance for people in the direst poverty?

Person A: 'We live in constant fear, starvation; there is a lack of government. Personal safety is crucial, so wellbeing is not there yet. Things are very difficult as people are living in despair.'

Person B: 'Before, we always talked of improving living standards, which mostly meant material needs. Now we talk of the importance of relationships among people and between people and the environment.'

Person C: 'The land looks after us. We have plenty to eat, but things are changing. There are no fish now, not like when my father was a boy.'

RESOURCES — ONLINE ONLY

Explore more with this weblink: Gauging interconnections

Deepen your understanding of this topic with related case studies and questions.
World population

9.3 SkillBuilder: Constructing and interpreting a scattergraph

WHAT IS A SCATTERGRAPH?

A scattergraph is a graph that shows how two or more sets of data, plotted as dots, are interconnected. This interconnection can be expressed as a level of correlation.

Go online to access:

- a clear step-by-step explanation to help you master the skill
- a model of what you are aiming for
- a checklist of key aspects of the skill
- a series of questions to help you apply the skill and to check your understanding.

FIGURE 1 Scattergraph showing the income and the Human Development Index for sub-Saharan Africa and the rest of the world, 2011

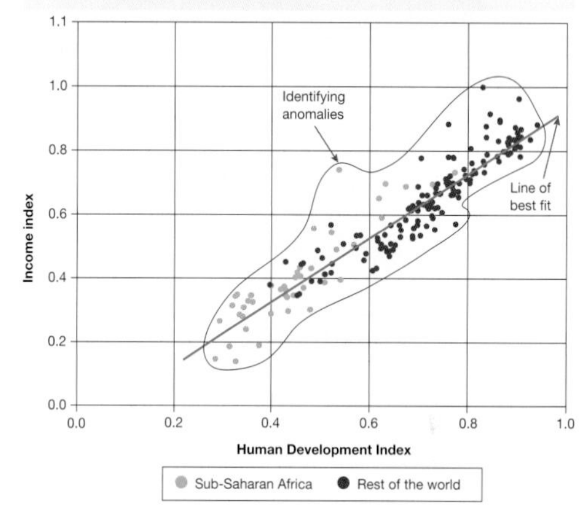

Source: © United Nations Development Programme

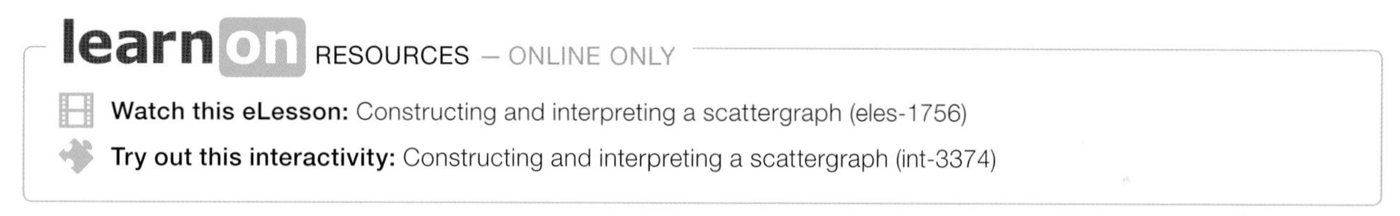

learnON RESOURCES — ONLINE ONLY

Watch this eLesson: Constructing and interpreting a scattergraph (eles-1756)

Try out this interactivity: Constructing and interpreting a scattergraph (int-3374)

9.4 How do we measure wellbeing?

9.4.1 How do we describe development?

Whichever method of measuring development or wellbeing we choose, it is important to understand the terms that have been used, the values that underpin it, and what perspective (often Western) we take. With an overwhelming amount of data available to us, the world is often divided simplistically into extremes such as 'rich' or 'poor'. Is this the best way?

The annotated classifications in figure 1 have been used in the past century, but they are very general and as such have been questioned by geographers for their accuracy (and sometimes offensiveness). Today, we use terminology such as more economically developed country (MEDC) and less economically developed country (LEDC) to describe levels of development — in the economic, social, environmental and political spheres. A newly **industrialised** country (NIC) is one that is modernising and changing quickly, undergoing rapid economic growth. Emerging economies (EEs) are places also experiencing rapid economic growth, but these are somewhat volatile in that there are significant political, monetary or social challenges.

FIGURE 1 World map showing various definitions of development

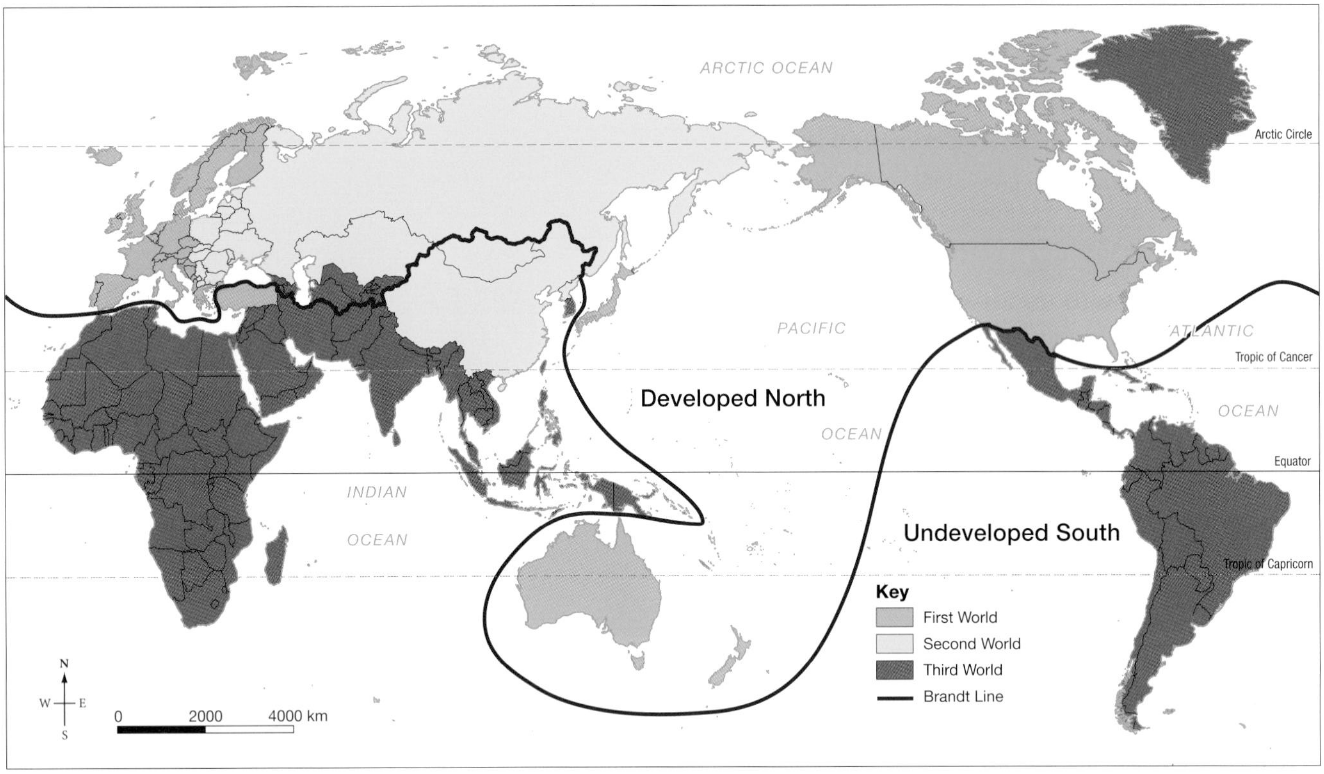

Source: United Nations Development Report

DEVELOPED OR DEVELOPING?

One of the most common ways of talking about the level of development in various places is to label them as 'developed' or 'developing'. These terms assume that development is a linear process of growth, so each country could be placed on a continuum of development. Countries that are developing are still working towards achieving a higher level of living standard or economic growth, implying that the country could ultimately become 'developed'.

NORTH OR SOUTH?

In 1980, the Chancellor of Germany, Willy Brandt, chaired a study into the inequality of living conditions across the world. The imaginary Brandt Line divided the rich and poor countries, generally following the line of the equator. The North included the USA, Canada, Europe, the USSR, Australia and Japan. The South represented the rest of Asia, Central and South America, and all of Africa. Once again, these terms have become obsolete as countries have developed differently and ignored these imaginary boundaries.

FIRST WORLD OR THIRD WORLD?

The terminology First, Second and Third Worlds was a product of the Cold War. The Western, industrialised nations and their former colonies (North America, western Europe, Japan and Australasia) were the First World. The Soviet Union and its allies of the Communist bloc (the former USSR, eastern Europe, China) were the Second World.

The Third World referred to all of the other countries. However, over time this term became more commonly used to describe the category of poorer countries that generally had lower **standards of living**.

The Second World ceased to exist when the Soviet Union collapsed in 1991.

9.4.2 What is poverty?

There is a strong interconnection between development and poverty. The United Nations defines poverty as 'a denial of choices and opportunities, a violation of human dignity ... It means not having enough to feed and clothe a family, not having a school or clinic to go to, not having the land on which to grow one's food or a job to earn one's living ... It means susceptibility to violence, and it often implies living in marginal or fragile environments, without access to clean water or sanitation.' However, poverty is most often measured using solely economic indicators. More than 1 billion people live in poverty, as shown in figure 2.

FIGURE 2 The proportion of the world's population (shown as a cartogram) living on less than US$1.25 per day

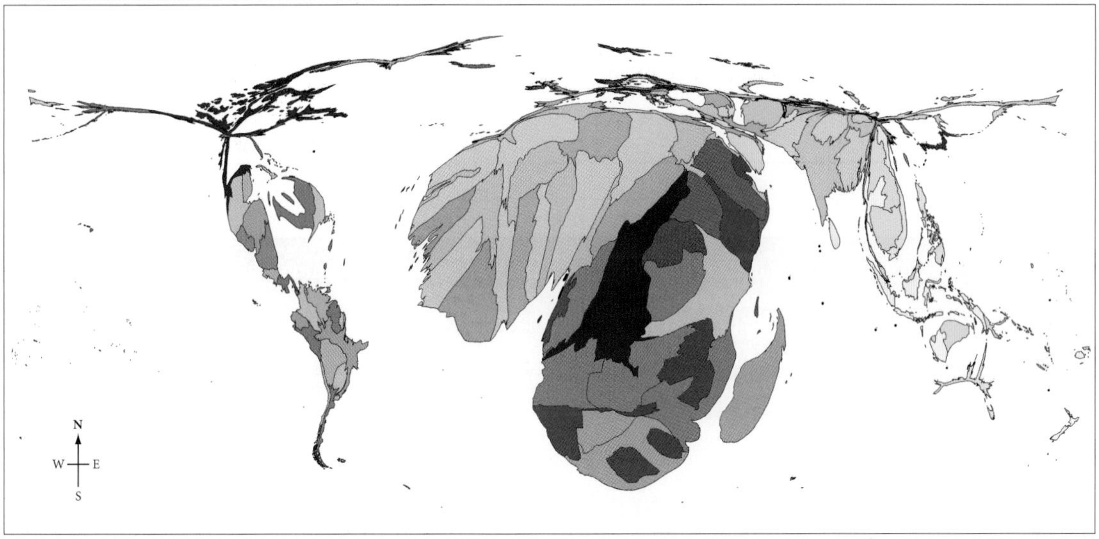

Source: The World Bank: Poverty headcount ratio at $1.25 a day PPP % of population: World Development Indicators

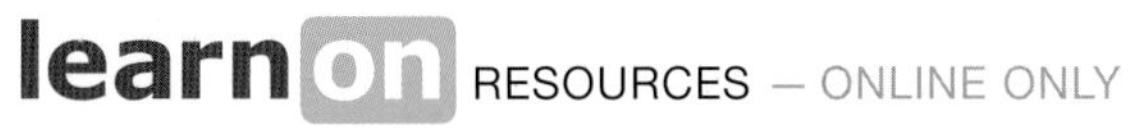

RESOURCES — ONLINE ONLY

 **Try out this interactivity:** Living on less than $1.25 a day (int-3304)

9.4 Activities

To answer questions online and to receive **immediate feedback** and **sample responses** for every question, go to your learnON title at www.jacplus.com.au. *Note*: Question numbers may vary slightly.

Remember

1. Identify two examples of ***places*** that would have been classified as 'developed North' and two that would have been classified as 'undeveloped South'.

Explain

2. What do you think about Australia being labelled a part of the 'developed North'?
3. Although indicators are measuring different aspects of quality of life, they are also ***interconnected***. For example, if a country goes through an economic recession, other indicators will be affected. Explain with examples (a flow chart may be useful to step out your thinking).

Think

4. How are MEDCs and LEDCs different? Complete table 1 (try to include your own explanations where possible).

TABLE 1 Comparison of MEDCs and LEDCs

	MEDC	LEDC
Birth rate		High — many children die so the birth rate increases to counteract fatalities
Death rate	Low — good medical care available	
Life expectancy	High — good medical care and quality of life	
Infant mortality rate		High — poor medical care and nutrition
Literacy rate	High — access to schooling, often free	
Housing type		Poor — often no access to fresh water, no sanitation, infrequent or no electricity

Deepen your understanding of this topic with related case studies and questions.

- **Indigenous Australians**
- **Wellbeing in Western Sydney**
- **Wellbeing in Sudan**
- **Child labour around the world**

9.5 SkillBuilder: Interpreting a cartogram

WHAT IS A CARTOGRAM?

A cartogram is a diagrammatic map. These maps use a single feature, such as population, to work out the shape and size of a country. Therefore, a country is shown in its relative location but its shape and size may be distorted. Cartograms are usually used to show information about populations and social and economic features.

Go online to access:

- a clear step-by-step explanation to help you master the skill
- a model of what you are aiming for
- a checklist of key aspects of the skill
- a series of questions to help you apply the skill and to check your understanding.

FIGURE 1 Cartogram showing estimated world population, 2050

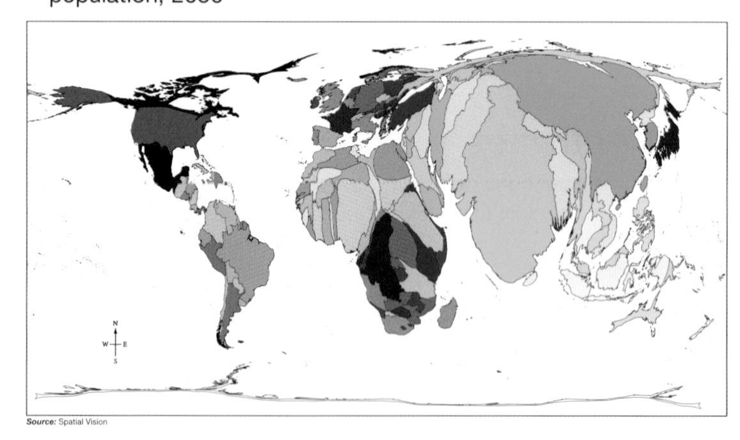

Source: Spatial Vision

RESOURCES — ONLINE ONLY

Watch this eLesson: Interpreting a cartogram (eles-1757)

Try out this interactivity: Interpreting a cartogram (int-3375)

9.6 Does wealth equal wellbeing?

9.6.1 What is a multiple component index?

A wellbeing approach to development takes into account a variety of quantitative and qualitative indicators. Some of these are a little more difficult to measure, such as the idea of happiness. Before you read on, make a list of 10 indicators that you think would give an accurate measure of a teenager's happiness in their country of residence.

A single indicator gives us only a narrow picture of the development of a country. A country may have a very high GDP but, if we dig a little deeper and look at each individual's share in that country's income or their **life expectancy**, we may not find what we expected. Inequalities may be revealed. A combination of many indicators will create a more accurate picture of the level of wellbeing in a particular place. Much like using our five senses to try a new cuisine, a combination of indicators will give us better insight into a country's wellbeing. The **Human Development Index (HDI)** is one such index. It was developed in 1990 to measure wellbeing according to four indicators (see figure 1).

FIGURE 1 Measuring quality of life encompasses many indicators.

The richest one per cent of adults worldwide owned 50 per cent of global assets in the year 2015, and the richest 10 per cent of adults accounted for 87.7 per cent of the world total. In contrast, the bottom half of the world adult population owned less than one per cent of global wealth.

Wealth is heavily concentrated in North America, western Europe and high-income Asia–Pacific countries. People in these countries collectively hold 90 per cent of total world wealth.

Source: Credit Suisse Wealth Report, 2015

9.6.2 Is this the best measure of wellbeing?

Over thousands of years, different societies have measured progress in different ways. A GDP-led development model focuses solely on boundless economic growth on a planet with limited resources — and this is not a balanced equation. The HDI has become one of the most common ways to measure wellbeing, but it has also attracted criticism for its narrow approach. These measures do not recognise some of the greatest environmental, social and humanitarian challenges of the twenty-first century, such as pollution or stress levels.

9.6.3 Measuring twenty-first century wellbeing

Which country is the happiest?

The new Happy Planet Index (HPI) results (see figure 2) map the extent to which 151 countries across the globe produce long, happy and sustainable lives for the people that live in them.

FIGURE 2 The happiest countries in the world according to the Happy Planet Index

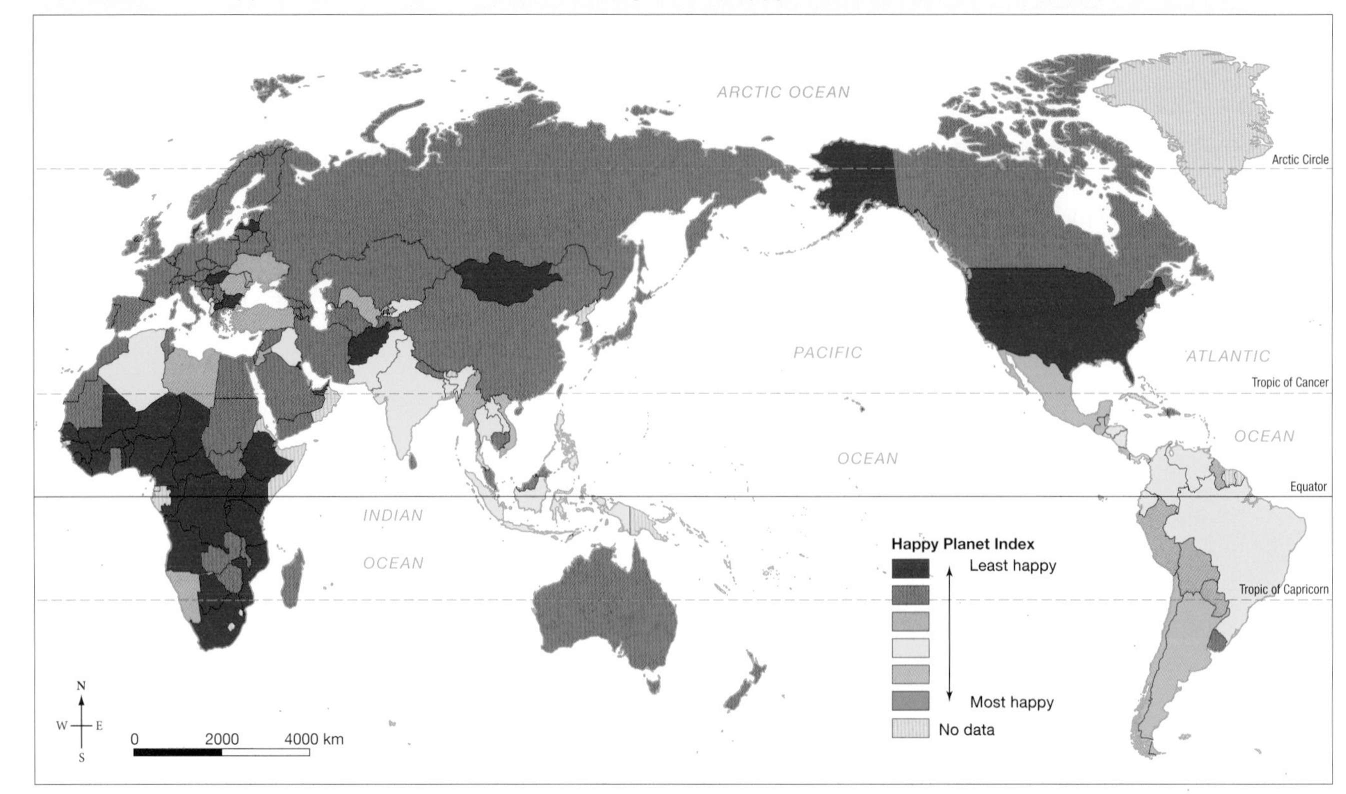

Source: The Happy Planet Index: 2012 Report. A global index of sustainable wellbeing (nef: London)

Each of the three component measures — life expectancy, experienced wellbeing and **ecological footprint** — is given a traffic-light score based on thresholds for good (green), middling (amber) and bad (red) performance. These scores are combined to an expanded six-colour traffic light for the overall HPI score. To achieve bright green (the best of the six colours), a country would have to perform well on all three individual components.

What is Gross National Happiness?

In 2011, the Prime Minister of Bhutan (Central Asia) demonstrated his country's commitment to its well-being by developing the world's first measure of national happiness, and he encouraged world economies to do the same. UN Secretary-General Ban Ki-moon supported this innovation: 'Gross national product (GNP) … fails to take into account the social and environmental costs of so-called progress … Social, economic and environmental wellbeing are indivisible. Together they define gross global happiness.'

$$\text{Happy Planet Index} = \frac{\text{experienced wellbeing} \times \text{life expectancy}}{\text{ecological footprint}}$$

How does Australia assess wellbeing?

The Australian National Development Index (ANDI), which was approved in 2015, incorporates 12 indicators measuring elements of progress including health, education, justice and **Indigenous** wellbeing. Measures such as this demonstrate a new direction in articulating wellbeing, recognising that happiness is not directly proportionate to our bank balance or how long we expect to live. This new measure of wellbeing will reflect what is important to Australians to feel happy as individuals, as well as the happiness of our communities. It will allow Australians to measure the future we want.

9.6 Activities

To answer questions online and to receive **immediate feedback** and **sample responses** for every question, go to your learnON title at www.jacplus.com.au. *Note*: Question numbers may vary slightly.

Remember

1. Without referring to figure 2, name three ***places*** you would expect to rank high on the Happy Planet Index and three you would expect to rank low. Now, check your predictions on the map. Were you correct?

Explain

2. Provide a detailed explanation of each of the four indicators used to calculate the HDI. Is the HDI the best indicator of a country's development? Give reasons for your answer.
3. Can you make any comments about the distribution of the happiest/unhappiest countries across the world according to the data in figure 2? What do you think would make a country unhappy?

Discover

4. Locate Bhutan on a world map. Describe its location. How does it rate on the Happy Planet Index?
5. It has been well documented that as people move beyond a certain income level, they do not become any happier. Try measuring your own happiness using the **HPI survey** weblink in the Resources tab. How does your happiness rate against that of your classmates?
6. A number of countries have already adopted a national measure of wellbeing. Either individually or in pairs, research the history of one of the following indexes, identify the indicators used to measure it and evaluate its success.
   - Gross National Happiness (Bhutan)
   - Key National Indicator System (USA)
   - Canadian Index of Wellbeing (Canada)

Think

7. Suggest why a range of indexes is being developed in the twenty-first century to measure wellbeing.
8. Should wellbeing or happiness be a core goal of a country's government? Debate this in a small group.
9. Suggest two indicators that might be used in ANDI.

Explore more with this weblink: HPI survey

9.7 How can we improve wellbeing?

9.7.1 How can we bridge the gap?

We have much to be thankful for. We live in a world where we live much longer than our ancestors, we have better nutrition and education, and we generally have a better outlook for our lives. But in an age where some are globally connected, educated, fed, clothed and medicated, it is easy to forget that many of our fellow human beings go without each and every day.

Have you ever given some loose change to a tin-shaker on the street or helped collect money for a fundraiser? If so, then you are already a part of the cycle of aid. Aid (also known as international aid, overseas aid or foreign aid) is the voluntary transfer of resources from one country to another, given at least partly with the aim of benefiting the receiving country.

Why do we give aid? Aid may be given by government, private organisations or individuals. **Humanitarianism** is still the most significant motivation for the giving of aid, but it may be motivated by other functions as well:

- as a sign of friendship between two countries
- to strengthen a military ally
- to reward a government for actions approved by the donor
- to extend the donor's cultural influence
- to gain some kind of business or commercial access to a country.

FIGURE 1 The Friendship Bridge across the Mekong River, which connects Thailand with Laos, was built with Australian aid.

9.7.2 What types of aid exist?

Bilateral aid is aid given by governments to donor countries. Multilateral aid is provided through international institutions such as UNICEF. **Non-government organisation** (NGO) or charity aid is voluntary, private, individual donations collected by organisations such as the Red Cross. Aid takes many forms: money, food, medicine, equipment, expertise, scholarships, training, clothing or military assistance (to name just a few). Large-scale aid (top-down aid) is usually given to the government of a developing country so that it can spend it on the projects that it needs. Small-scale aid projects (bottom-up aid) target the people most in need of the aid and help them directly, without any government interference. Aid from NGOs tends to be bottom-up aid.

There are positive and negative impacts of aid (see figure 2). Aid can increase the dependency of LEDCs on donor countries. Sometimes aid is not a gift but a loan, and poor countries may struggle to repay the money. Aid may also be used to put political or economic pressure on a country, which may leave its people feeling like they owe their donors a favour. There is always the threat that corruption among politicians and officials will prevent aid from reaching the people who need it most. If aid does not provide for and empower citizens, then wellbeing will not be improved.

FIGURE 2 Advantages and disadvantages of aid

Bilateral aid	Multilateral aid	NGO/charity aid
+ Helps expand infrastructure: roads, railways, ports, power generation.	+ The organisations have clear aims around what they are trying to achieve (e.g. WHO combats disease and promotes health).	+ Usually targeted at long-term development within a country
+ Aid which directly supports economic, social or environmental policies can result in successful programs.	+ Leading experts in their field work to help achieve multilateral aid program objectives.	+ Raises awareness of specific situations in a country or region
– 'Tied aid' obliges the country receiving aid to spend it on goods and services from the donor country (may be expensive).	– Sometimes directed only towards specific areas or organisations, leaving many without benefit.	– The greatest source of need may not be prioritised (e.g. the 2006 tsunami devastation received many donations, but areas in sub-Saharan Africa on a daily basis were just as much in need).
– Inappropriate technology may be given (e.g. tractors are of little use if there are no spare parts or fuel).	– May come with conditions to make big changes to structures, which can be difficult to manage once aid has 'finished'.	– Up to 30% of donations may be 'eaten up' by administration costs.

9.7.3 How does Australia help?

The Australian Government's official development assistance (ODA) is designed to promote prosperity, reduce poverty and enhance stability in the developing countries of the Indo-Pacific region (90 per cent of its aid allocation). In 2015–16 Australia provided $4051.7 billion worth of official development assistance. It focuses on strengthening private sector development and enabling human development. Specifically, it will promote aid for trade; investment in infrastructure, agriculture, fishing, health and education; and effective governance.

FIGURE 3 Australian aid helps these primary school students in north-western Laos.

9.7 Activities

To answer questions online and to receive **immediate feedback** and **sample responses** for every question, go to your learnON title at www.jacplus.com.au. *Note*: Question numbers may vary slightly.

Remember

1. What is the difference between the three types of aid?
2. List the various forms of aid mentioned in this section. Can you add any more types to this list?

Discover

3. Think of a charity that you or your family have supported in the past. Find out more about the charity. Where is your money going, and who are the beneficiaries?
4. Using the internet, review some of the Australian Government programs currently in operation. In which ***places*** are most of these programs focused?
5. Study figures 1 and 3. What benefits would each of these aid projects bring to the recipients?
6. Discuss in a small group what limitations might exist in administering an aid program in (a) a developing country and (b) a country that has been devastated by a natural disaster (e.g. an earthquake). Suggest possible ways of overcoming the problems you identify.
7. Do you think the Australian Government's focus will shift in 10 years' time? In 50 years' time? Which region do you think we might have to shift our focus to?

Think

8. Reflect on what you have studied so far in this topic. Why are some people 'poor' and some people 'rich'?
9. Think about the challenges that might be faced by someone delivering emergency aid to a LEDC. How might they be affected by the physical and emotional conditions of their work?
10. Is aid ever inappropriate? Discuss this in a small group.

RESOURCES — ONLINE ONLY

Try out this interactivity: Helping others (int-3305)

9.8 Are we on track?

9.8.1 Did we succeed?

Recognising the need to assist impoverished nations more actively, the United Nations (UN) held a series of summits dating from the year 2000. In 2001, the UN member states formally adopted the United Nations Millennium Development Declaration and its eight goals, which set out targets aimed at improving social and economic conditions in the world's poorest countries by 2015.

In his foreword to the publication *The Millennium Development Goals Report 2015*, United Nations Secretary-General Ban Ki-moon wrote, 'The global mobilization behind the Millennium Development Goals has produced the most successful anti-poverty movement in history. The landmark commitment entered into by world leaders in the year 2000 — to "spare no effort to free our fellow men, women and children from the abject and dehumanizing conditions of extreme poverty" — was translated into an inspiring framework of eight goals and, then, into wide-ranging practical steps that have enabled people across the world to improve their lives and their future prospects.'

TABLE 1 Global progress of the Millennium Development Goals measured against 1990 data

Millennium Development Goal	Indicator for developing countries	1990	2015
1. Eradicate extreme poverty and hunger	• Living on less than $1.25 a day • People undernourished	47% 23.3%	14% 12.9%
2. Achieve universal primary education	• Primary school enrolment • Literacy rate youth aged 15–24	83% (year 2000) 83%	91% 91%
3. Promote gender equality and empower women	• Women's paid employment in non-agriculture sector • Gender equality in education: girls per 100 boys	35% 74 (Southern Asia)	41% 103 (Southern Asia)
4. Reduce child mortality	• Under-five mortality rate (deaths per 1000 live births)	90	43
5. Improve maternal health	• Maternal mortality (deaths per 100 000 live births) • Births assisted by health personnel	440 59%	240 75%
6. Combat HIV/AIDS, malaria and other diseases	• New infection rate of HIV • Malaria incidence rate (per 1000 at risk)	3.1 million (year 2000) 146 (year 2000)	2 million 91
7. Ensure environmental sustainability	• Using an improved drinking water source • Sanitation • Slum living	75% 36% 39.4% (year 2000)	90% 68% 29.7%
8. Global partnership for development	• Development aid • Internet penetration	$81 billion 6% (2000)	$135.2 billion 43%

Over the past two decades we have made huge progress in improving the quality of life throughout the developing world. Progress has varied from place to place, but overall there has been a reduction in poverty levels and increased access to health, education, water and other essential services. The Millennium Development Goals (MDGs) have been an important motivational force and have allowed us to take accurate measurements through quantitative data. Many countries have achieved a significant number of the MDG targets. This has transformed the quality of life of hundreds of millions of people who have been given hope and incentive for change. However, progress has been uneven across regions and countries, in particular in countries affected by conflict.

Sustainable Development Goals

The Sustainable Development Goals (SDGs) came into being on 25 September 2015 to replace the expired Millennium Development Goals (MDGs). The three overarching themes are to end poverty, protect the planet and ensure prosperity for everyone over the next 15 years. Each of the 17 goals has a number of targets to be met. Indicators will be used to assess each target. The SDGs apply to all countries, which was also the intention of the MDGs; however, the MDGs became targets for poor countries to achieve with finance from wealthier countries. The SDGs apply to all countries.

Former New Zealand Prime Minister and now United Nations Development Programme Administrator Helen Clark commented, 'This agreement marks an important milestone in putting our world on an inclusive and sustainable course. If we all work together, we have a chance of meeting citizens' aspirations for peace, prosperity, and wellbeing and to preserve our planet'.

TABLE 2 A brief outline of the Sustainable Development Goals

	Sustainable development goal	Targets (SUSTAINABLE DEVELOPMENT GOALS)
1 NO POVERTY	**Goal 1:** End poverty in all its forms everywhere	Major target: • By 2030 no-one should live on less than $1.25 per day.
2 ZERO HUNGER	**Goal 2:** End hunger, achieve food security and improved nutrition and promote sustainable agriculture	Significant targets include: • By 2030 ensure access by all people to safe, nutritious and sufficient food all year round. • By 2030 end all forms of malnutrition.
3 GOOD HEALTH AND WELL-BEING	**Goal 3:** Ensure healthy lives and promote well-being for all at all ages	Targets for 2030 include: • Reduce the global maternity mortality ratio to less than 70 per 100 000 live births. • Attain an under-five mortality of, at most, 25 per 1000 live births. • End the epidemics of AIDS, tuberculosis and malaria.
4 QUALITY EDUCATION	**Goal 4:** Ensure inclusive and quality education for all and promote lifelong learning	Targets to achieve include: • By 2030 all boys and girls can complete free, equitable and quality primary and secondary schooling with effective outcomes. • All women and men have equal access to affordable and quality ongoing educational opportunities.
5 GENDER EQUALITY	**Goal 5:** Achieve gender equality and empower all women and girls	Targets include: • End discrimination against all women and girls everywhere and eliminate violence towards them too.

(Continued)

TABLE 2 A brief outline of the Sustainable Development Goals (*Continued*)

	Sustainable development goal	**Targets** SUSTAINABLE DEVELOPMENT GOALS
6 CLEAN WATER AND SANITATION	**Goal 6:** Ensure access to water and sanitation for all	Targets by 2030 include: • Achieve universal and equitable access to safe and affordable drinking water for all. • Achieve access to adequate and equitable sanitation and hygiene for all.
7 AFFORDABLE AND CLEAN ENERGY	**Goal 7:** Ensure access to affordable, reliable, sustainable and modern energy for all	Targets for 2030 include: • Provide access to affordable, reliable and modern energy services, especially renewable energies.
8 DECENT WORK AND ECONOMIC GROWTH	**Goal 8:** Promote inclusive and sustainable economic growth, employment and decent work for all	Major targets: • Sustain economic growth and productivity aiming to achieve by 2030 full and productive employment and decent work for all. • By 2025 eliminate child labour in all its forms, including forced labour, modern slavery, human trafficking and child soldiers.
9 INDUSTRY, INNOVATION AND INFRASTRUCTURE	**Goal 9:** Build resilient infrastructure, promote sustainable industrialisation and foster innovation	General targets include: • Develop quality, reliable, sustainable and resilient infrastructure to support economic development and human wellbeing. • Promote inclusive and sustainable industries that raise industries' share of employment and GDP. • Provide universal and affordable internet access to least developed countries by 2020.
10 REDUCED INEQUALITIES	**Goal 10:** Reduce inequality within and among countries	Key target: • By 2030 achieve and sustain income growth of the bottom 40 per cent of the population.
11 SUSTAINABLE CITIES AND COMMUNITIES	**Goal 11:** Make cities inclusive, safe, resilient and sustainable	Targets include: • By 2030 ensure adequate, safe, affordable and sustainable housing and transport for all. • Protect and safeguard the world's cultural and natural heritage.
12 RESPONSIBLE CONSUMPTION AND PRODUCTION	**Goal 12:** Ensure sustainable consumption and production patterns	2030 targets include: • Achieve the sustainable management and efficient use of natural resources • Halve per capita global food waste by consumers and during production. • Ensure all people have the relevant information and awareness for sustainable development and lifestyles in harmony with nature.
13 CLIMATE ACTION	**Goal 13:** Take urgent action to combat climate change and its impacts	Targets include: • Strengthen resilience and adaptive capacity to climate-related hazards and natural disasters in all countries. • Integrate climate change measures into national policies, strategies and planning.
14 LIFE BELOW WATER	**Goal 14:** Conserve and sustainably use oceans, seas and marine resources	Significant targets: • By 2025 prevent and reduce marine pollution of all kinds. • By 2020 sustainably manage and protect marine and coastal ecosystems. • By 2020 regulate and end overfishing.

(*Continued*)

TABLE 2 A brief outline of the Sustainable Development Goals (*Continued*)

Sustainable development goal		Targets
15 LIFE ON LAND	**Goal 15:** Sustainably manage forests, combat desertification, halt and reverse land degradation, halt biodiversity loss	Targets: • By 2020 protect inland freshwater ecosystems and all types of forests. • By 2020 prevent the introduction of invasive alien spaces and prevent the extinction of threatened species. • By 2030 combat desertification and protect mountain ecosystems.
16 PEACE, JUSTICE AND STRONG INSTITUTIONS	**Goal 16:** Promote just, peaceful and inclusive societies	Targets: • End abuse, exploitation, trafficking and all forms of violence against children. • Reduce bribery and corruption in all forms. • By 2030 provide legal identity for all, including birth registration.
17 PARTNERSHIPS FOR THE GOALS	**Goal 17:** Revitalise the global partnership for sustainable development	Target: • Consider finance, technology, capacity building, trade and systemic issues.

9.8 Activities

To answer questions online and to receive **immediate feedback** and **sample responses** for every question, go to your learnON title at www.jacplus.com.au. *Note*: Question numbers may vary slightly.

Remember

1. When were the MDGs conceived? In which year was it hoped the goals would be achieved?
2. What is the basis of the SDGs?

Discover

3. Search the internet for 'Millennium Development Goals Progress Report 2015'.
   (a) List the regions in the summary report. Suggest why the report does not include all regions of the world.
   (b) Which regions seemed to make most progress? Support your answer with evidence.
   (c) Which of the eight MDGs seemed to show least improvement? Quote evidence for your response.
4. Using the three overarching themes of the SDGs, draw up a table to show where each of the 17 goals is aligned. Is your table the same as a classmate's? Discuss any differences.

Think

5. Within your class, divide into groups and assign the SDGs across the class. Using the internet, research the targets of each goal and provide a tick or a cross depending on whether you think the world will meet each target as set down. Be prepared to argue your point of view in a class debate on 'Are there too many goals and targets to be met by all countries in the world?'

9.9 Can we help the bottom billion?

Access this subtopic at **www.jacplus.com.au**

9.10 What are human rights?

9.10.1 The basis of human rights

Human rights are so much a part of our daily lives here in Australia that we tend to take them for granted. Many principles that have been adopted in international human rights practices have their roots in traditions and religions that are thousands of years old. Different countries, societies and cultures have come up with their own definitions over time to suit their particular environment or context.

In some societies, human rights may be enshrined in law and legislation, whereas in others they may simply exist as guidelines that reflect the values of that particular community. In short, the concept of human rights stems from the belief that there is an instinctive human ability to distinguish right from wrong.

Human rights can be defined in different ways. The Australian Human Rights Commission notes that definitions may include:

- the recognition and respect of people's dignity
- a set of moral and legal guidelines that promote and protect a recognition of our values
- our identity and ability to ensure an adequate standard of living
- the basic standards by which we can identify and measure inequality and fairness
- those rights associated with the Universal Declaration of Human Rights.

FIGURE 1 Origins of human rights

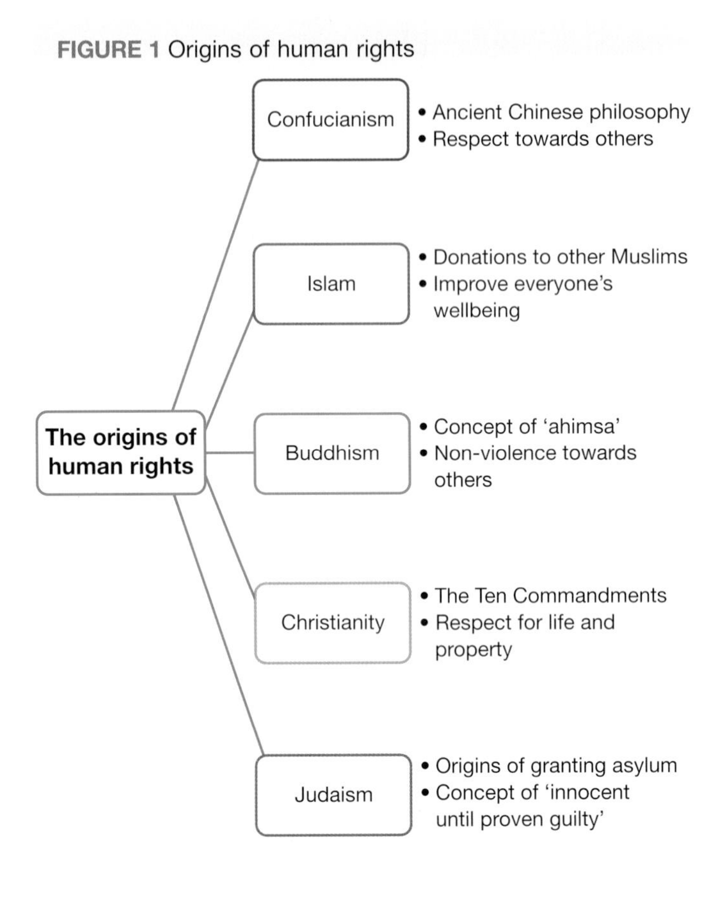
The origins of human rights
Confucianism
• Ancient Chinese philosophy
• Respect towards others
Islam
• Donations to other Muslims
• Improve everyone's wellbeing
Buddhism
• Concept of 'ahimsa'
• Non-violence towards others
Christianity
• The Ten Commandments
• Respect for life and property
Judaism
• Origins of granting asylum
• Concept of 'innocent until proven guilty'

9.10.2 What role does the United Nations play?

The UN was formed in the aftermath of World War II on 24 October 1945 by countries committed to preserving peace through international cooperation and security. Today, nearly every nation (currently 193 countries) in the world belongs to the UN. One of the main aims of the UN Charter is to promote respect for human rights. The **Universal Declaration of Human Rights**, proclaimed by the UN General Assembly in 1948, sets out basic rights and freedoms to which all women and men are entitled, including:

- the right to life, liberty and nationality
- the right to freedom of thought, conscience and religion
- the right to work and to be educated
- the right to food and housing
- the right to take part in government.

These rights are legally binding by virtue of two **International Covenants**, to which most states are parties. One covenant deals with economic, social and cultural rights and the other with civil and political rights (see figure 2). Together with the Declaration, they constitute the **International Bill of Human Rights**.

FIGURE 2 Political rights around the world, 2015. A free country is one where political rights are available and protected. A country that is not free is one where basic political rights are absent, and basic civil liberties are widely and systematically denied.

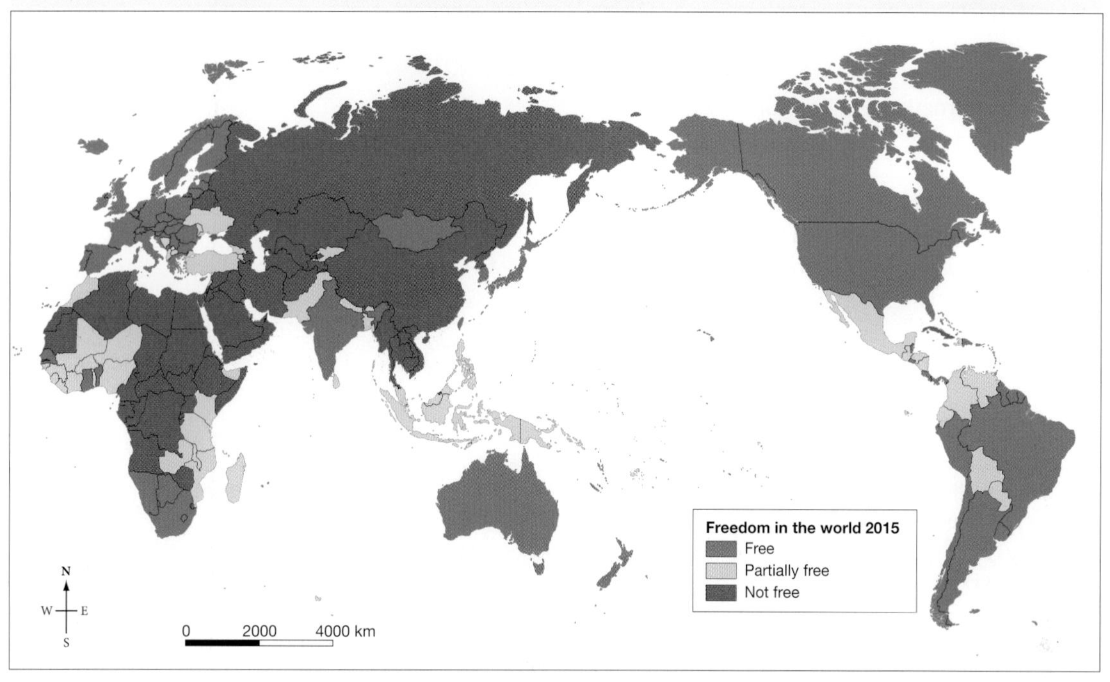

Source: Freedom House

9.10.3 Who protects our human rights?

While Australia has agreed to be bound by these major international human rights treaties, they do not form part of Australia's domestic law unless they have been specifically written into Australian law through legislation. The Australian Human Rights Commission is the national organisation that advocates for promotion and protection of human rights. In addition to monitoring economic, social, cultural, civil and political rights, other areas of human rights include peacekeeping, eradication of poverty and the humanitarian tribunals (for example, the International Criminal Court that deals with mass human rights violations, such as genocide). The death penalty (and capital punishment) is one area in which Amnesty International focuses its human rights advocacy work, a contentious issue on the global political stage (see figure 3).

FIGURE 3 The death penalty violates the right to life as proclaimed in the Universal Declaration of Human Rights.

At least **20 292** people worldwide were under sentence of death at the end of 2015.	
Top five countries where most executions happened in 2015 (1634 people, excluding those from China)	Top five countries where most people were sentenced to death in 2015 (1998 in total worldwide)
China (unknown)	China (unknown)
Iran 977 +	Egypt 538 +
Pakistan 326 +	Bangladesh 197 +
Saudi Arabia 158 +	Nigeria 171
USA 28	Pakistan 121 +
140 countries worldwide, more than two-thirds, are abolitionist in law or practice.	

9.10 Activities

To answer questions online and to receive **immediate feedback** and **sample responses** for every question, go to your learnON title at www.jacplus.com.au. *Note*: Question numbers may vary slightly.

Remember

1. What is the International Bill of Human Rights?

Explain

2. Study figure 1. Which philosophies have influenced your understanding of human rights?
3. Define the term *human rights* in your own words.
4. Who does the Universal Declaration of Human Rights apply to?

Discover

5. Use the internet to find out when Human Rights Day occurs each year and why the date was chosen.

Think

6. Observe figure 2.
   (a) What does this map illustrate?
   (b) Which ***places*** around the world are 'free', and which are 'not free'?
   (c) Identify one of the countries that is not free and conduct additional research. What violations of this area of human rights have contributed to this rating? You may want to use the **Human Rights Watch** weblink in the Resources tab as one source of information.
7. Some of the basic human rights are outlined in this section. In pairs or small groups, develop a 'Teenager's Bill of Rights' (include at least 10 rights) that you believe would provide for a better existence for all teenagers.

Explore more with this weblink: Human Rights Watch

9.11 Who are the unprotected?

9.11.1 Do you know your rights?

International human rights organisations recognise that children have special human rights because of their vulnerability to exploitation and abuse. The United Nations General Assembly adopted the Convention on the Rights of the Child (the CRC) in November 1989. How are your rights protected? And what are some of the big issues for children's rights today?

Some of the rights and protections that a **child** is entitled to according to the CRC include:

- the right to life
- the right to a name and a nationality
- the right to live with their parents
- the right to freedom of thought, conscience and religion
- the right to privacy
- protection from abuse and neglect
- the right to education
- the right to participate in leisure, recreation and cultural activities
- protection from economic exploitation
- protection from or prevention of abduction, sale or trafficking.

Two key areas that are currently a focus for rights are the use of children in conflict and for labour.

9.11.2 Too young to serve?

The issue of children in armed conflict has become a pressing one over the past few decades because of the serious risks of involving children in war or conflict situations. Approximately 300 000 children are believed to be combatants in conflicts worldwide. **Child soldiers** have gone to battle in a range of countries, including Burundi, Colombia, the Democratic Republic of the Congo, Liberia, Myanmar and Uganda.

The Child Labour Index 2014 evaluated the frequency and severity of reported child labour incidents in 197 countries. Worryingly, nearly 40 per cent of all countries were classified as 'extreme risk' in the index, with conflict-torn and authoritarian states topping the ranking (see figure 1).

FIGURE 1 Risk of child labour worldwide

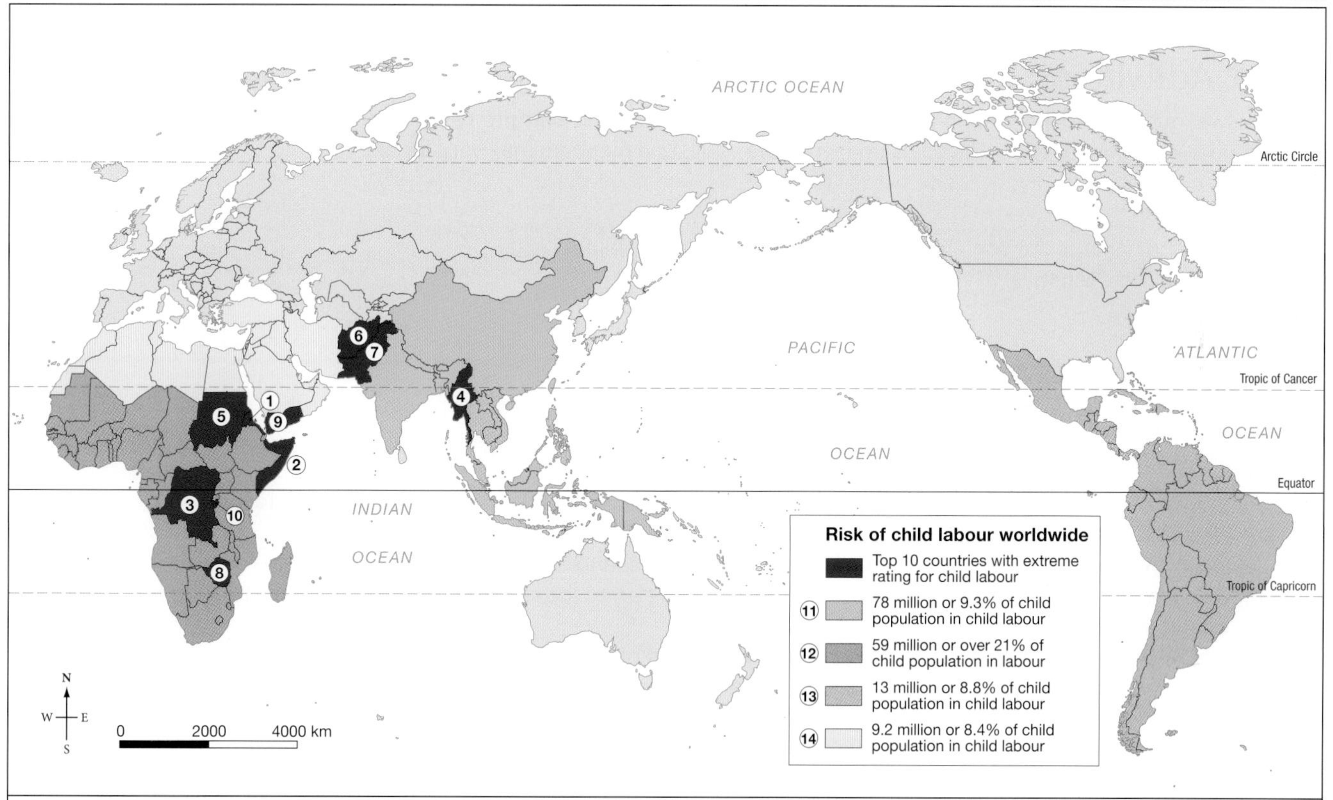

1 **Eritrea:** Ranked equal first in the world for countries where child labour is most prevalent. Forced child labour in agriculture and military recruitment are key issues.
2 **Somalia:** Ranked equal first in the world for countries where child labour is most prevalent. The use of child soldiers is a major issue.
3 **Democratic Republic of Congo:** Ranked equal third in the world for countries where child labour is most prevalent. The recruitment of child soldiers and the use of forced child labour in mining are key concerns.
4 **Myanmar:** Ranked equal third in the world for countries where child labour is most prevalent. The use of child labour in various areas, particularly agriculture, is an issue.
5 **Sudan:** Ranked equal third in the world for countries where child labour is most prevalent. The use of child soldiers is of particular concern here.
6 **Afghanistan:** Ranked equal sixth in the world for countries where child labour is most prevalent. Child labour exists in a range of areas such as agriculture, the production of carpets and bricks, and for military purposes.
7 **Pakistan:** Ranked equal sixth in the world for countries where child labour is most prevalent. Of particular concern is the sexual exploitation of children and their use in illegal activities and in the production of various goods including bricks, carpets, coal, glass bangles and leather.
8 **Zimbabwe:** Ranked number eight in the world for countries where child labour is most prevalent. Children's labour is used in agricultural and mining activities.
9 **Yemen:** Ranked number nine in the world for countries where child labour is most prevalent. Children are used in the fishing industry and in armed conflict as child soldiers.
10 **Burundi:** Ranked equal tenth in the world, along with Nigeria, for countries where child labour is most prevalent. Concerns include the commercial sexual exploitation of children, and their use in agricultural work and armed conflict.
11 **Asia and the Pacific:** Almost 78 million, 9.3% of the child population, are in child labour.
12 **Sub-Saharan Africa:** 59 million, over 21% of the child population, are in child labour.
13 **Latin America and the Caribbean:** 13 million, 8.8% of the child population, are in child labour.
14 **Middle-East and North Africa:** 9.2 million, 8.4% of the child population, are in child labour.

Source: Based on data from International Labour Organization, Maplecroft and United States Department of Labor

9.11.3 Growing up too quickly?

Recent figures from the International Labour Organization (ILO) show that:

- globally, 1 in 10 children work (168 million children between the ages of 5 and 17)
- 85 million children work in hazardous conditions
- the highest number of child labourers are in the Asia–Pacific region (78 million children)
- the highest proportion of child labourers is in sub-Saharan Africa, where 27 per cent of children (59 million) work.

In many countries, poor girls are put to work as domestic servants for richer families. In many places, children (especially girls) perform unpaid work for their families. In all cases, children are exploited, and in many cases, they are excluded from attending school (denying them their right to education).

9.11 Activities

To answer questions online and to receive **immediate feedback** and **sample responses** for every question, go to your learnON title at www.jacplus.com.au. *Note*: Question numbers may vary slightly.

Remember

1. What is the name of the document that sets out the rights of children?
2. How is a *child* defined?

Explain

3. Why do children need a separate declaration outlining their rights?
4. In which areas or industries is a child's 'right to be protected from economic exploitation' most at risk?

Discover

5. Find out more about the items that may be produced with child labour. Using the **Products of slavery** weblink in the Resources tab, explore the world map.
   (a) In which ***spaces*** across the world are children most exposed to the risk of child labour?
   (b) Select a ***place*** of interest and write a short paragraph detailing the level of risk and items most likely to be produced there by child labour.
6. The ILO's Convention 182 works to prevent the worst forms of child labour worldwide.
   (a) Research details of Convention 182 and what types of exploitation it is trying to prevent.
   (b) Research countries that have and have not **ratified** Convention 182 and create a simple world map displaying this information. Use the **ILO** weblink in the Resources tab to search for 'Ratifications of C182 — Worst Forms of Child Labour Convention'.
   (c) In 2012, it was alleged that the maker of AFL footballs was using child labour, with Indian children working up to 10 hours a day, seven days a week hand-stitching footballs for 12 cents each. Conduct some additional research online to find out more about the allegation, and what resulted from the investigation.

Predict

7. Since the 2008 Global Financial Crisis, the situation for child labourers has ***changed*** for the worse. Why do you think this might be the case? Justify your explanation.

Think

8. Only a small selection of the rights outlined by the CRC is provided in this section. In a small group or in pairs, answer the following questions.
   (a) How would you rank the 10 rights listed in this section in order of importance (1 being the most important)? Justify your choices.
   (b) Do you think someone in Myanmar or the other countries profiled in figure 1 would agree with your choices? Would you add any different rights or protections to your list now?
9. If you had to stay home and babysit your younger siblings, then your right to an education may be compromised. How might simple daily events prevent you from achieving your rights or protections as outlined by the CRC?

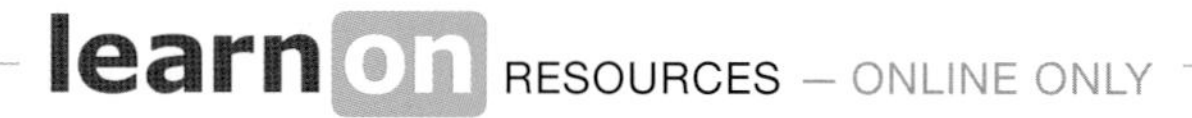

RESOURCES — ONLINE ONLY

Try out this interactivity: The two worlds (int-3306)

Explore more with these weblinks: Products of slavery

9.12 Review

9.12.1 Review

The Review section contains a range of different questions and activities to help you revise and recall what you have learned, especially prior to a topic test.

9.12.2 Reflect

The Reflect section provides you with an opportunity to apply and extend your learning.

Access this subtopic at **www.jacplus.com.au**

TOPIC 10 Human wellbeing and change

10.1 Overview

Numerous **videos** and **interactivities** are embedded just where you need them, at the point of learning, in your learnON title at www.jacplus.com.au. They will help you to learn the content and concepts covered in this topic.

10.1.1 Introduction

As the world's living standards have improved, so too has our population grown. In October 2011, the world's population reached seven billion. It is expected that the number of people on the planet will continue to grow. In 2050, it is estimated that our population will be between 8 and 10.6 billion. It is not just a matter of how many people we can fit in a particular place, but also the manner in which we live (our ecological footprint) that affect our wellbeing. Our wellbeing is clearly interconnected with our population characteristics.

Global population diversity

Starter questions

1. Why would the estimates for the world population for 2050 vary so widely?
2. Think back to the definition of wellbeing you covered in topic 9.
   (a) How might the number of people in a given ***place*** be ***interconnected*** with their wellbeing?
   (b) How could the wellbeing of a particular ***place*** have an impact on the number of people living at that location?

INQUIRY SEQUENCE

10.2 Where is everybody?

10.2.1 Global population distribution

Whether you have travelled within Australia or to another country, you would be aware that there is considerable variation in the number of people found in one place compared to another. The seven billion people on our planet are not spread evenly across space (continents, countries, and rural or urban areas). We may feel crowded in or feel we have plenty of space; we may feel isolated within or embraced as part of a community. This has a major impact on our wellbeing.

The world's population is distributed unevenly across the globe. Although there is an average of 47 persons per square kilometre, as shown in figure 2, the **population density** varies considerably. Places well below this figure include large regions of most continents, particularly in the inland sections, such as central Asia and Australia, with well below 10 persons per square kilometre. Regions of highest density are clustered in Europe, East and South-East Asia and in the eastern half of the United States of America. For example, Germany has 232 persons per square kilometre and Japan has 349 persons per square kilometre.

FIGURE 1 View of Dhaka, Bangladesh — an area of high density

Figure 2 also shows that most regions of high density are dominated by large cities. The majority of the world's population lives in urban environments. Within the largest cities, population density may be considerably higher than the country average. Dhaka, the capital city of Bangladesh (see figure 1), is considered to be the most densely populated place in the world, with an estimated 50 000 persons per square kilometre.

FIGURE 2 Global population density and the world's largest cities

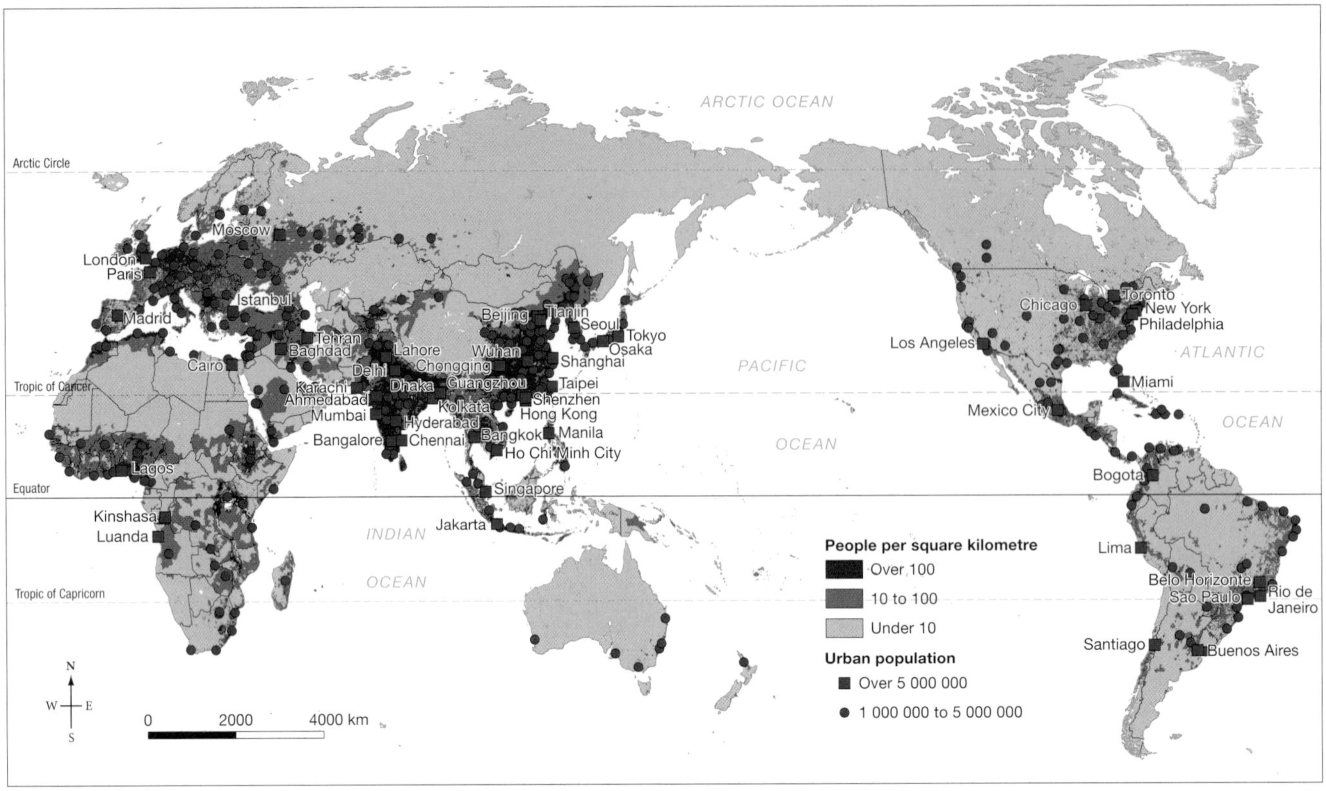

Source: Spatial Vision.

10.2.2 Why does population distribution vary so much?

Physical factors play a large part in determining global **population distribution**. Characteristics of the natural environment that favour human settlement include the availability of freshwater resources, fertile soil, moderate climate and sea ports. Inhospitable features such as mountains, jungles and deserts tend to deter high population densities.

Of course, human factors also influence population distribution. Urban places around the world are attracting an increasing number and percentage of people due to the availability of employment, particularly in the manufacturing and service sectors (for example, the industrial areas of Mumbai, India and Yokohama, Japan, and the coastal ports of Rotterdam in the Netherlands and Rio de Janiero in Brazil). Population density is closely interconnected with energy demands as reflected in the pattern of lights visible in figure 3. Not all regions of high density are urban places. Rural environments such as in parts of central Europe and South-East Asia may contain large numbers of people per square kilometre; for example, rural locations in Vietnam have over 900 persons per square kilometre.

Government policies may also affect population distribution. Examples include migration of people from one country to another, such as from Mexico to the United States of America; encouraging settlement in a particular location, such as the movement of Han Chinese into Tibet; and establishing service centres based around resource development, as in the Pilbara in Western Australia.

FIGURE 3 Population distribution by night

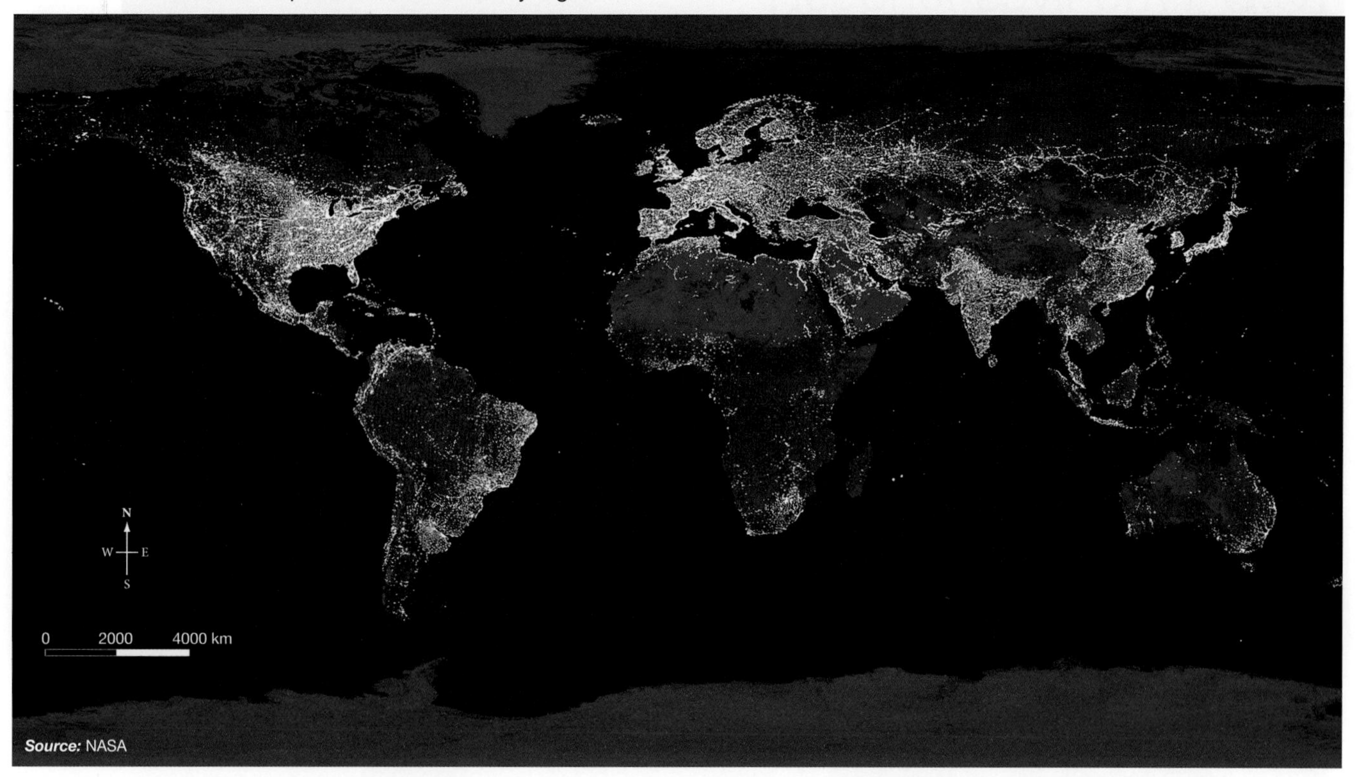

10.2 Activities

To answer questions online and to receive **immediate feedback** and **sample responses** for every question, go to your learnON title at www.jacplus.com.au. *Note*: Question numbers may vary slightly.

Remember

1. Name the continents with the highest and lowest population density.
2. Use the **Population concepts** weblink in the Resources tab to distinguish between population density and population distribution.

Explain

3. Account for (give reasons for) the uneven distribution of global population.
4. Give reasons for the uneven distribution of global population with reference to specific ***environments***.

Discover

5. Refer to the Australian Bureau of Statistics website or the website for your local government area. Find out the population density of your local government area, or calculate this (by dividing the size of the area by the number of people). How does it compare to Dhaka's population density?

Predict

6. How might global population distribution ***change*** in the next 20 years? Justify your answer.

Think

7. Why would some governments want to redistribute their populations away from existing large cities?
8. What disadvantages might there be for those people moving from an area of dense population?

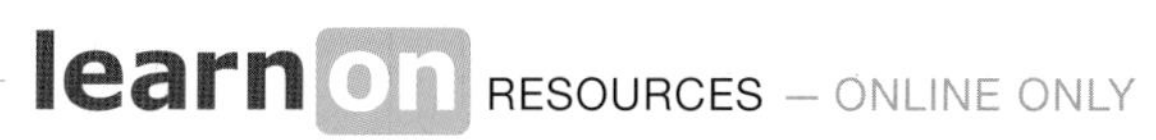

RESOURCES — ONLINE ONLY

Explore more with this weblink: Population concepts

10.3 Who lives longest?

10.3.1 Life expectancy

One of the major indicators of wellbeing is life expectancy. Globally, on average, people are expected to live longer than at any previous time in history. Sayings such as '60 is the new 50' reflect our changing expectations in Australia as to how long we expect to lead active lives. However, with the variation in living conditions around the world, the answer to the question 'how long can we expect to live?' also varies considerably.

How long we can expect to live when we are born is referred to as our **life expectancy** and is calculated according to the conditions in a particular country in that year. A child born in Japan in 2014 can expect to live 83 years, while one born in the African country of Sierra Leone can expect to live only 50 years. Figure 1 shows variation in life expectancy worldwide.

FIGURE 1 Global life expectancy

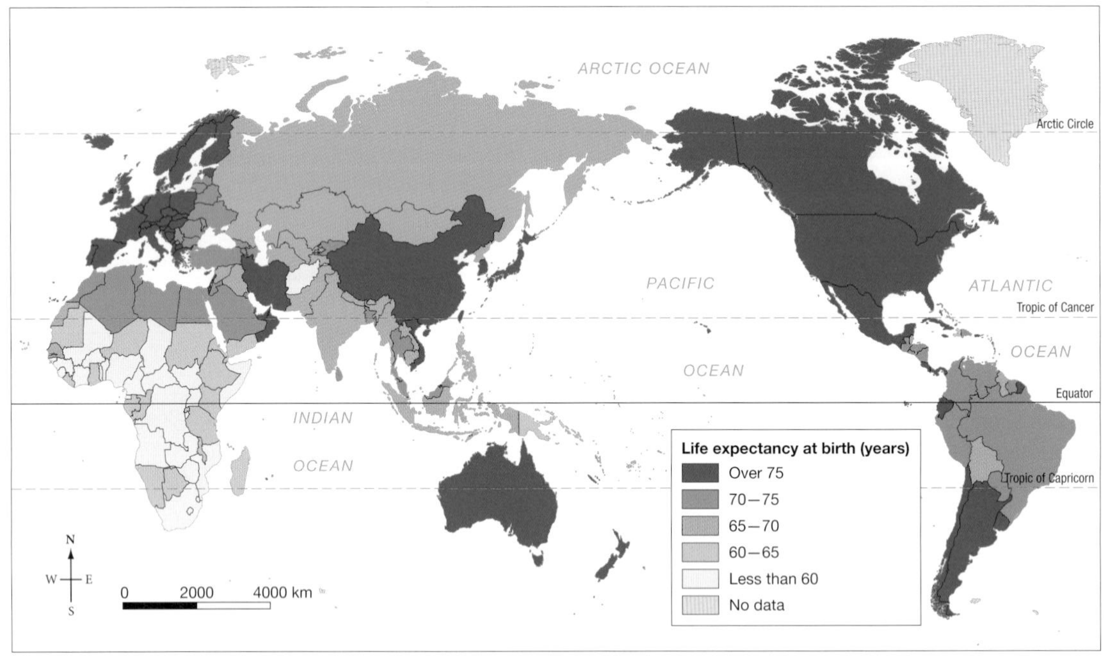

Source: © United Nations Publications

Life expectancy around the world started to increase in the mid 1700s due to improvements in farming techniques, working conditions, nutrition, medicine and hygiene. There is a clear interconnection between wealth and life expectancy: wealthier people in all countries can expect to live longer than poorer people. In general, women outlive men. A higher income enables people to have better access to education, food, clean water and health care. One region where life expectancy is decreasing rather than increasing is sub-Saharan Africa, where many countries have been affected by HIV and AIDS.

10.3.2 Child mortality

Life expectancy is closely interconnected to child mortality: countries with high death rates for children under five years of age have low life expectancy. This is well above the United Nations Sustainable Development Goal 3 target of 25 deaths per 1000. Young children are particularly vulnerable to infectious diseases due to their lower levels of immunity. Major causes of death include pneumonia, diarrhoea, measles and malnutrition. In wealthier households, child deaths are lower as these children are more likely to have better nutrition and to be immunised, and parents are more likely to be educated and aware of how to prevent disease.

Under the United Nations' Millennium Development Goals program, which operated from 1990 to 2015, child mortality was reduced considerably. The number of deaths of children under the age of five declined from 12.7 million in 1990 to 6 million in 2015 — the equivalent of nearly 17 000 fewer children dying each day. The greatest success occurred in northern Africa and eastern Asia. The highest levels of under-five mortality continue to be found in sub-Saharan Africa and southern Asia where four out of five children die. However, substantial improvements are being made. For example, since 2000 increased measles vaccination programs have prevented nearly 15.6 million child deaths.

Life expectancy and child mortality allow us to measure and compare human wellbeing in different places.

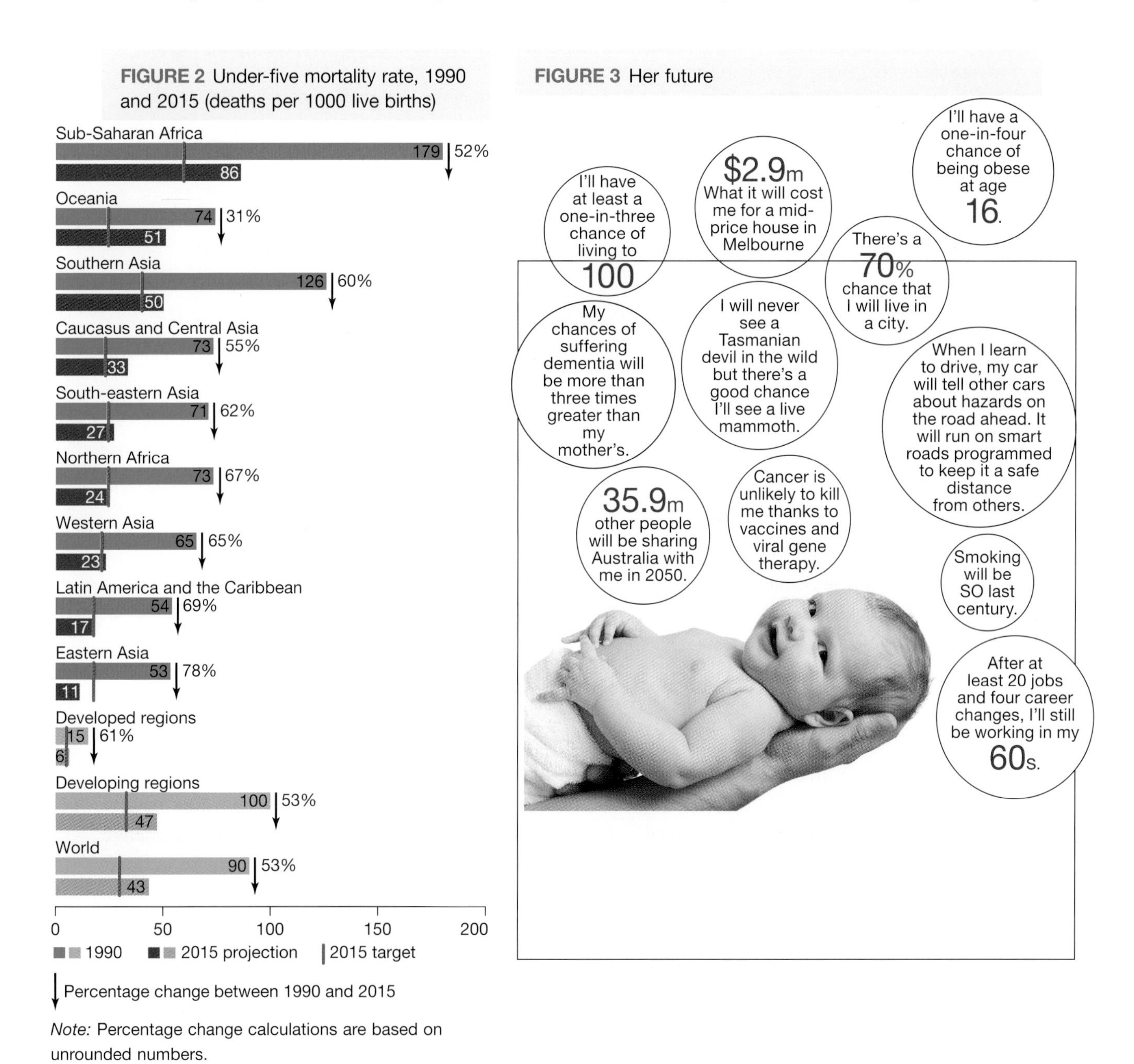

Sub-Saharan Africa
179
52%
86
Oceania
74
31%
51
Southern Asia
126
60%
50
Caucasus and Central Asia
73
55%
33
South-eastern Asia
71
62%
27
Northern Africa
73
67%
24
Western Asia
65
65%
23
Latin America and the Caribbean
54
69%
17
Eastern Asia
53
78%
11
Developed regions
15
61%
6
Developing regions
100
53%
47
World
90
53%
43
0
50
100
150
200
1990
2015 projection
2015 target
Percentage change between 1990 and 2015
Note: Percentage change calculations are based on unrounded numbers.
I'll have at least a one-in-three chance of living to 100
$2.9m What it will cost me for a mid-price house in Melbourne
I'll have a one-in-four chance of being obese at age 16.
There's a 70% chance that I will live in a city.
My chances of suffering dementia will be more than three times greater than my mother's.
I will never see a Tasmanian devil in the wild but there's a good chance I'll see a live mammoth.
When I learn to drive, my car will tell other cars about hazards on the road ahead. It will run on smart roads programmed to keep it a safe distance from others.
35.9m other people will be sharing Australia with me in 2050.
Cancer is unlikely to kill me thanks to vaccines and viral gene therapy.
Smoking will be SO last century.
After at least 20 jobs and four career changes, I'll still be working in my 60s.

FIGURE 2 Under-five mortality rate, 1990 and 2015 (deaths per 1000 live births)

FIGURE 3 Her future

10.3 Activities

To answer questions online and to receive **immediate feedback** and **sample responses** for every question, go to your learnON title at www.jacplus.com.au. *Note*: Question numbers may vary slightly.

Remember

1. Describe the distribution of life expectancy shown in figure 1.
2. What age groups are likely to have the highest mortality rates? Why?

Discover

3. Refer to the **MDGs Report** weblink in the Resources tab. Construct a table to show statistics for the highest and lowest life expectancies in each region of the world.
4. Use the **Demographic indicators** weblink in the Resources tab to examine interactive maps on life expectancy and other demographic indicators.
5. Use the **Gauging interconnections** weblink in the Resources tab to examine the correlation between life expectancy and other indicators such as infant mortality. What do you find?

Predict

6. Predict how life expectancies in sub-Saharan Africa may ***change*** if a cure for HIV/AIDS is discovered.
7. Will increased incomes always lead to increased life expectancy? Justify your answer.

Think

8. Write your own version of 'Her future' (figure 3), imagining what your life will be like when you are 35 years old.
9. What implications does an increase in life expectancy have on the provision for health care?

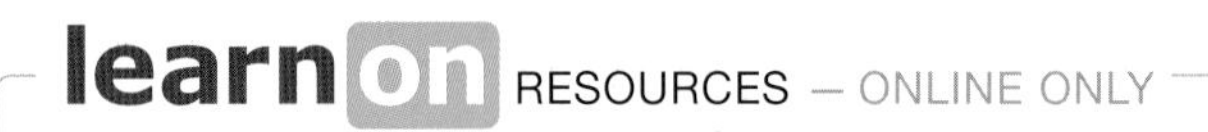

RESOURCES — ONLINE ONLY

Try out this interactivity: Long life, short life (int-3307)

Explore more with these weblinks: MDGs Report, Demographic indicators, Gauging interconnections

10.4 How do hatches and dispatches vary?

10.4.1 Births and deaths

Every minute there are an estimated 278 births and 109 deaths worldwide. The difference between these two figures gives us a growth rate of 1.08 per cent. While this does not sound like a big number, this natural increase equates to an extra 169 people at a global scale every minute. However, the rate of population change varies considerably across the world, with some places experiencing a decline rather than an increase in numbers of people. Rates of population change have an impact on wellbeing, both now and in the future.

Figure 1 shows the global distribution of birth rates. The continent of Africa clearly stands out here with the highest figures. According to the United Nations, in 2014 nearly half of the sub-Saharan African countries had an estimated **fertility rate** above five children per woman. Countries such as Niger and Burundi have fertility rates as high as 7.6 and 6.2 respectively. In contrast, Europe has very low birth rates, with the average birth rate in 2014 being 1.6 children per woman.

As figure 2 illustrates, sub-Saharan Africa is again at the high end of the spectrum, with many places in that region experiencing death rates above 10 per 1000. However, high death rates are more dispersed, with many European countries, such as Bulgaria and Ukraine, included. Low death rates are widely distributed across the regions of the Americas, much of Asia and Oceania.

Whether a population increases or decreases is largely dependent on variations in births and deaths producing a **natural increase**. Where a fertility level is well above the **replacement rate** of 2.1 children, population growth will occur. Conversely, fewer births over a period of time will ultimately result in a declining population. Figure 3 indicates the rate of natural population change, which ranges from over 3 per cent growth primarily in African nations (this would result in a doubling of population in approximately 23 years) to negative growth primarily in Europe. It should be noted that on a national scale, population change is also affected by migration.

FIGURE 1 Global distribution of birth rates

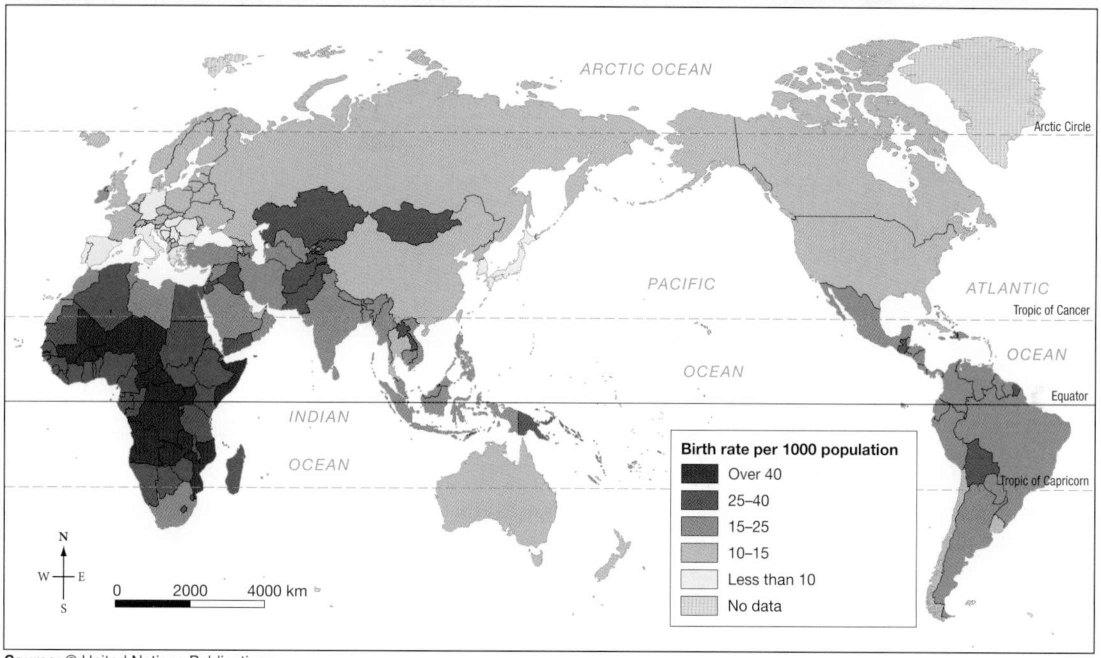

ARCTIC OCEAN
Arctic Circle
PACIFIC
OCEAN
ATLANTIC
OCEAN
Tropic of Cancer
Equator
INDIAN
OCEAN
Tropic of Capricorn
Birth rate per 1000 population
Over 40
25–40
15–25
10–15
Less than 10
No data
N
W
E
S
0 2000 4000 km

Source: © United Nations Publications

FIGURE 2 Global death rates

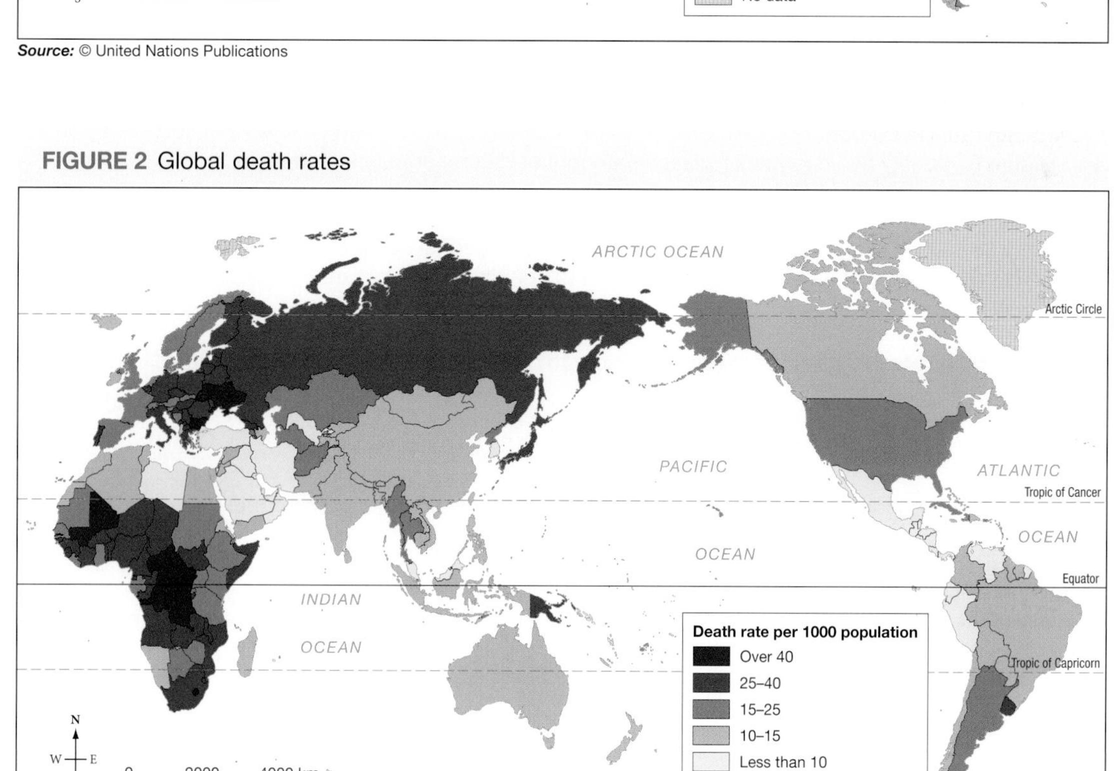

Source: © United Nations Publications

FIGURE 3 Population change

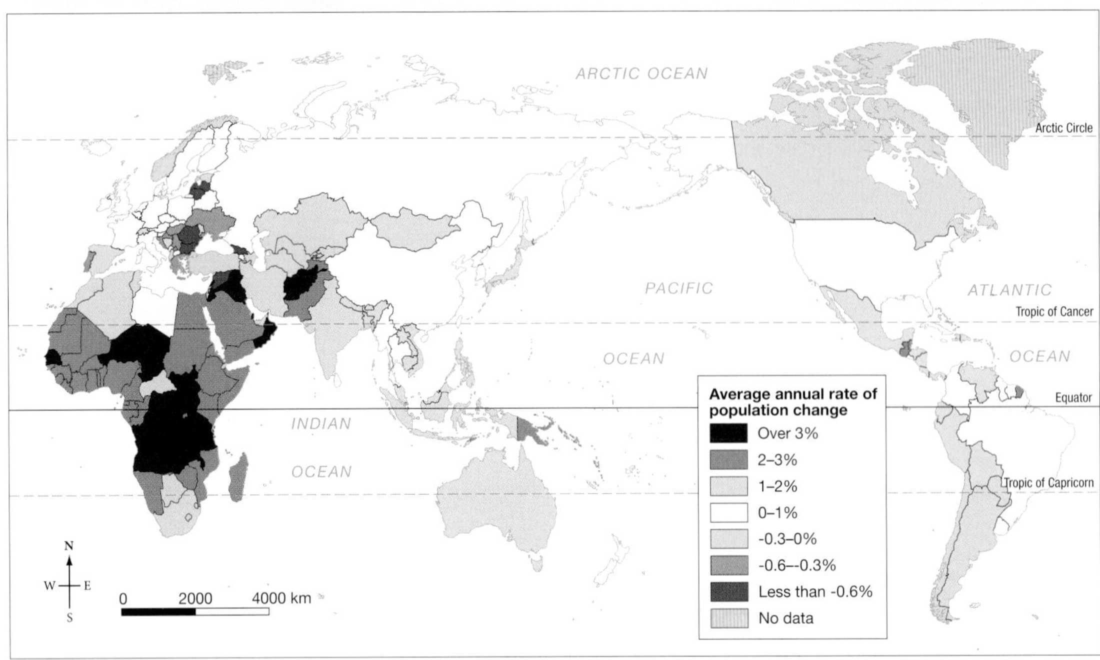

Source: © United Nations Publications

10.4 Activities

To answer questions online and to receive **immediate feedback** and **sample responses** for every question, go to your learnON title at www.jacplus.com.au. *Note*: Question numbers may vary slightly.

Remember

1. Using figures 1 and 2 and an atlas, identify countries that are exceptions to these patterns:
   - high birth rates in Africa
   - low birth rates in Europe
   - low death rates in Asia
   - low death rates in Europe.
2. Using figure 3, identify two countries for each category of natural population ***change***.

Explain

3. What are the major contributors to population ***change*** at a national ***scale***?
4. Explain the significance of a national fertility level of 2.1 children per woman.

Discover

5. (a) Use the **Birth rates** weblink in the Resources tab to discover and list the top 10 countries in terms of birth rates.
   (b) Find out three geographical characteristics that these countries have in common.

Predict

6. What are the implications for a country if its fertility rate is below the replacement rate?

Think

7. Suggest reasons to explain why some countries have higher birth rates than others.
8. How would the reasons for some countries in Europe experiencing a high death rate differ from those African countries with similar statistics?

RESOURCES — ONLINE ONLY

Try out this interactivity: Births and deaths (int-3308)

Explore more with this weblink: Birth rates

10.5 What is the link between population and wellbeing?

10.5.1 Population growth over time

From our examination of birth and death rates, we have seen that subsequent population change varies considerably across the world. To generalise, the more developed countries of the world tend to have lower birth and death rates and therefore lower population growth, while developing nations experience higher rates of births and deaths and higher growth. What are the reasons for such a large variation? What impacts does this have on wellbeing?

Global population growth has been rapid. From approximately one billion people in the year 1800, our planet now supports over seven billion people. Annually, global population is growing by some 88 million, although the rate of growth is now declining. These changes have been due to gains in wellbeing. Improvements in food production, education, medicine and hygiene have resulted in rapidly decreasing death rates, especially in infants and young children, and increased life expectancy. The **demographic transition model** (figure 1) attempts to explain changes in population growth by examining the interconnection between population characteristics and changes in wellbeing.

FIGURE 1 The demographic transition model

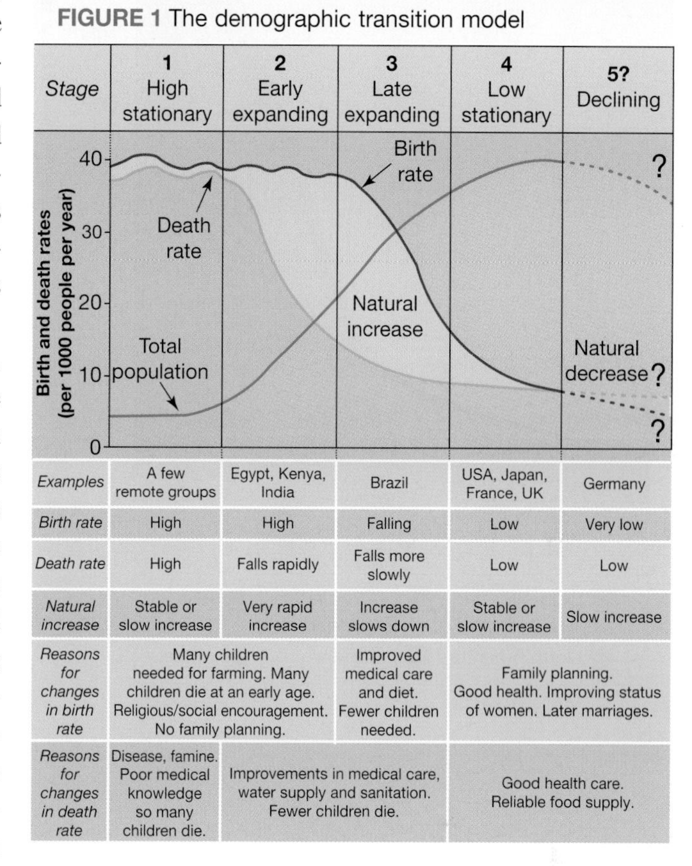

Examples	A few remote groups	Egypt, Kenya, India	Brazil	USA, Japan, France, UK	Germany
Birth rate	High	High	Falling	Low	Very low
Death rate	High	Falls rapidly	Falls more slowly	Low	Low
Natural increase	Stable or slow increase	Very rapid increase	Increase slows down	Stable or slow increase	Slow increase
Reasons for changes in birth rate	Many children needed for farming. Many children die at an early age. Religious/social encouragement. No family planning.		Improved medical care and diet. Fewer children needed.	Family planning. Good health. Improving status of women. Later marriages.	
Reasons for changes in death rate	Disease, famine. Poor medical knowledge so many children die.	Improvements in medical care, water supply and sanitation. Fewer children die.		Good health care. Reliable food supply.	

Most of the global population growth is taking place in developing countries, particularly the poorest nations (see figure 2). By 2050, with an estimated nine billion people in the world, some eight billion people (86 per cent) will be in developing countries, with two billion of those in the least developed nations. Despite continued global population growth, global fertility rates are falling (see figure 3). Declines in fertility have coincided with improvements in living conditions, greater access to education (particularly for women), improved health care and access to contraception. It is anticipated that fertility rates in developing regions will continue to fall, particularly with increasing rural–urban migration. In the cities, a child is more likely to be an economic burden than an asset, and there is better access to health services and family planning programs.

10.5.2 Population structure and wellbeing

The level of wellbeing of a population in terms of its health and life expectancy is reflected in its **population structure**. Increases in life expectancy and a decrease in the number of children being born has resulted in an increasing proportion of people in the older age groups. This ageing is expected to occur on a global scale in both developed and developing countries, with the rate of change being faster in the latter.

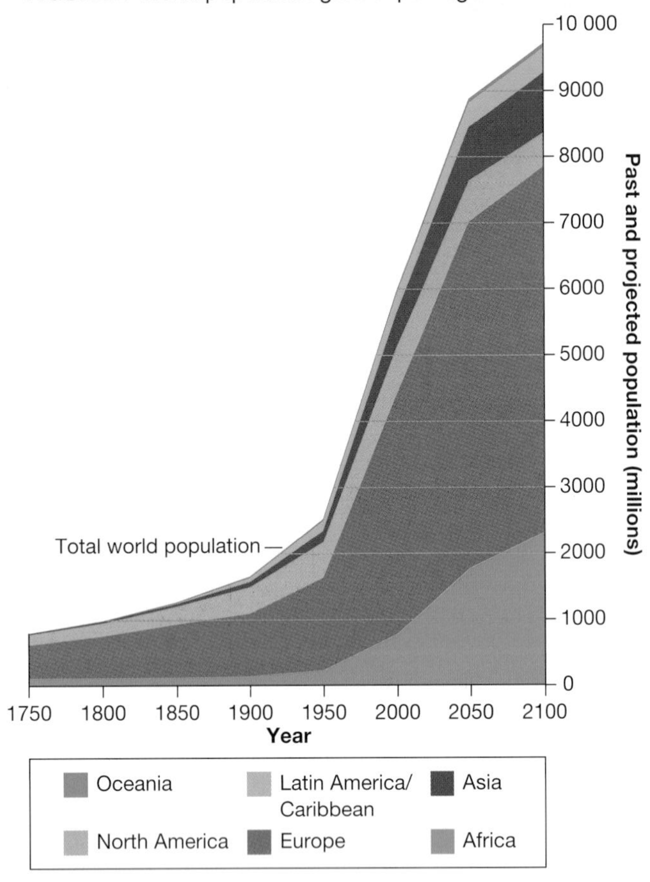

FIGURE 2 World population growth per region

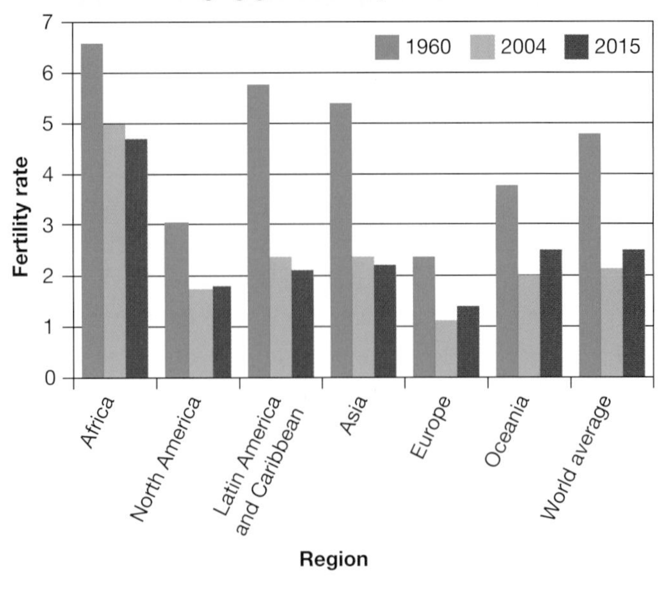

FIGURE 3 Changing global fertility rates

The proportion of the population in the **dependent population** affects the wellbeing of a country. A youthful population, as in Niger and Kenya, has implications in terms of future provision of infrastructure, education and employment. In addition, a high proportion of youth means that the population has momentum to cause large future growth, placing stress on a country's resources. However, provided those young people are healthy and well educated, they provide a potential skilled workforce in future years. An **ageing population** and a high percentage of elderly population, as found in Germany and Japan, has implications in terms of a decreasing workforce and tax base and increased demands on health services. On the positive side, the aged population does make a significant economic contribution, often in terms of voluntary labour such as caring for grandchildren and assisting with community projects.

FIGURE 4 The expanding urban environment of Seoul, South Korea. From 1970 to 2014, South Korea's urban population increased from 28 to 82 per cent, and its fertility rate fell to 1.2.

10.5 Activities

To answer questions online and to receive **immediate feedback** and **sample responses** for every question, go to your learnON title at www.jacplus.com.au. *Note*: Question numbers may vary slightly.

Remember

1. Which regions of the world have the highest and lowest fertility rates?
2. Describe how world population growth has ***changed*** over time.

Explain

3. Explain why fertility is likely to decrease with an increasing proportion of people living in cities.
4. What do you think is meant by the statement 'Children can be an economic burden or an economic asset'?

Predict

5. In which stage of the demographic transition model (refer to figure 1) is Kenya likely to be in 10 years? Justify your answer.
6. If a country reaches stage 5 of the demographic transition model, is it likely to remain there? Justify your answer.

Discover

7. Refer to the **Population growth** weblink in the Resources tab. What is the relationship between population growth and wellbeing, and how is this likely to ***change*** in the future?
8. Refer to the **HDI over time** weblink in the Resources tab. How has wellbeing, as measured by the Human Development Index, ***changed*** over time?

RESOURCES — ONLINE ONLY

Explore more with these weblinks: Population growth, HDI over time

10.6 How have governments responded to population and wellbeing issues?

The number of children in a family directly affects their wellbeing. In Australia, parents may comment on the cost of raising children. In other countries, children may make a contribution to family income by completing simple jobs such as collecting firewood. At a national scale, the numbers of children also affect the wellbeing of the country as a whole. Concerns about too few or too many children have resulted in a variety of government responses. Two are outlined below.

TABLE 1 Selected demographic characteristics for Japan and Kenya

Demographic characteristic	Kenya	Japan
Population mid-2015	44.3 million	126.9 million
Life expectancy at birth	62 years	83 years
Fertility rate	3.9	1.4
Natural increase	2.3%	–0.2%
Infant mortality	39 per 1000	2.1 per 1000
Projected population 2050	81.4 million	96.9 million
Population under 15 years	41%	13%
Population 60+ years	3%	26%
Percentage urban	24%	93%
Gross National Income per capita (US$)	2890	37 930

Source: PRB, Data Sheet, 2015

10.6.1 Kenya: response to a youthful population

FIGURE 1 Youthful population in Kenya

While Kenya's fertility rate has fallen substantially in the past 30 plus years from over seven children to just over four, the country still has a relatively high rate of population growth. Its population structure has a high proportion of young people (see table 1 and figure 2(a) and (b)), so by 2030 it is estimated that there will be over 65 million people. This increase will put pressure on Kenya's resources in terms of providing food, services and employment. With a predominantly rural population, the amount of arable land per person is falling.

Under Kenya's Vision 2030, a national framework for development, population management is an essential component of achieving wellbeing goals for health, poverty reduction, gender equality and environmental sustainability. The United Nations Population Fund (UNFPA), has been working with the Kenyan Government since the 1970s to help improve wellbeing in the country. Between January 2009 and December 2013 it provided $96.25 million. This financed a range of services including family planning with free contraceptives provided, increased availability of maternal and newborn health services, services to prevent the contraction of HIV and sexually transmitted infections, advocation for the education of girls and elimination of gender-based violence. Unfortunately, despite this work, there is still a huge unmet need for family planning in Kenya, particularly among the poorest women, where almost half report they have unplanned pregnancies.

FIGURE 2 Population pyramid for Kenya (a) 2014 and (b) 2050

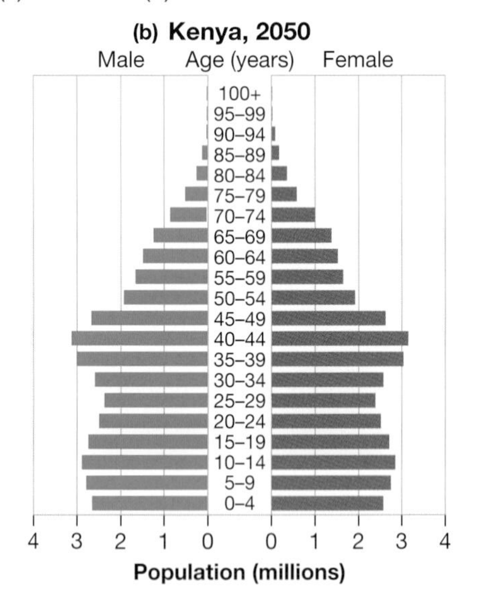

10.6.2 Japan: response to an ageing population

Japan has one of the highest life expectancies in the world, and this, combined with a very low fertility rate, has led to an ageing population, with more than one quarter of Japan's population in the 60 plus age group (see table 1 and figure 4(a) and (b)). Fertility in Japan has been consistently below replacement level since the 1970s. A high standard of living, increased participation of women in the workforce, high costs of raising children and lack of supporting childcare facilities have all contributed to this. Japan's total population is expected to decline from almost 127 million to an estimated 111 million in 2032.

The workforce is expected to fall 15 per cent over the next 20 years and halve in the next 50 years. This means that in 2025, three working people will have to support two retirees. The Japanese Government also faces rising pension and healthcare costs. These economic concerns led to the Japanese Government implementing a number of measures in 1994 such as subsidised child care and bonus payments for childbirth via a policy known as the Angel Plan (revised in 1999). The policy has been largely ineffective: although the fertility rate rose slightly initially, it has remained well below replacement level. The Japanese Government has historically been reluctant to use immigration to fill labour shortages, and although this may change slowly, improving female workforce participation rates, particularly after marriage, may be a more viable option. Use the **Ageing population: Japan** weblink in the Resources tab to discover more about the problems of an ageing population in Japan and the impact on wellbeing.

FIGURE 3 Ageing population, Japan

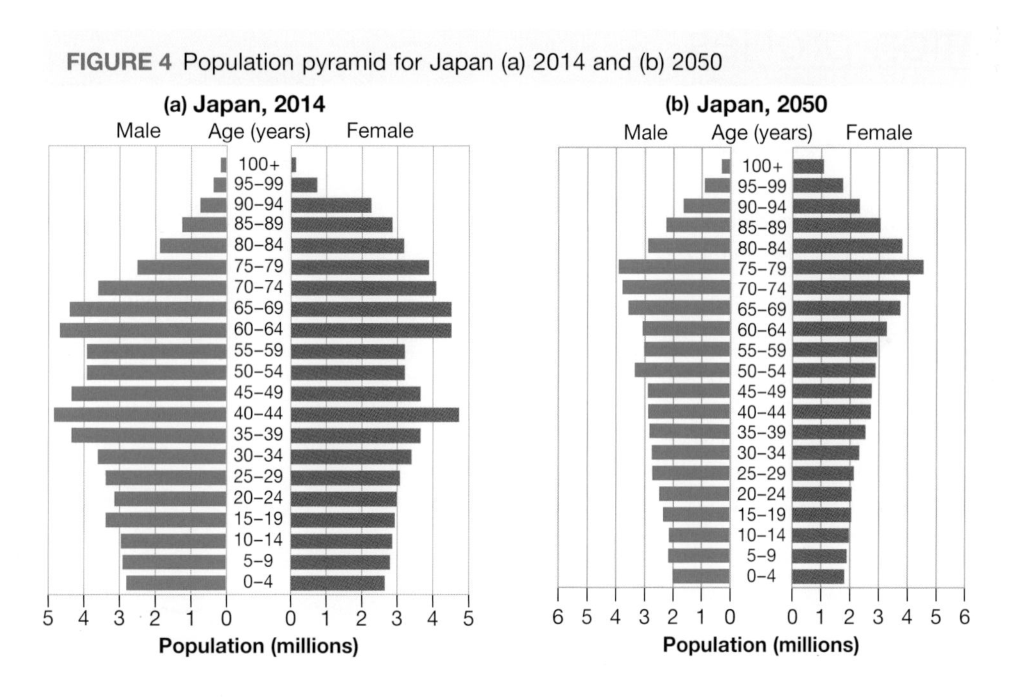

FIGURE 4 Population pyramid for Japan (a) 2014 and (b) 2050

10.6 Activities

To answer questions online and to receive **immediate feedback** and **sample responses** for every question, go to your learnON title at www.jacplus.com.au. *Note*: Question numbers may vary slightly.

Remember

1. How has an improvement in living conditions led to a ***change*** in population structure?

Explain

2. Account for (give reasons for) the variation in shape of the population pyramids for Japan and Kenya in 2012 and 2050.

Discover

3. Describe the ***changing*** percentage of aged population between 2012 and 2050 in both Kenya and Japan.

Predict

4. What problems does the Kenyan Government face with a large proportion of young population?
5. What problems does Japan face with a large proportion of aged population?
6. How do these issues affect the wellbeing of people in those countries?

Think

7. Of the problems you listed above, which do you consider more serious? Why?

RESOURCES — ONLINE ONLY

Explore more with this weblink: Ageing population: Japan

Try out this interactivity: Revealing population pyramids (int-3309)

10.7 SkillBuilder: Using Excel to construct population profiles

WHY DO WE USE EXCEL TO CONSTRUCT POPULATION PROFILES?

When constructing population profiles, there is a large amount of data and large numbers to handle. The use of an Excel spreadsheet simplifies the process.

Go online to access:

- a clear step-by-step explanation to help you master the skill
- a model of what you are aiming for
- a checklist of key aspects of the skill
- a series of questions to help you apply the skill and check your understanding.

FIGURE 1 Population pyramid, or profile, created from the Excel spreadsheet of population statistics for the United States, 2010

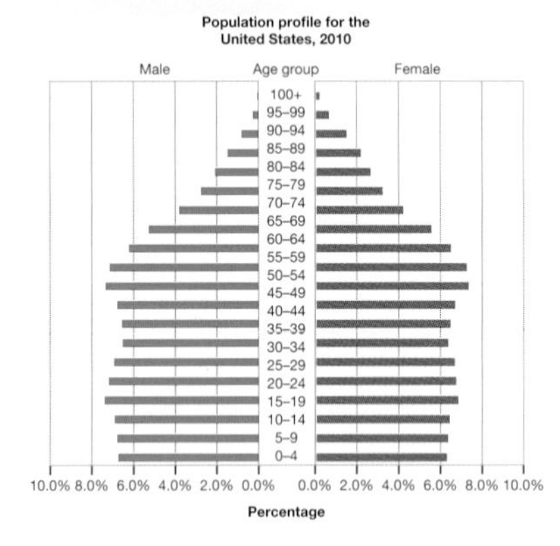

RESOURCES — ONLINE ONLY

Watch this eLesson: Using Excel to construct population profiles (eles-1758)

Try out this interactivity: Using Excel to construct population profiles (int-3376)

10.8 How does wellbeing vary in India?

10.8.1 How and why is India's population changing?

You probably know that China has the biggest population in the world with a population of 1.37 billion. With some 1.32 billion people in 2015, India is set to surpass China's population by 2025, when its population will reach an estimated 1.4 billion. With a predicted 1.6 billion by 2050, what happens to India's population will have major implications in terms of the wellbeing of the people in that country.

India's population is growing at a rate of 1.4 per cent. Improvements in water supply, a decrease in infectious diseases and an increase in education levels have resulted in a reduced death rate since the 1950s, while the birth rate has not declined to the same extent. Infant mortality remains high as over two-thirds of the population are rural dwellers who may not have ready access to health and reproductive services. Children remain a vital part of the family's labour force both on farms (as shown in figure 1) and for old age support, so it is essential for families to have more children to improve the chance of them surviving to adulthood. Twenty-nine per cent of the population is under 15 years of age, creating huge momentum for future growth (see figure 2).

FIGURE 1 Indian children assisting with rice planting

FIGURE 2 Population pyramid for India (a) 2014 and (b) 2050

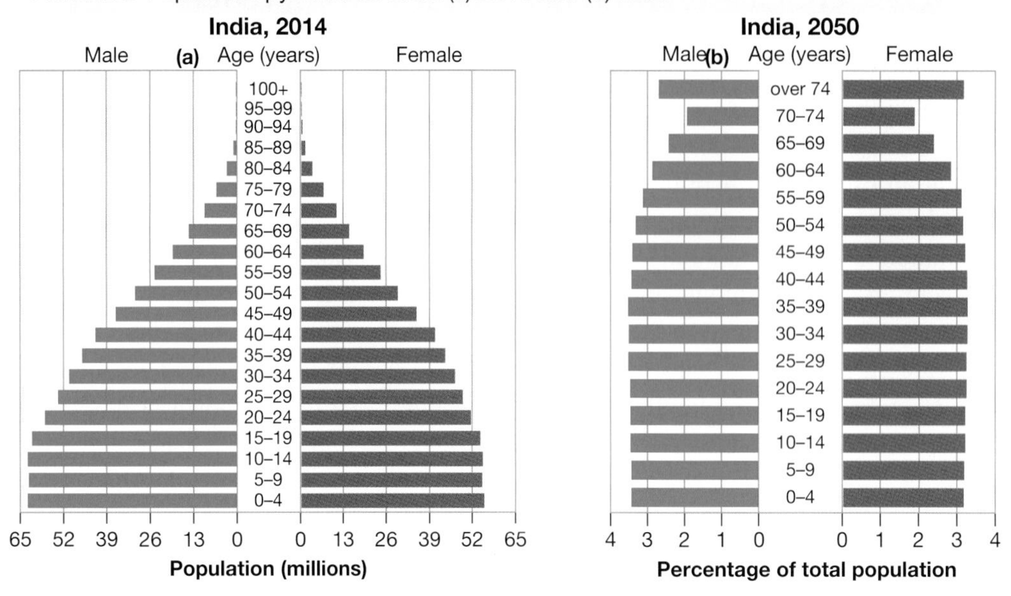

10.8.2 Regional variation in wellbeing and population

The number of children per woman in India has declined substantially from 5 in the 1970s to 2.3 in 2014. There is considerable regional variation. The levels of literacy and poverty shown in figures 4 and 5 reflect a varying distribution of wellbeing in India. For information about the Indian Government's moves to reduce poverty and improve wellbeing across the country, use the **Poverty challenge** weblink in the Resources tab.

FIGURE 3 Proportion of children 0–6 years to total population, India, 2011

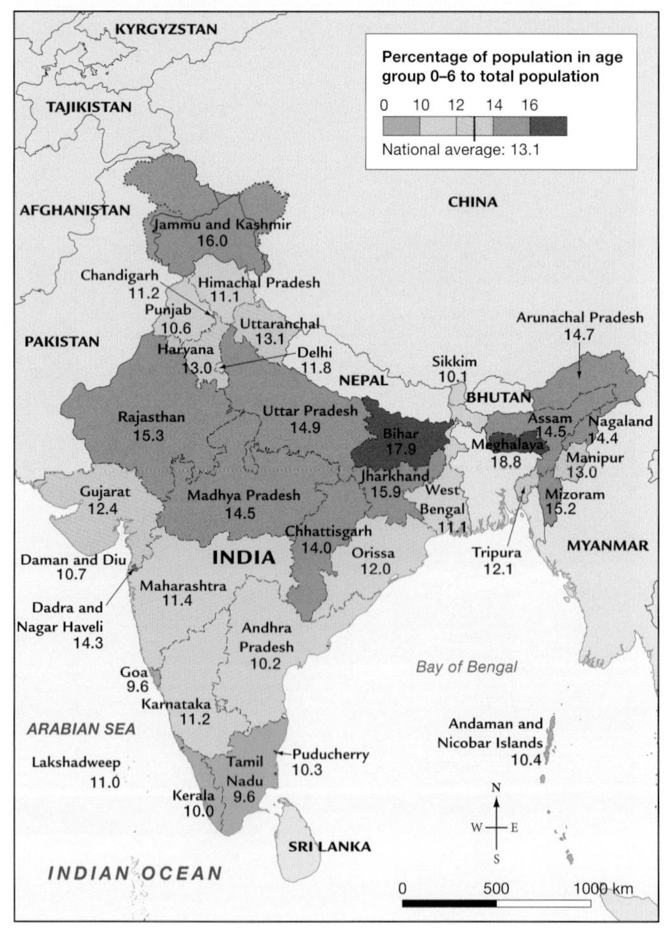

Source: Spatial Vision

FIGURE 4 Poverty levels in India

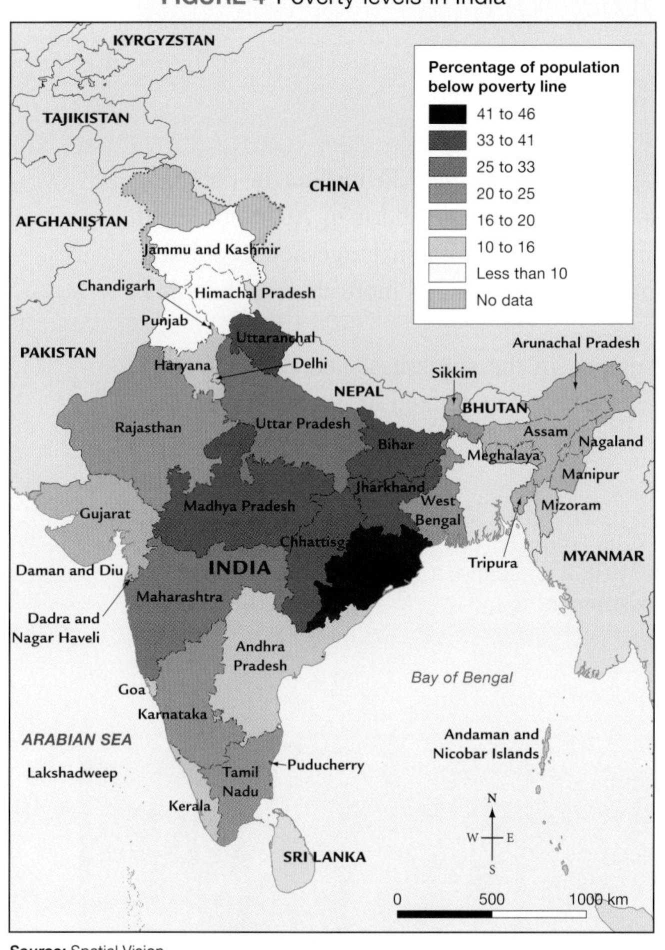

Source: Spatial Vision

FIGURE 5 Literacy rates (percentage) in India, 2011

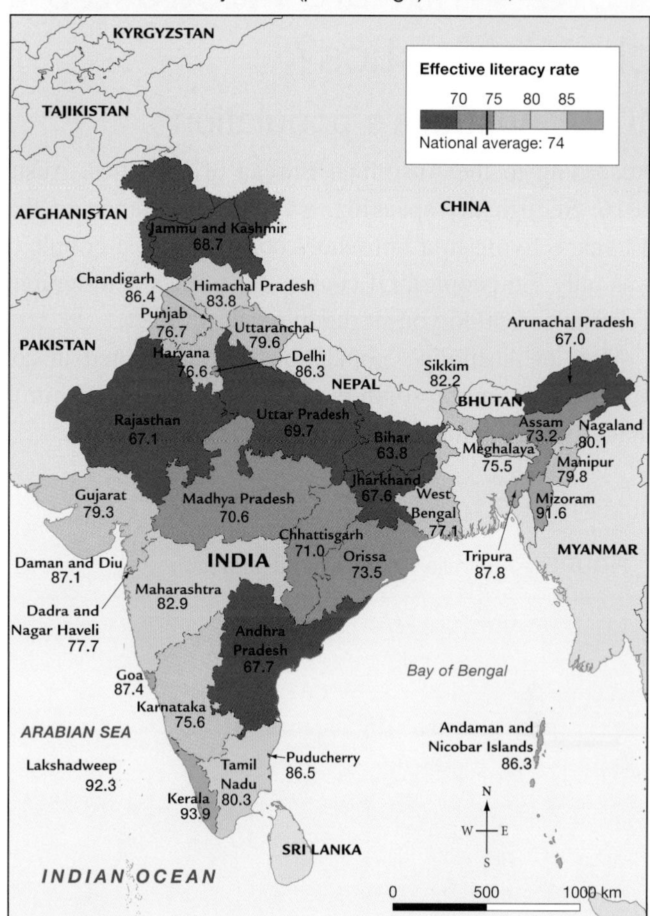

Source: Spatial Vision

10.8 Activities

To answer questions online and to receive **immediate feedback** and **sample responses** for every question, go to your learnON title at www.jacplus.com.au. *Note*: Question numbers may vary slightly.

Remember

1. With reference to the population pyramids shown in figure 2, account for India's ***changing*** population growth.

Explain

2. Explain why India is set to overtake China in terms of total population.
3. Using the data provided in this section, describe and account for the variation in wellbeing in India. You can also use the **India: contrasting wellbeing** weblink in the Resources tab to learn more about the contrasting conditions in wellbeing within the country.

Discover

4. (a) Using figure 4, pick two states that fall into different categories on the map. Use the **Census India** weblink in the Resources tab to compare the 2011 demographic characteristics for those two states.
   (b) Share your findings with other members of your class who selected different states.

RESOURCES — ONLINE ONLY

Explore more with these weblinks: Poverty challenge, India: contrasting wellbeing, Census India

10.9 What are Australia's population characteristics?

10.9.1 Australia's population

According to the Australian Bureau of Statistics, Australia's population reached 24 million in February 2016. Statistically speaking, a typical Australian in that year would be female, born in Australia, aged 37 years, living in a household consisting of a couple and children (although the average household size was only 2.6 people). Of course, Australia's demographic characteristics are much more diverse than this. To what extent do you fit the 'typical' profile?

Most of Australia's population is concentrated in coastal regions in the south-east and east and, to a lesser extent, in the south-west. The population within these regions is concentrated in urban centres, particularly the capital cities (see figure 1).

FIGURE 1 Australia's population distribution

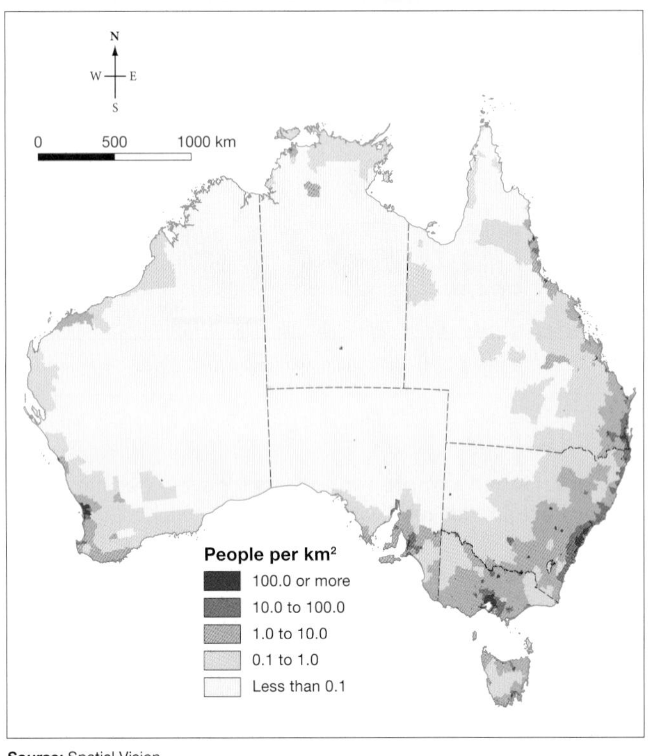

Source: Spatial Vision

FIGURE 2 Brisbane, a typical Australian urban environment

FIGURE 3 Innamincka: an outback town

FIGURE 4 Population pyramid for Indigenous and non-Indigenous Australians

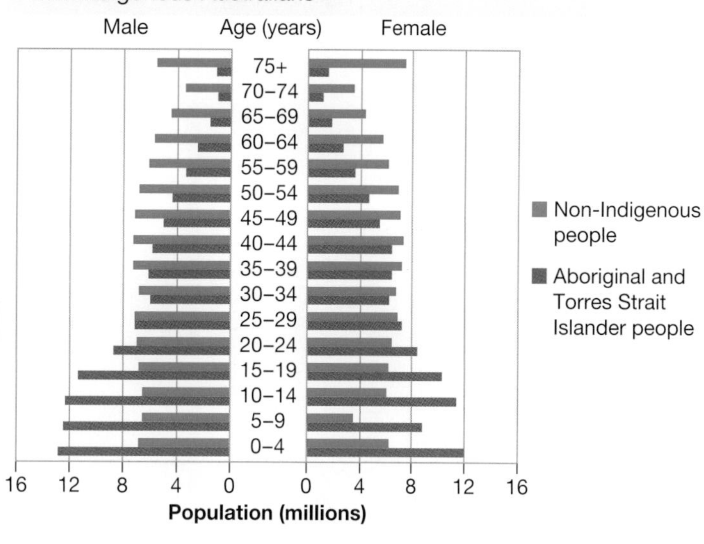

Australia's population has increased considerably over time and is continuing to grow. Between 2005 and 2015, the population increased by over three million people, at an average rate of 1.8 per cent per year (see figure 5).

FIGURE 5 Australia's changing population growth

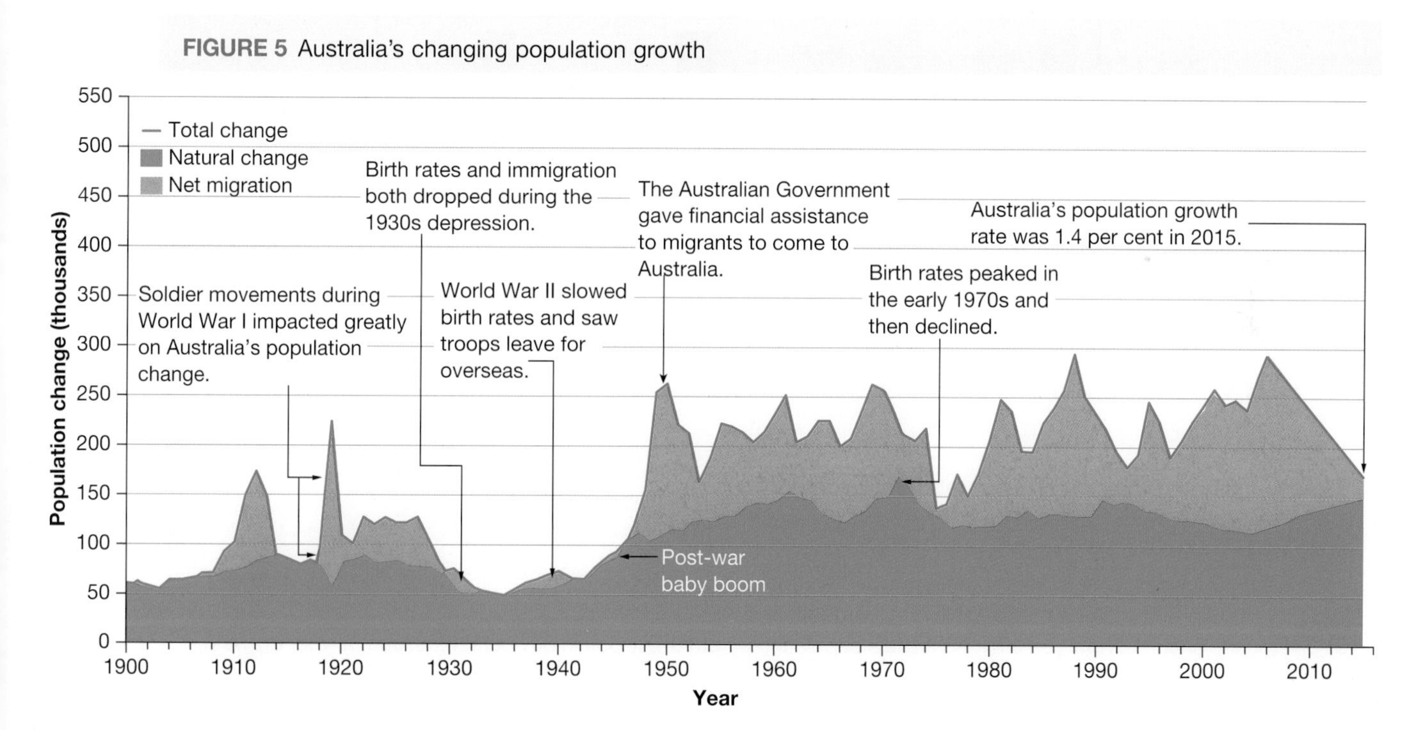

Our population growth is due to immigration rather than natural increase. The level of migration is set annually by the Federal Government and is currently about 170 000 per year.

Our rate of fertility has declined steadily since the 1970s and is now well below replacement rate. Despite attempts to increase the number of children via a Federal Government baby bonus of approximately $5000 per baby, which was in place between 2003 and 2013, our fertility rate was 1.9 in 2015.

The decline in fertility and increased life expectancy has resulted in an ageing population (see figure 6). The proportion of the population aged 65 years and over increased from 11.3 per cent to 15 per cent between 30 June 1991 and 30 June 2015.

FIGURE 6 Australia's changing population structure

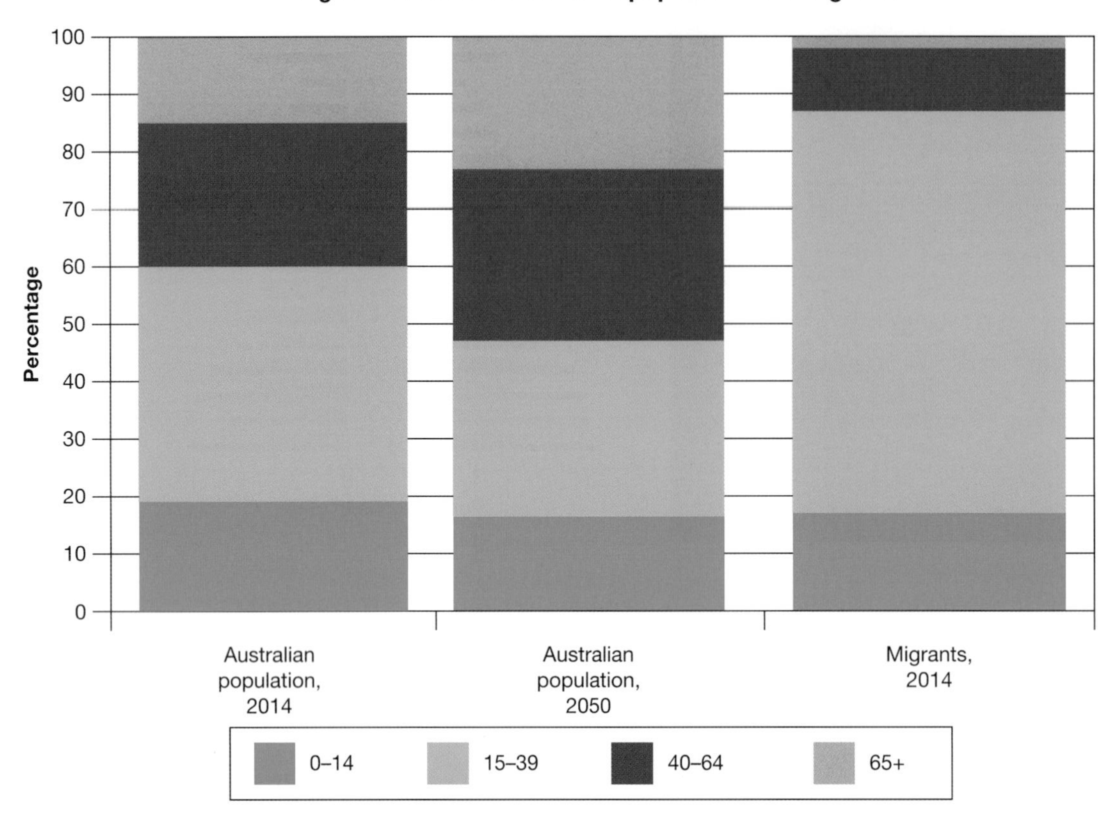

10.9 Activities

To answer questions online and to receive **immediate feedback** and **sample responses** for every question, go to your learnON title at www.jacplus.com.au. *Note*: Question numbers may vary slightly.

Remember

1. What factors have accounted for Australia's ***changing*** population growth over time?
2. How is Australia's population structure expected to ***change*** in the future?

Explain

3. Account for the variation in Australia's population distribution.

Predict

4. Predict the impact of Australia's ageing population on our demand for different facilities. Use the **Lateline** weblink in the Resources tab to assist you with this answer.
5. Sketch the shape of how you think Australia's population pyramid will look in 50 years' time. Justify your drawing.

Think

6. What other methods could the Australian Government use in order to encourage population growth in Australia?
7. What are the advantages and disadvantages of a 'big Australia' and a projected future population of 35 million? Use the **Population puzzle** weblink in the Resources tab to find out about one side of the argument.

Discover

8. (a) Use the Australian Bureau of Statistics website to access statistics on four demographic characteristics of your Local Government Area.
   (b) How do the statistics for your Local Government Area compare to those for Australia as a whole and those of your state?

learn on RESOURCES — ONLINE ONLY

 **Explore more with these weblinks:** Lateline, Population puzzle

10.10 SkillBuilder: How to develop a structured and ethical approach to research

WHAT IS A STRUCTURED AND ETHICAL APPROACH TO RESEARCH?

A structured and ethical approach to research involves organising your work clearly and meeting research standards without pressuring anyone into providing material and without destroying environments while gathering the data. Your work must also be your own, and anything that is someone else's work must be referenced in the text and included in the reference list.

Go online to access:

- a clear step-by-step explanation to help you master the skill
- a model of what you are aiming for
- a checklist of key aspects of the skill
- a series of questions to help you apply the skill and check your understanding.

FIGURE 1 A sample of a contribution form

Name:
School:
The research that I have undertaken has contributed to my understanding of the topic. At all times I have acted in such a way as to not harm the feelings of people or destroy the environment. This research is presented in my own words and is my understanding of the topic.
I, ____________________[name], certify the accuracy of this statement of contribution.
Signature: **Date:**

learn on RESOURCES — ONLINE ONLY

Watch this eLesson: How to develop a structured and ethical approach to research (eles-1759)

Try out this interactivity: How to develop a structured and ethical approach to research (int-3377)

10.11 The biggest child killer?

Access this subtopic at **www.jacplus.com.au**

10.12 HIV no longer a death sentence?

Access this subtopic at **www.jacplus.com.au**

10.13 Review

10.13.1 Review

The Review section contains a range of different questions and activities to help you revise and recall what you have learned, especially prior to a topic test.

10.13.2 Reflect

The Reflect section provides you with an opportunity to apply and extend your learning.

Access this subtopic at **www.jacplus.com.au**

TOPIC 11
Is life the same everywhere?

11.1 Overview

Numerous **videos** and **interactivities** are embedded just where you need them, at the point of learning, in your learnON title at www.jacplus.com.au. They will help you to learn the content and concepts covered in this topic.

11.1.1 Introduction

All of us have travelled to different places during our lives. These places may be within our own suburb, within our town or city, in another state of Australia, or, if we are very fortunate, in another country. While we tend to be more conscious of differences between our own country and others, variations also occur at local and regional scales. Variation may be between urban and rural environments or even within the one city or town. Think about how the various spaces near where you live might reflect differences or similarities in wellbeing and the reasons for these characteristics.

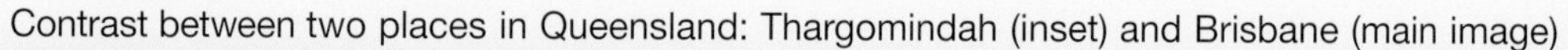

Contrast between two places in Queensland: Thargomindah (inset) and Brisbane (main image)

Starter questions

1. What are the characteristics of the particular rural or urban ***environment*** in which you live (your suburb)?
2. In what ways do these characteristics vary from those of the neighbouring ***environment***, whether it is a farm, town or suburb?
3. What ***interconnection*** is there between these characteristics and the wellbeing of the people in these ***places***?
4. Why do similarities or variations in wellbeing occur at a local or regional ***scale***?

INQUIRY SEQUENCE

11.2 Does gender affect wellbeing?

11.2.1 Women's health and wellbeing

Rarely would women in Australia consider that pregnancy and giving birth could be one of the most life-threatening activities in which they could engage. For the large majority, having children is something that fits into our busy lifestyles without health complications to either mother or baby. Sadly, this is not the case for a huge number of women worldwide, for whom child-bearing has a negative impact on their wellbeing.

Every day approximately 800 women die from complications in relation to pregnancy or childbirth. Most of these deaths are from preventable complications: severe bleeding, infections and complications from unsafe abortions. The incidence of **maternal mortality** and related illness is interconnected to poverty and lack of accessible and affordable quality health care. Use the **Maternal mortality** weblink in the Resources tab to learn more about global maternal mortality and how it can be addressed.

Figure 1 shows the global distribution of maternal mortality. Eighty-six per cent of maternal deaths are in sub-Saharan Africa and southern Asia, with the former accounting for two-thirds of all these deaths. Highest maternal mortality rates are recorded in Sierra Leone and Chad, where mothers have a 1 in 17 and 1 in 18 risk of dying. At a national scale, two countries account for one-third of total global maternal deaths: Nigeria at 19 per cent (58 000) followed by India at 15 per cent (45 000).

Millennium Development Goal (MDG) 5 set the targets of reducing by three-quarters the maternal mortality rate between 1990 and 2015, and the achievement of universal access to reproductive health by 2015. While maternal mortality fell by 45 per cent between 1990 and 2015, globally the target was not met, particularly in countries in southern Africa where AIDS had a major impact.

FIGURE 1 Global scale distribution of maternal mortality

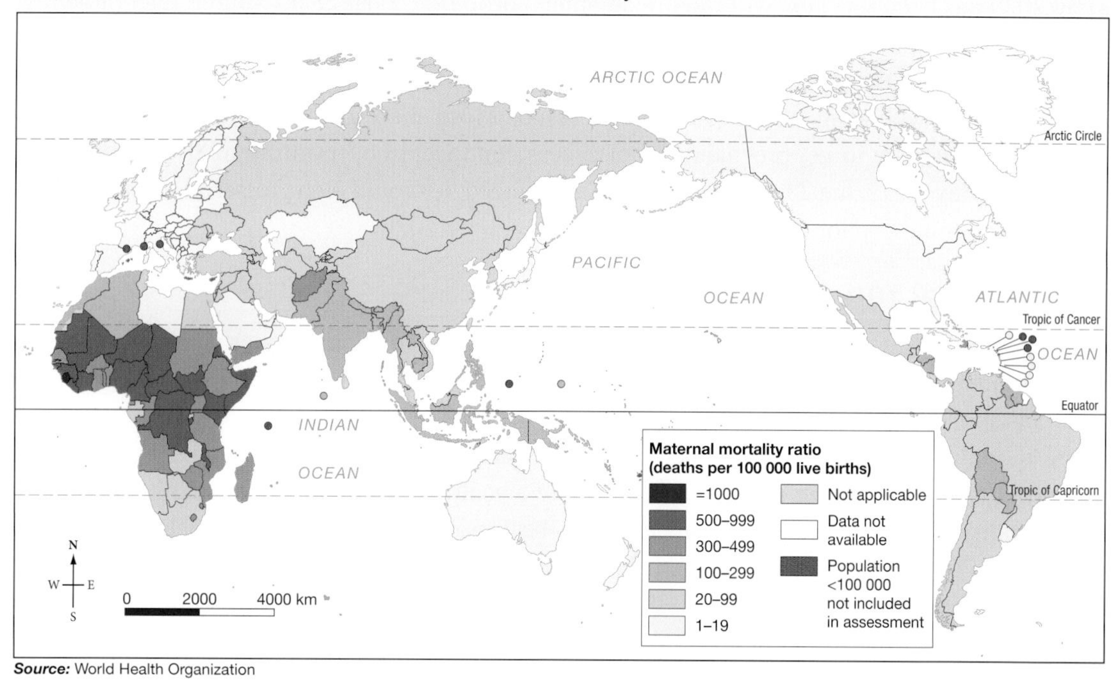

Source: World Health Organization

Figure 3 indicates progress in terms of access to reproductive health. Most indicators fell well short of universal access (considered to be 80 per cent). While contraceptive use has increased substantially, wealthier women in urban areas have the best access. Unmet need among the world's poor women is substantial. Providing access to contraception is a means of empowering women to make choices about family size.

FIGURE 2 Child-bearing: a threat to wellbeing

FIGURE 3 Reproductive health indicators, 1990–2015

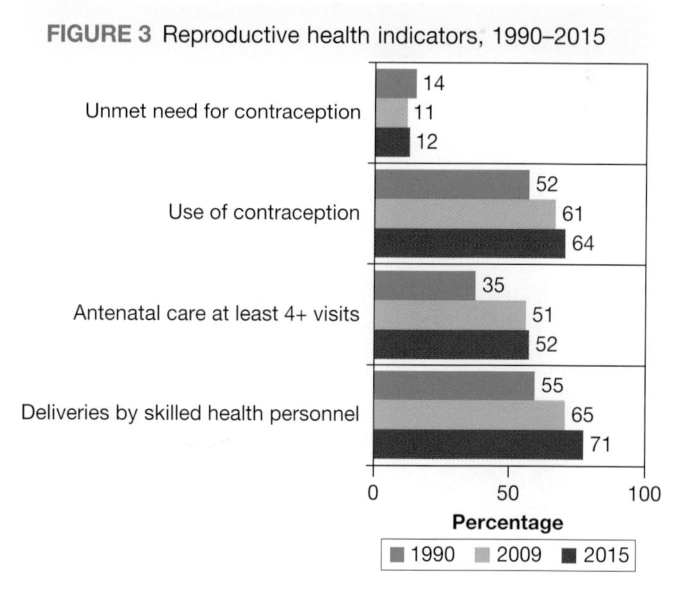

Under the Sustainable Development Goals (SDGs) the target of a maternal mortality rate below 70 per 100 000 by 2030 has been set. This will require an annual drop of 7.5 per cent — more than three times the reduction that occurred under the MDGs. However, Cambodia, Rwanda and Timor-Leste all achieved this type of reduction rate in the past 15 years, proving that this goal is achievable. The Gates Foundation sees women's health and wellbeing as a social justice issue. More than 220 million women in developing countries who do not want to get pregnant cannot access contraception, resulting in 80 million unintended pregnancies. In May 2012 the Gates Foundation announced it would help raise $4 billion in order to increase access to contraception to 120 million more women by 2020.

11.2.2 Maternal mortality in India

As previously mentioned, India accounts for a large percentage of global maternal deaths. On average, maternal mortality rates in that country declined 68 per cent between 1990 and 2015. However, there is substantial variation within India, as figure 4 indicates.

Maternal mortality is strongly interconnected with poverty in both rural areas and urban slums: places with poor provision of **sanitation** and a lack of affordable health services are associated with high levels of maternal mortality. In addition, women are likely to be less well-nourished than males in a household. According to the 2011 Indian Census, women also have much lower literacy levels — a 65 per cent literacy rate compared to 82 per cent for men — so they are less likely to be able to access information on health and contraception. Use the **In silence** weblink in the Resources tab to watch a video regarding women dying in childbirth in India.

FIGURE 4 Maternal mortality rates in India

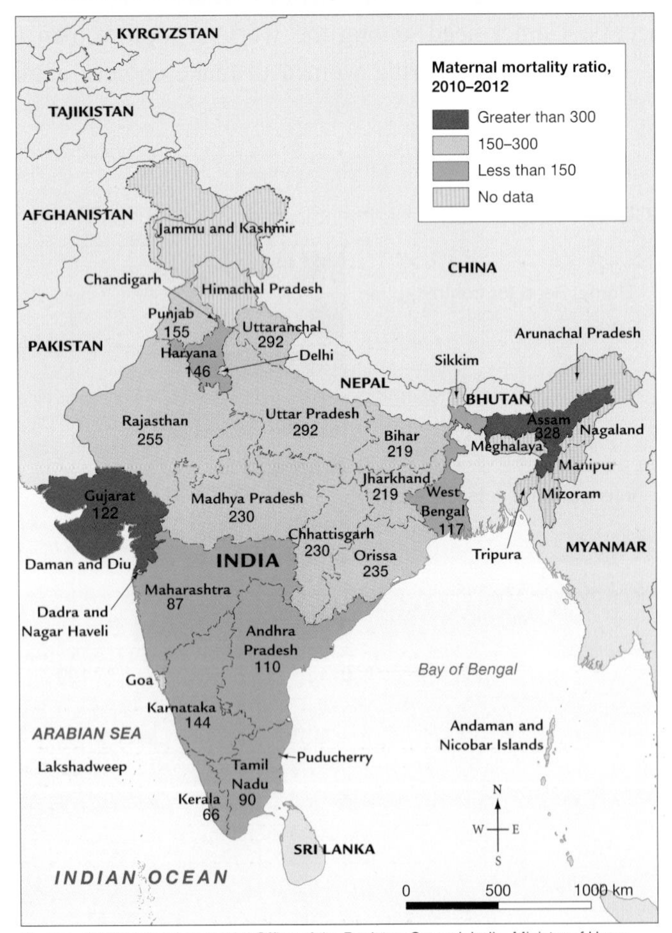

Source: Published and issued by Office of the Registrar General, India, Ministry of Home Affairs http://www.censusindia.gov.in/vital_statistics/SRS_Bulletins/MMR_Bulletin-2010-12.pdf

FIGURE 5 This Indian mother survived childbirth.

The government of India launched the National Rural Health Mission in 2005, with a specific focus on maternal health. This was reinforced in their 2013 Call to Action. Efforts have been focused on those districts that account for 70 per cent of all infant and maternal deaths. Under this program, community workers have been trained to deliver babies, and 10 million women have been provided with a cash incentive to enable them to give birth in clinics rather than at home. Maternal mortality has fallen, but Human Rights Watch reports that many women are being charged for services as they are unaware of these entitlements.

A related issue for pregnant women in India is the pressure to produce a son. Census data in 2011 revealed the number of female children (0–6 years) has decreased from 927 to 914 girls per 1000 boys in the past decade, despite some overall improvement in the **sex ratio** across all age groups (see figure 6). Males are traditionally preferred over female children: sons are seen as the breadwinners who carry the family name, while daughters are often perceived as an economic burden. Although **female infanticide** is illegal, use of ultrasound for sex-determination tests has led to sex-selective abortions, with an estimated 500 000 girls aborted each year (although sex selective abortion is also illegal). The pressure to produce a son means that many Indian women have multiple pregnancies, thereby increasing their risk of maternal mortality over their reproductive years. Figures 4 and 6 indicate a strong interconnection between the places of high maternal mortality and those with a large imbalance in the sex ratio.

FIGURE 6 Variation in sex ratio within India

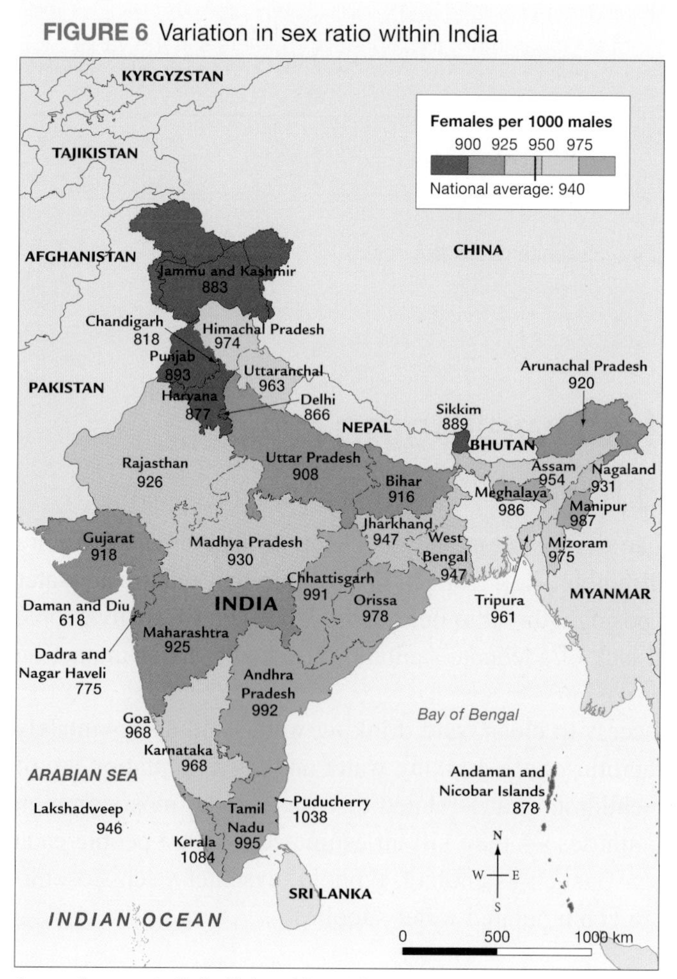

Source: Government of India, Ministry of Home Affairs, Office of Registrar General Made with Natural Earth. Map by Spatial Vision GAT-40

FIGURE 7 Son preference has resulted in an imbalance in India's sex ratio.

11.2 Activities

To answer questions online and to receive **immediate feedback** and **sample responses** for every question, go to your learnON title at www.jacplus.com.au. *Note*: Question numbers may vary slightly.

Remember

1. Refer to figure 1. Describe the variation in maternal mortality on a global ***scale***.

Explain

2. Explain why women living in developing ***places*** are more likely to have lower levels of wellbeing than their male counterparts.
3. Refer to figure 4 in topic 10, subtopic 10.8 'How does wellbeing vary in India?', showing the distribution of poverty in India. To what extent is the ***interconnection*** between poverty and maternal mortality (as shown in figure 4 in section 11.2.2) evident?

Discover

4. Use the **Women's health** weblink in the Resources tab to learn more about the UN Global Strategy for women and children's health. Note any key information provided.

Predict

5. Predict the shape of India's population pyramid if the trends in India's sex ratio continue.

Think

6. Suggest measures that could be introduced by the Indian Government to help Indian parents see the value of female babies equally with that of boys.

learnON RESOURCES — ONLINE ONLY

Explore more with these weblinks: Maternal mortality, In silence, Women's health

Try out this interactivity: His and hers (int-3310)

11.3 How does water affect wellbeing?

11.3.1 Water supply, sanitation and health

In Australia most of us take it for granted that we can turn on a tap and get clean, drinkable water. We are confident that when we drink our tap water we will not get sick from it. We assume that our waste water and sewage will be treated and disposed of without posing a threat to our health. However, for many people around the world, lack of access to clean water and lack of adequate sanitation has had a major impact on their health and therefore their wellbeing.

Approximately 663 million people do not have access to clean, safe drinking water, and approximately 2.4 billion people are without adequate sanitation facilities. Safe drinking water and basic sanitation are of crucial importance to human health, especially for children. Water-related diseases are the most common cause of death among the poor in less developed countries — they kill an estimated 842 000 people each year, especially children under the age of five. Diseases such as cholera, typhoid, dysentery, schistosomiasis and worm infestations are directly attributable to contaminated water supplies.

Global progress

Remarkable progress was made in regard to Target 10 of MDG 7, which was to halve the proportion of people without access to safe drinking water by 2015. This target was met more than five years ahead of schedule with 91 per cent of the global population using an improved drinking water source in 2015 compared with 76 per cent in 1990. Fifty-eight per cent of the world's people now have access to piped drinking water. Rural and urban coverage does vary however, as shown in figures 1 (a) and (b), 2 and 3. Use the **WHO/UNICEF** weblink in the Resources tab, which shows the interconnection between water, sanitation and other indicators.

FIGURE 1 (a) Proportion of the population using improved drinking water sources, 2015 (b) Trends in drinking water coverage (%), by rural and urban residency

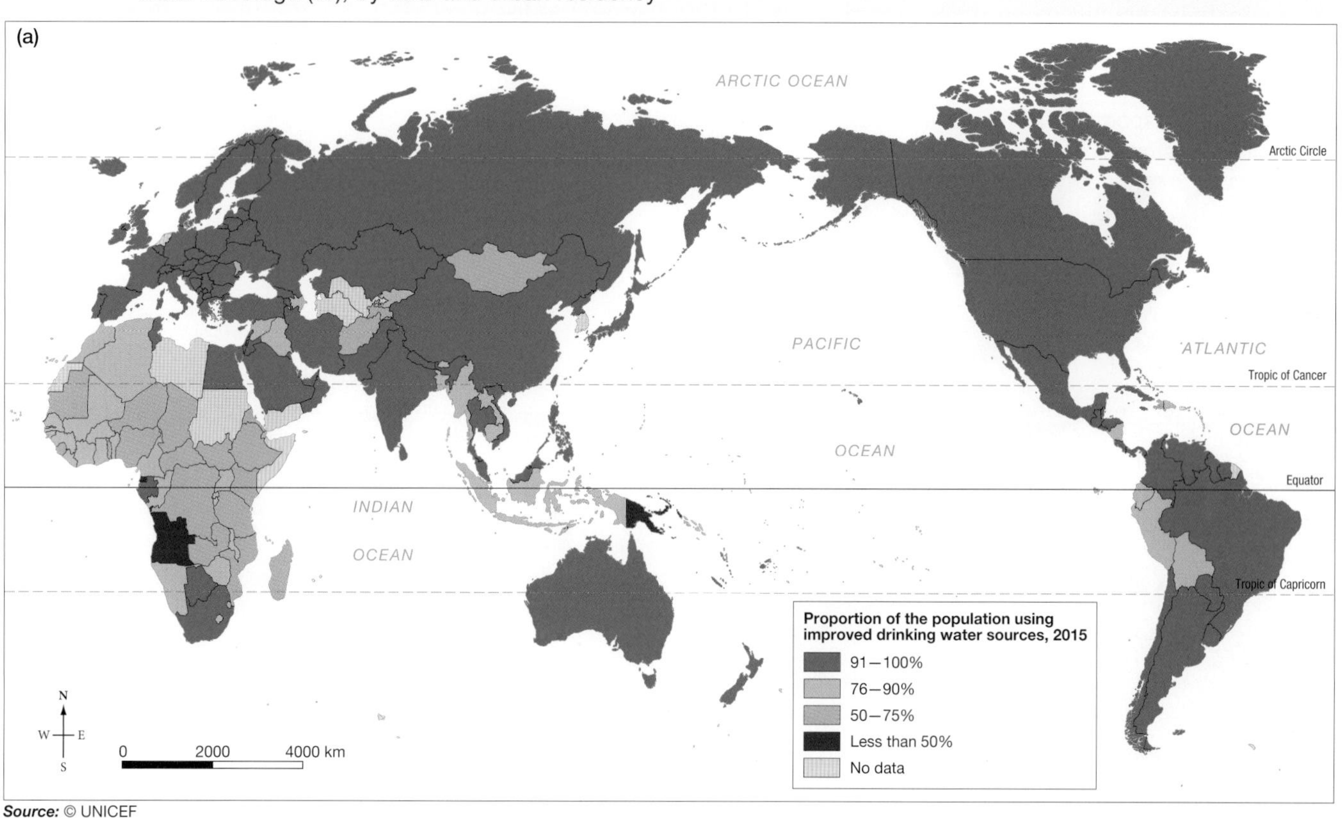

Source: © UNICEF

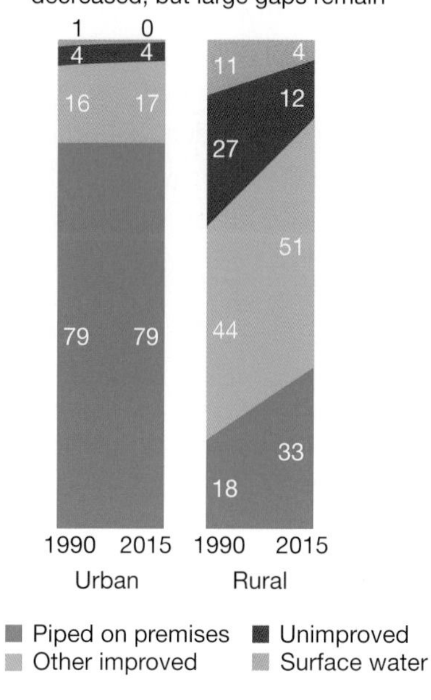

FIGURE 2 Transporting water in a rural area in Kenya

Unfortunately, progress towards the MDG sanitation target of 75 per cent was much slower and was not met overall, although 95 countries did succeed. Only 68 per cent of the world's population has access to improved sanitation. (see figure 4). Improving access to sanitation for all now forms part of SDG 6.

There are huge regional variations across the globe as well as within countries. For example, 90 per cent or more of people in Latin America, Northern Africa and much of Asia have improved water supply, in contrast with only 68 per cent in sub-Saharan Africa. In terms of sanitation, most people who lack access are again rural dwellers, even within locations where water supply has improved. For example, 590 million people in India practise open defecation. Use the **India: Untouchables** weblink in the Resources tab to learn more about the issues facing the poorest members of Indian society in terms of sanitation and health.

FIGURE 3 Transporting water to urban areas in Tasmania, Australia

FIGURE 4 Changes in provision of sanitation

Year	1990	1995	2000	2005	2010	2015
Unimproved target	51	48	44	40	37	32
Improved sanitation	49	52	56	60	63	68

Percentage (0–100); Year

■ Improved sanitation ■ Unimproved target
- - - MDG target

Success stories

Many non-government organisations (NGOs) have been involved in successful projects to improve access to water supply and sanitation. The success of these projects hinges not just on provision of clean water and toilets, but also on community involvement and education. Oxfam is working with communities in rural Cambodia. It has provided families with access to water filters, safe and clean toilets, water pumps and rainwater harvesting systems for collecting and storing water. These facilities have negated the need for girls and mothers to spend hours carrying water or finding additional firewood in order to boil unsafe drinking water. The mothers now have more time to work on their farms, potentially improving food availability and income, and girls have more time free from essential chores, so they can instead attend school.

11.3 Activities

To answer questions online and to receive **immediate feedback** and **sample responses** for every question, go to your learnON title at www.jacplus.com.au. *Note*: Question numbers may vary slightly.

Remember

1. What impact does poor sanitation have on health and wellbeing?
2. Refer to figure 1 (a) and (b). How has the distribution of ***places*** with clean water ***changed*** over time?

Explain

3. Explain the ***interconnection*** between poor sanitation, unclean water, health and wellbeing.
4. What impact does poor sanitation have on the natural ***environment***?

Discover

5. Undertake research on one of the following to find out the role unclean water plays in its spread: cholera, typhoid, dysentery, schistosomiasis, worm infestations.
6. Go to the website for either Oxfam or World Vision. Gather the following information on one of their current projects involving improvements to the provision of clean water and sanitation: location of project, ***scale*** of project, problems the project aimed to fix, what the project involves.

Predict

7. Which countries and regions do you expect to make greatest progress in terms of improving access to clean water and sanitation? Justify your answer.
8. What is the likely impact of improved water provision on literacy for girls in places such as Cambodia?

Think

9. 'Access to adequate sanitation is increasingly a problem in urban ***places*** in the developing world.' Evaluate this statement.
10. If you were travelling to a ***place*** that does not have clean water or the level of sanitation you are used to, what steps could you take to ensure you did not become ill during your visit?

Explore more with these weblinks: WHO/UNICEF, India: Untouchables

11.4 How does poverty affect wellbeing?

Access this subtopic at **www.jacplus.com.au**

11.5 The great divide?

11.5.1 Rural–urban variation in Australia

Wellbeing varies considerably from one place to another within the one country and also within urban and rural environments. Sometimes these variations are quite distinct, but at other times they may be quite minor. As you will have seen in earlier chapters, the particular indicators used may give a different picture of these places.

According to the Australian Institute of Health and Welfare (AIHW), people living in rural places tend to have shorter lives and higher levels of some illnesses than those in major cities. The level of health of Australians is much lower in **regional and remote areas** than in major cities. The location of these categories is shown in figure 1. People from regional and remote areas tend to be more likely than their major cities counterparts to smoke and drink alcohol in harmful or hazardous quantities. This is reflected in higher mortality rates than for those living in major cities.

FIGURE 1 Australia's population by remoteness classification

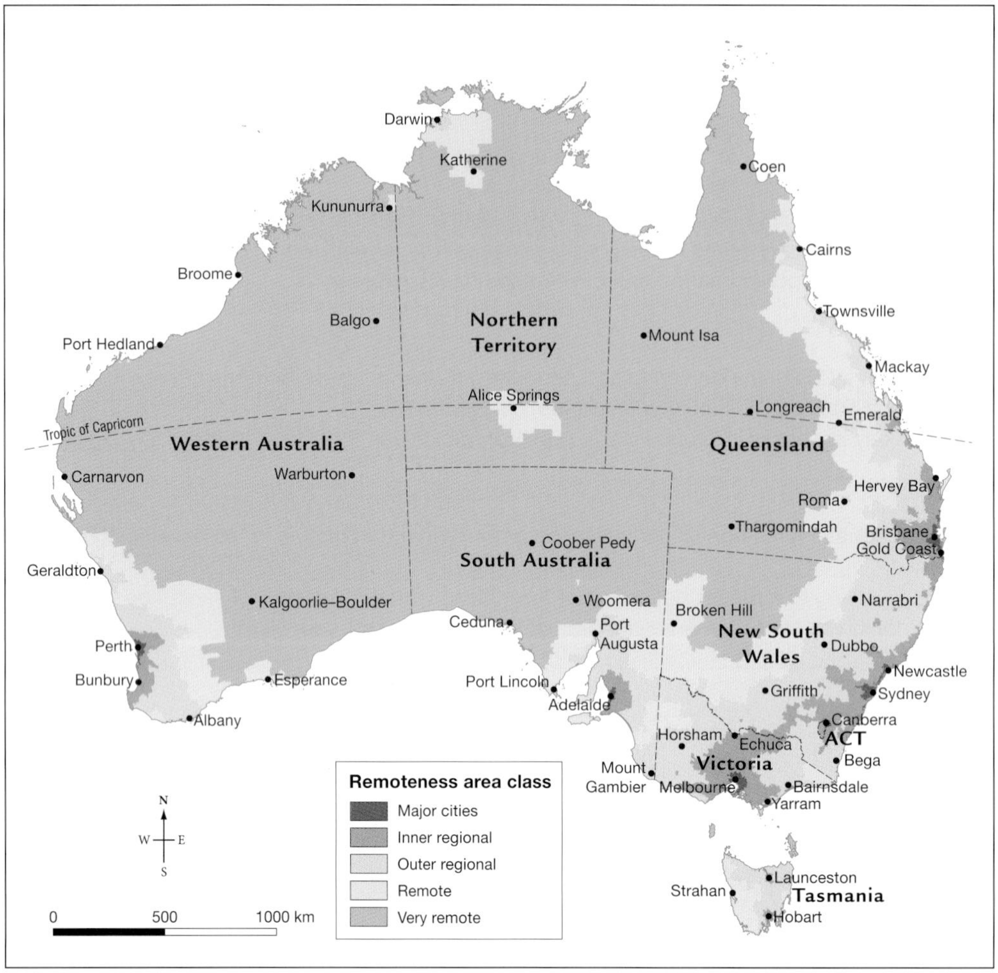

Source: Australian Bureau of Statistics

These higher death rates may relate to differences in access to services, increased risk factors and the regional/remote environment. More physically dangerous occupations in rural areas lead to higher accident rates. Factors associated with driving, such as long distances, greater speed and animals on roads, contribute to elevated road accident rates in country areas.

People with disability living outside major cities are significantly less likely to access disability support services. Where health services are provided, regional and remote residents also face higher out-of-pocket expenses. Use the **National Health Survey** weblink in the Resources tab to find out more about the health of the Australian population.

In general, people living in rural Australia do not always have the same opportunities for good health as those living in major cities. Residents of more inaccessible areas of Australia are generally disadvantaged in their access to health facilities with skilled personnel. Figure 3 shows access to services is at least partially affected by the number of available health workers per population. Medical personnel in rural areas have a higher average age and face longer hours than their city counterparts. Recruitment difficulties in rural areas also affect the sustainability of such services.

Higher costs and more limited availability of products such as fresh fruit and vegetables also impact on health and wellbeing. A government survey found that absence of competition in remote areas led to mark-ups of up to 500 per cent on some foods, particularly fresh fruit and vegetables, which took up to two weeks to reach their destination. A typical packet of pasta cost approximately five times more than in metropolitan stores. Fewer educational and employment opportunities are other challenges faced by those in regional and remote places.

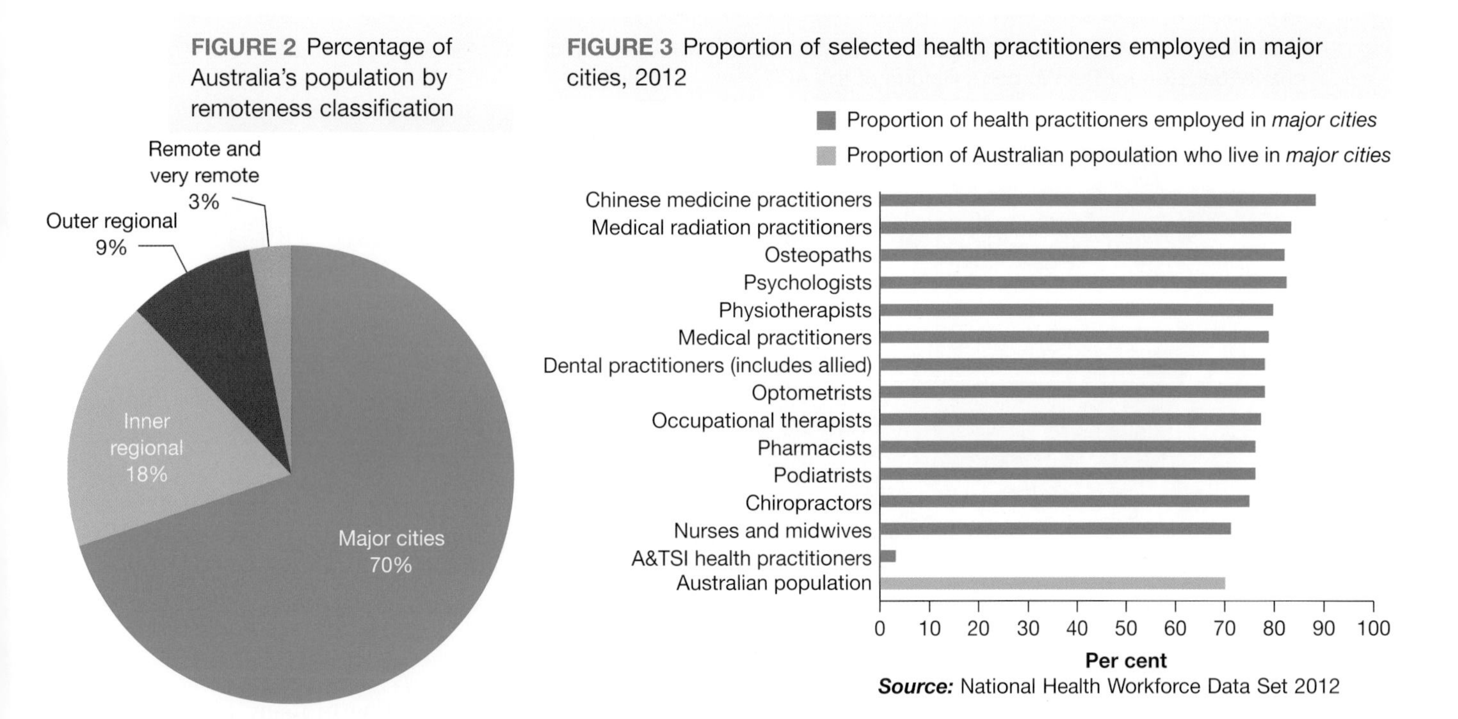

FIGURE 2 Percentage of Australia's population by remoteness classification

FIGURE 3 Proportion of selected health practitioners employed in major cities, 2012

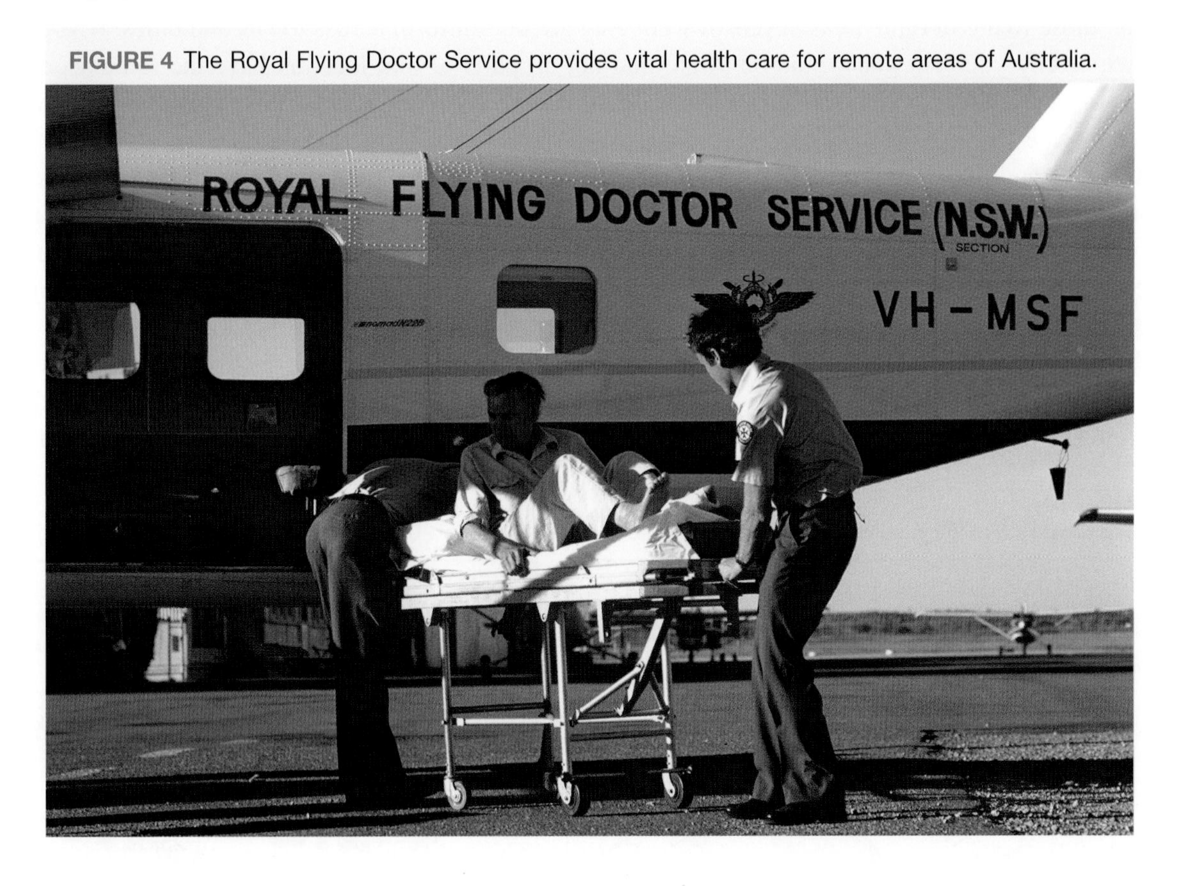

FIGURE 4 The Royal Flying Doctor Service provides vital health care for remote areas of Australia.

On the positive side, in terms of wellbeing, rural Australians tend to have higher levels of social cohesiveness, as reflected in higher rates of participation in volunteer work and feelings of safety in their community. The Country Women's Association is one such volunteer organisation (see figure 5).

FIGURE 5 Members of the Canberra Evening Branch of the Country Women's Association held a charity bake-off for the Women's Refuge of the ACT.

11.5.2 Variation in wellbeing within cities

Although people in urban places generally have a higher standard of wellbeing according to many measures than those living in rural areas, levels of wellbeing are not uniform across towns and cities. If you live in a town or city yourself, you would be aware that not all parts of that location have the same access to facilities or the same types of housing. Variations in wellbeing occur on a local scale as well as at national and global scales.

Figure 6 (a) and (b) shows **housing affordability** as a measure of wellbeing in terms of the median price for houses and units in Melbourne. The cost of housing is a major expenditure for people so its affordability directly impacts on people's living standards. Based on this information, it is difficult for many people to purchase a home as only 2 per cent of Melbourne suburbs have a median house value of under $300 000.

Located in the city of Mumbai, India, the slum of Dharavi (see figure 7) contains an estimated one million people in an area of 175 hectares. Although this location has minimal formal infrastructure and has poor drainage, its cheap rent, manufacturing activities such as leather tanning and its location between two

FIGURE 6 (a) House affordability and (b) unit affordability in Melbourne, 2013

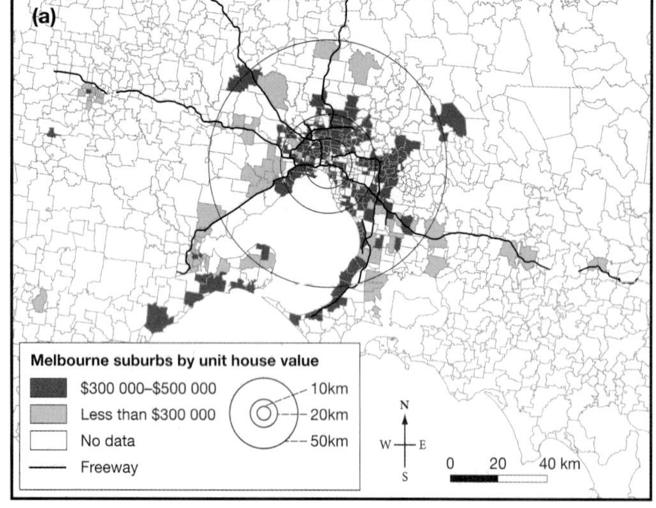

Source: © Core Logic RP Data

(b)
Melbourne suburbs by median house value
$300 000–$500 000
Less than $300 000
No data
Freeway
10km
20km
50km
0 20 40 km

Source: © Core Logic RP Data

major suburban railway lines mean that it continues to grow. In contrast, figure 8 shows a new residential development in the eastern part of the same city. Mumbai's population of more than 22 million people includes India's richest person, Mukesh Ambani, whose US$21 billion fortune, made via the textile industry, also ranks him among the richest people in the world. The contrasts in wellbeing in India are clearly seen by these two vastly different environments.

FIGURE 7 Dharavi, a large slum in the city of Mumbai, India

FIGURE 8 New residential development for the wealthy in Mumbai, India

11.5 Activities

To answer questions online and to receive **immediate feedback** and **sample responses** for every question, go to your learnON title at www.jacplus.com.au. *Note*: Question numbers may vary slightly.

Remember

1. Draw a table to show the advantages and disadvantages of rural versus urban areas in terms of wellbeing in Australia.
2. Describe the distribution of housing and unit affordability shown in figure 6 (a) and (b).

Explain

3. Refer to figures 2 and 3 to explain why people living in rural Australia do not always have the same opportunities for good health as those living in major cities.
4. Explain why the information shown in figure 6 (a) and (b) may be considered a measure of wellbeing.

Predict

5. Suggest the positive and negative impacts of the mining boom on levels of wellbeing in remote communities in Australia.
6. What is the long-term potential outcome of the contrast in wellbeing shown in figures 7 and 8?

Think

7. Why would it be difficult to measure wellbeing in a slum area such as Dharavi in Mumbai?
8. Suggest an alternative measure of wellbeing not mentioned in this section that could highlight the variation in wellbeing within an urban area.

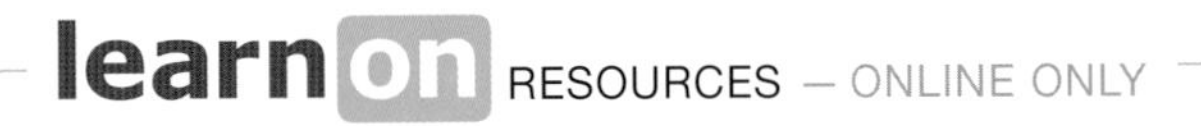

RESOURCES — ONLINE ONLY

Explore more with this weblink: National Health Survey

Try out this interactivity: Call the doctor! Call the nurse! (int-3311)

11.6 Is everybody equal?

11.6.1 Are all Australians equal?

Indigenous Australian culture has often been acknowledged as the world's oldest surviving culture. Sixty thousand years of history has resulted in rich traditions, strong spiritual beliefs and complex social structures. Since European settlement much has changed, but culture and the bonds within **Indigenous** communities remain strong. Why then, do Aboriginal and Torres Strait Islander peoples not enjoy the level of wellbeing experienced by the wider Australian community?

FIGURE 1 Closing the gap for Indigenous Australians will take generations of commitment.

In a just society like Australia, we would expect that everyone is able to experience a similar standard of living. It would be unfair for one sector of a community to experience significant disadvantage when the rest of the community enjoys the privileges of a 'good life'. Many Indigenous people consistently experience lower levels of health, education, employment and economic independence than those enjoyed by most Australians. These **socioeconomic** factors inhibit the ability of Indigenous Australians, who make up 2.5 per cent of the Australian population, to contribute to and benefit from all that Australia has to offer.

11.6.2 Why does disadvantage exist?

The inequalities may be attributed to three main causes:

- the dispossession of land
- the displacement of people
- discrimination.

Many generations of Indigenous people have experienced difficulties in accessing the same services and opportunities as other Australians. Disadvantage in one area, for example, poor access to health services, may affect a student's ability to attend school, which may in turn alter their employment prospects. Compared with other Australians, Indigenous people (as a group) remain disadvantaged (see figure 3).

11.6.3 How can we measure Indigenous wellbeing?

Indigenous peoples are culturally and linguistically diverse, but Indigenous culture differs markedly from non-Indigenous Australian culture. Concepts of family structure and community obligation, language, obligations to country and the passing down of traditional knowledge are all viewed and practised very differently by Indigenous cultures in comparison to non-Indigenous cultures. These are important factors that contribute to both identity and wellbeing, yet as indicators, they may be difficult to measure.

FIGURE 2 Better and more frequent access to health care will help close the gap. The Close the Gap Campaign is Australia's biggest ever public movement for health equality. Led by a coalition of leading Australian health and human rights organisations, it campaigns for long-term and sustainable change to close the gap in life expectancy and health standards for Indigenous Australians and to ensure that they are able to have direct control over their own health.

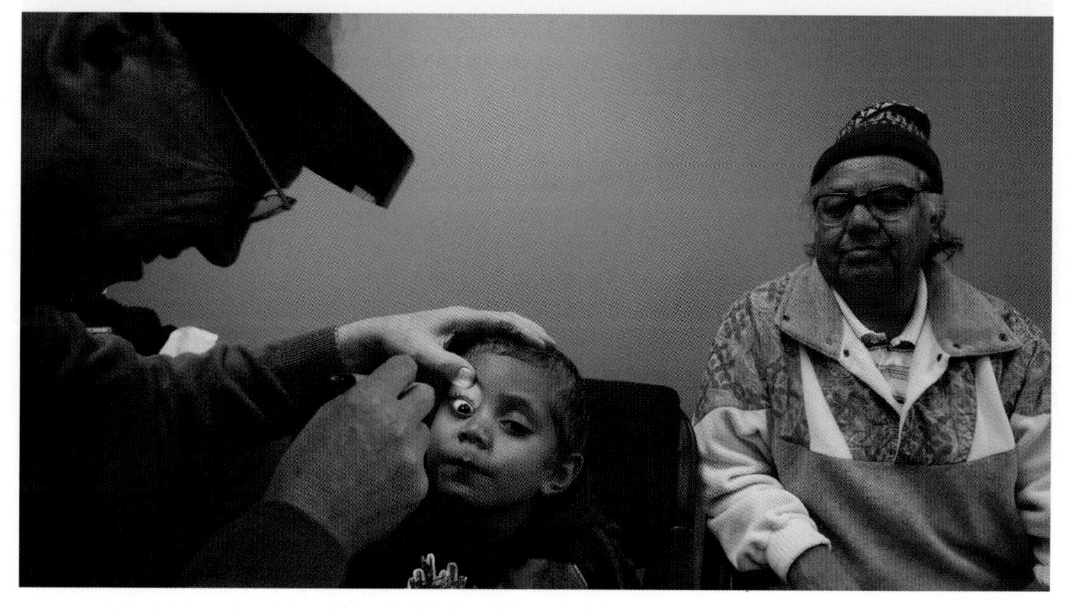

The National Aboriginal and Torres Strait Islander Health Survey conducted by the government aimed to measure the emotional and social health of Indigenous adults. In this, more than half the adult Indigenous population reported being happy (71 per cent), calm and peaceful (56 per cent), and/or full of life

(55 per cent) all or most of the time. Just under half (47 per cent) said they had a lot of energy all or most of the time. Indigenous peoples in remote areas were more likely to report having had these positive feelings all or most of the time than Indigenous peoples living in non-remote areas.

A note on statistics: collecting data for Indigenous people in Australia can be difficult, and the Australian Bureau of Statistics concedes that it encounters many difficulties in obtaining accurate figures such as those shown in figure 3.

FIGURE 3 Indicators of Indigenous wellbeing

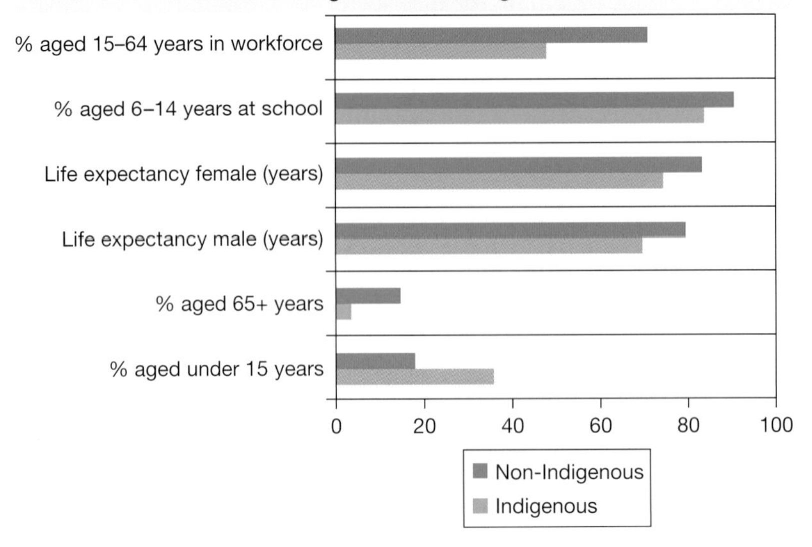

11.6 Activities

To answer questions online and to receive **immediate feedback** and **sample responses** for every question, go to your learnON title at www.jacplus.com.au. *Note*: Question numbers may vary slightly.

Remember

1. What are some of the reasons disadvantages exist for Indigenous Australians?
2. Refer to figure 3. What is the average life expectancy for Indigenous Australians and non-Indigenous Australians? What is the difference (in years) between these average life expectancies?

Explain

3. Explain how the National Aboriginal and Torres Strait Islander Health Survey might give us more insight into the wellbeing of Indigenous Australians.

Discover

4. Did any of the statistics about Australia's Indigenous people surprise you? Explain your reaction to them, and how they may have either ***changed*** or reinforced your own opinions or beliefs.

Think

5. Another factor contributing to disadvantage may be the remoteness of Indigenous communities. What innovative solutions can you come up with to try to solve these accessibility problems? Consider a range of socioeconomic areas.
6. 'Social justice' means fair and equitable access to a community's resources. Do you think Indigenous people experience social justice in Australia? Explain your answer.

11.7 How can wellbeing be improved for Indigenous Australians?

11.7.1 Programs to close the gap

Recognising the divides that exist at home, Australian governments and other agencies such as Oxfam are continuing to push initiatives aimed at improving some of the problems that many Indigenous communities face. Ultimately, all Australians benefit from a united effort to address Aboriginal and Torres Strait Islander disadvantage. When disadvantage is overcome, the need for government expenditure is decreased. At the same time, Aboriginal and Torres Strait Islander peoples will be better placed to fulfil their cultural, social and economic aspirations.

Under the Closing the Gap program, six specific targets were set relating to:

1. Indigenous life expectancy
2. infant mortality
3. early childhood development
4. reading, writing and numeracy
5. Year 12 attainment rates
6. employment.

The two specifically relating to health are to close the life expectancy gap within a generation (by 2030) and to halve the gap in mortality rates for Indigenous children aged under five years within a decade (by 2018). Child mortality rate targets are currently on track to be met due to improvements in **antenatal care**, sanitation and public health, better **neonatal intensive care**, and the development of immunisation programs. Meeting the life expectancy target will be challenging, particularly as overall life expectancy for the population as a whole is increasing. To meet the life expectancy target, average Indigenous life expectancy gains of 0.6 and 0.8 years per year are needed — that is almost 21 years by 2031 to close the gap.

11.7.2 Examples of programs to improve Indigenous wellbeing

The following initiatives provide examples as to how both government and non-government agencies are working to improve the health of the Indigenous population in Australia.

FIGURE 1 New food store at Ngukurr, Northern Territory

- *The National Partnership Agreement on Closing the Gap in Indigenous Health Outcomes*. For example, the Many Rivers Aboriginal Medical Service Alliance in northern New South Wales brings together 10 Aboriginal-controlled health organisations who share resources and programs servicing 35 000 people.
- *The Australian Government licensing scheme for community stores in the Northern Territory*. This scheme requires store managers to offer a range of healthy food and drinks and to make these attractive to customers. Prior to this, people in remote Indigenous communities often had little choice. Goods and food were of poor quality and basic consumer protection was lacking. More than 90 Northern Territory stores, such as that pictured in figure 1, are now licensed, with reported improvements in management, hygiene and employment of Indigenous staff.
- *Oxfam's Indigenous Health and Wellbeing Program*. Oxfam works with Indigenous organisations to hold governments to account over the Closing the Gap program. It also supports the Fitzroy Stars Football club, which competes in Melbourne's Northern Football League. This club brings together 300 Indigenous men with the aim of nurturing a culture that promotes a healthy lifestyle, fitness, nutrition and self-esteem. It also aims to build bridges between Indigenous and non-Indigenous communities.

Lombadina Indigenous community program

Indigenous communities themselves are also working hard to improve their wellbeing. Lombadina is an Indigenous community inhabited by the Bardi people. It is located on the north-western coast of Western Australia (see figure 2). Lombadina and the neighbouring Djarindjin community are home to approximately 200 Indigenous people. The Lombadina community is working towards self-sufficiency through ventures that include tourism operations, a general store, an artefact and craft shop, a bakery and a garage. The tourist ventures centre on sharing knowledge of an Indigenous lifestyle. In addition to providing serviced accommodation, many tours are offered, including cultural tours, fishing charters, kayaking and bushwalking. Lombadina has received a number of tourism awards. The considerable success of these businesses has contributed substantially to the wellbeing of this community. Use the **Lombadina** weblink in the Resources tab to experience a kayaking tour led by a Lombadina community member.

Lombadina is also involved in the EON Thriving Communities Project. EON is a non-government organisation operating by invitation in Indigenous communities in Western Australia. It aims to close the gap in terms of health; for example, via the provision of practical knowledge about growing and preparing healthy food in schools and communities. The project has community ownership and is designed to be sustainable, thus improving wellbeing in the long term.

FIGURE 2 Location of Lombadina, Western Australia

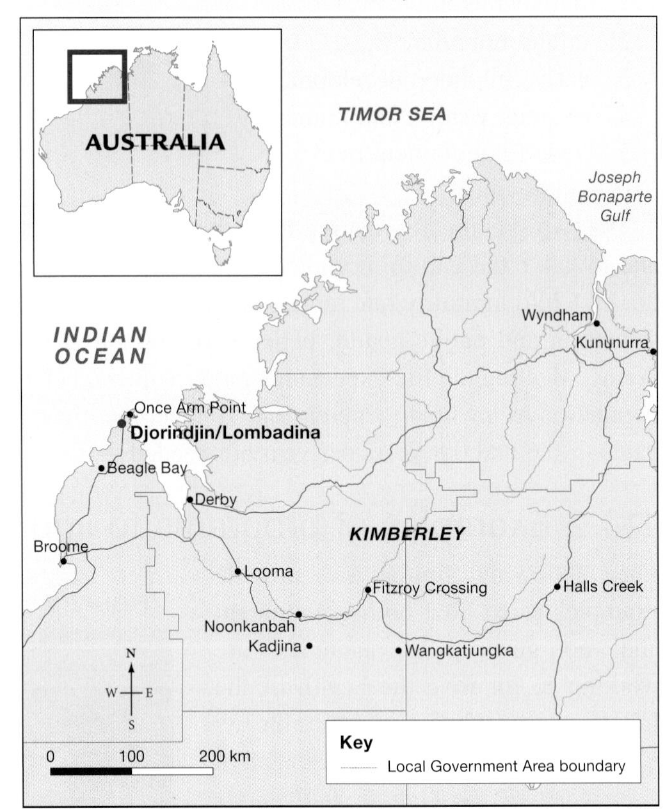

Source: Spatial Vision GAT-45; © Commonwealth of Australia Geoscience Australia 2013. © Commonwealth of Australia Australian Bureau of Statistics 2013.

11.7 Activities

To answer questions online and to receive **immediate feedback** and **sample responses** for every question, go to your learnON title at www.jacplus.com.au. *Note*: Question numbers may vary slightly.

Remember

1. What areas are being addressed by the Federal Government's Close the Gap program?

Explain

2. Why is the Close the Gap program necessary?

Discover

3. National Close the Gap Day is held in March each year to improve community awareness of the issue of Indigenous disadvantage and to publicise Federal Government action. Use the internet to find out what activities are taking place in your state and/or local area for Close the Gap Day.
4. Using the internet, research one of the following organisations that have experienced success in combating some of the health, social or educational disadvantages experienced by Indigenous Australians. Why have they been successful? What outcomes will be ***changed*** for Indigenous people?
   - Aboriginal Women Against Violence (NSW)
   - MPower — Family Income Management Plan (Qld)
   - Indigenous Enabling Program at Monash University (Vic.)

Think

5. How might Indigenous tourism initiatives such as those run by the Lombadina community improve the wellbeing of people beyond that community?

learnON RESOURCES — ONLINE ONLY

Explore more with this weblink: Lombadina

11.8 SkillBuilder: Understanding policies and strategies

onlineonly

WHAT ARE POLICIES AND STRATEGIES?

Policies are principles and guidelines that allow organisations to shape their behaviour and decisions, and to clarify future directions. Strategies ensure that the key components of a plan are implemented. Policies and strategies are particularly useful in large organisations, where information needs to be spread to all employees.

Go online to access:

- a clear step-by-step explanation to help you master the skill
- a model of what you are aiming for
- a checklist of key aspects of the skill
- a series of questions to help you apply the skill and to check your understanding.

Policy

The Metcalfe Boys' High School must maximise student numbers in order to remain viable and to offer a broad range of subjects.

Strategies

Long term

- The Metcalfe Boys' High School is to become a co-educational school. It will do this over a six-year period, beginning with Year 7.
- It must achieve a gender balance in the classes within six years.
- Awards for girls should be developed.
- Associations for past students should be created.

Short term

- Design and create a uniform for the girls.
- Create facilities such as toilets and gymnasium change rooms.
- Consider school camp facilities.
- Consider the need to join other sporting organisations such as softball and girls' competitions.

learnON RESOURCES — ONLINE ONLY

Watch this eLesson: Understanding policies and strategies (eles-1760)

Try out this interactivity: Understanding policies and strategies (int-3378)

11.9 How is teen pregnancy linked to wellbeing?

11.9.1 How does the teen pregnancy rate vary?

If you are a student, you are probably a teenager. You are likely to be thinking ahead about further study, employment or your next holiday. You are probably not thinking about starting a family. However, if you were a teenager in the United States, statistically you would have a higher chance of becoming a parent than teenagers in any other developed country. Why is this the case?

Although the **teenage pregnancy rate** has declined substantially in the past 25 years, teenagers in the United States have the highest birth rate of any developed country (see figure 1(a)). A teenager in the United States is nine times more likely to have a baby than a teenager in Switzerland. Roughly one in four teen girls in the United States will get pregnant at least once by age 20.

As figure 2 shows, there is substantial variation within the country, with the state of Mississippi having a rate of teen births four times that of New Hampshire.

FIGURE 1 How do teen pregnancy rates in the United States compare with other countries? (a) Teenage fertility rates and (b) relative child poverty rates

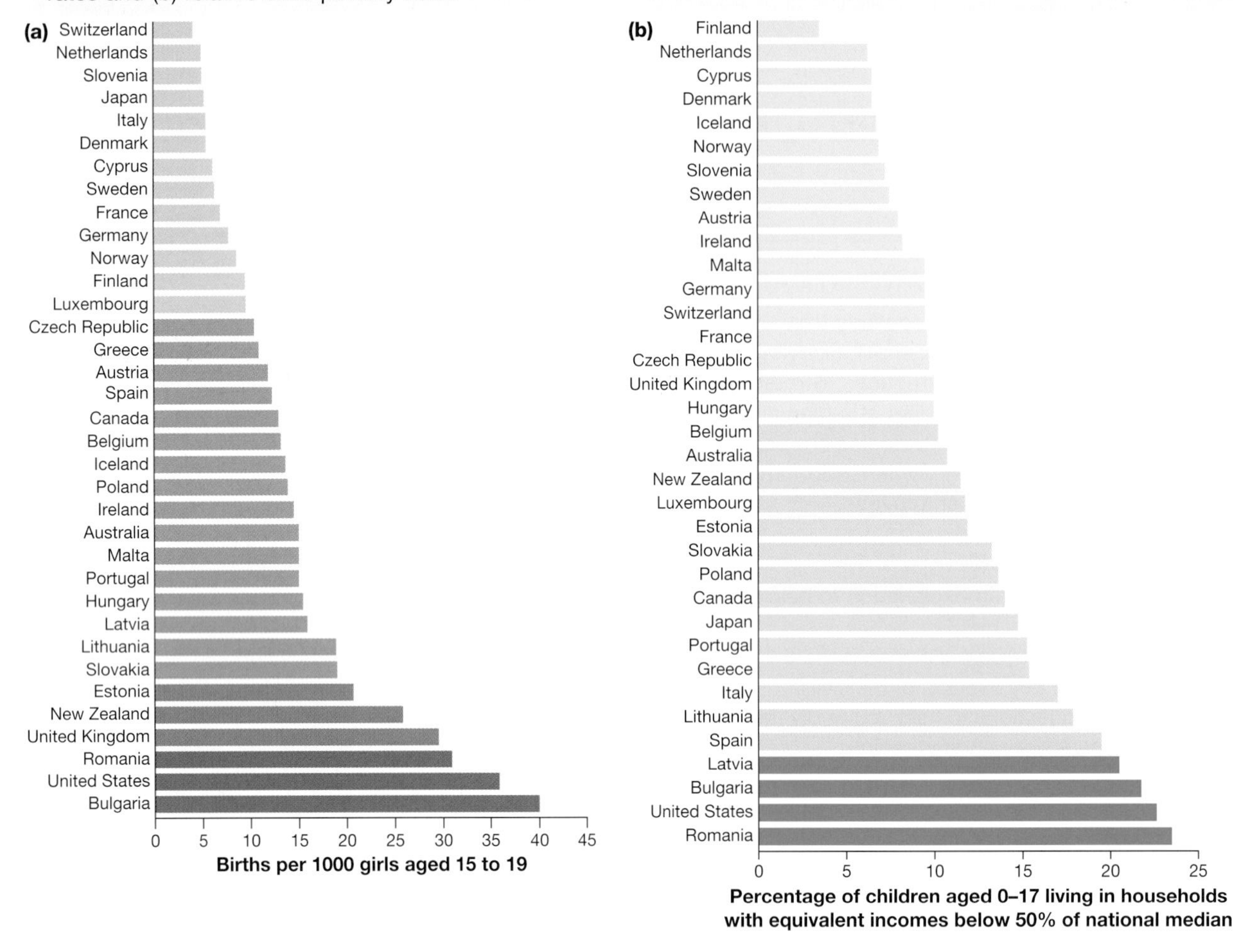

FIGURE 2 The distribution of the teen birth rate across the United States

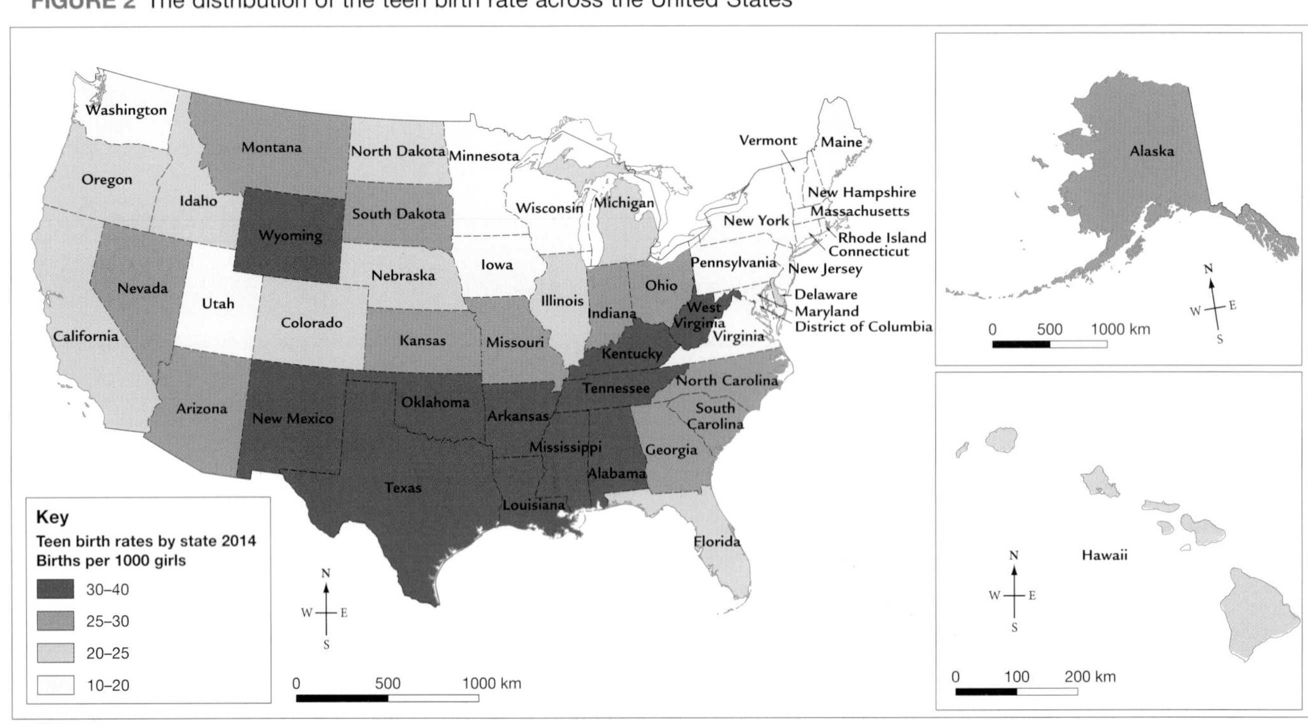

Source: © Public Domain http://www.census.gov/hhes/www/poverty/data/incpovhlth/2014/index.html

11.9.2 What factors account for the variation in teen pregnancy rates?

Sex education and access to contraception have been promoted across the United States, although rates of abortion are relatively low when compared with those in European countries. The majority of teenage girls who get pregnant welcome their pregnancy and decide to have their baby. It has been argued that the generous welfare system allows for this; however, European nations generally provide more support.

Recent studies in the United States have concluded that underlying social and economic problems have a major impact on teenage pregnancy rates and account for the variation across states. Teenage child-bearing is both a cause and effect of poverty. Teenage girls in areas of disadvantage as shown in figure 3 are choosing extramarital motherhood rather than continuing their education and pursuing economic success. They feel there is little chance of advancement through these channels.

FIGURE 3 The distribution of poverty in the United States of America has a strong interconnection to teen birth rates.

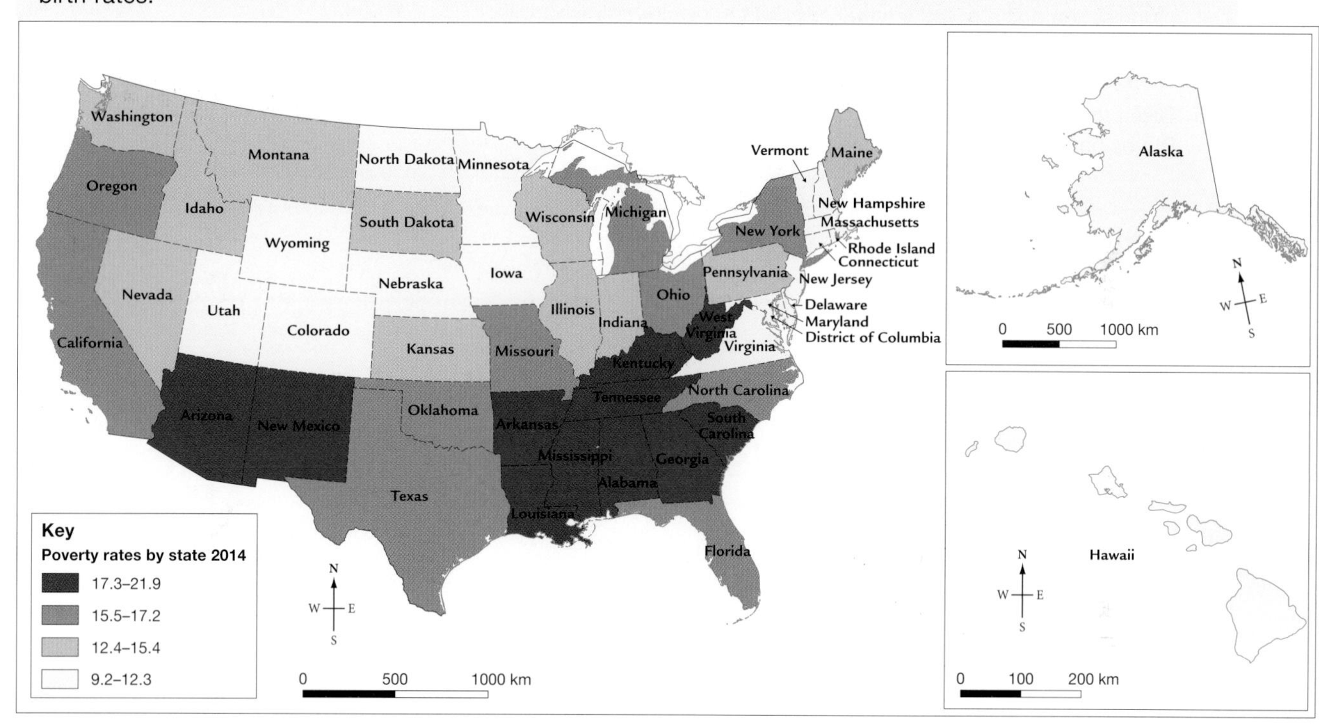

Source: © Public Domain http://www.census.gov/hhes/www/poverty/data/incpovhlth/2014/index.html

Both the short- and long-term social and economic costs of teen pregnancy and child-bearing can be high. For example, teen pregnancy and childbirth contribute significantly to drop-out rates among high school girls. This reduces their potential income. This cycle has a major impact on wellbeing. Teen pregnancy and childbirth costs US taxpayers an estimated $9.4 billion per year, taking into account welfare payments and indirect costs such as lost tax revenue.

11.9.3 Reducing teen pregnancy

Prevention of teen pregnancy is of paramount importance to the wellbeing of young people and communities throughout the United States and is one of six public health priorities. President Obama's Teen Pregnancy Prevention Initiative (TPPI) aims to reduce teen pregnancy and address disparities in teen pregnancy and childbirth rates. TPPI is focused on communities with the highest rates: African American and Hispanic and Latino youth. To date, this program has been partially successful.

11.9 Activities

To answer questions online and to receive **immediate feedback** and **sample responses** for every question, go to your learnON title at www.jacplus.com.au. *Note*: Question numbers may vary slightly.

Explain

1. Use map evidence to evaluate the extent of the ***interconnection*** between poverty and teen birth rates in the United States.
2. How do high teenage birth rates contribute to lower levels of wellbeing in both the short and long term for both individuals and the United States as a whole?

Discover

3. Use the internet to find out the current teenage pregnancy rate in Australia. How does this figure compare to that of the United States?
4. Use the **TheNext.Org** weblink in the Resources tab to explore this site. What is this non-government organisation doing to reduce teen births in the United States?

Predict

5. Given the information provided, what impact would an improvement in economic outlook for the United States have in the distribution of teenage birth rates?

Try out this interactivity: Teen mums (int-3312)

Explore more with this weblink: TheNext.Org

11.10 SkillBuilder: Using multiple data formats

online only

WHAT ARE MULTIPLE DATA FORMATS?

Multiple data formats are varied forms of data presentation, used when a range of data needs to be shown. All the information must be read before the data can be interpreted.

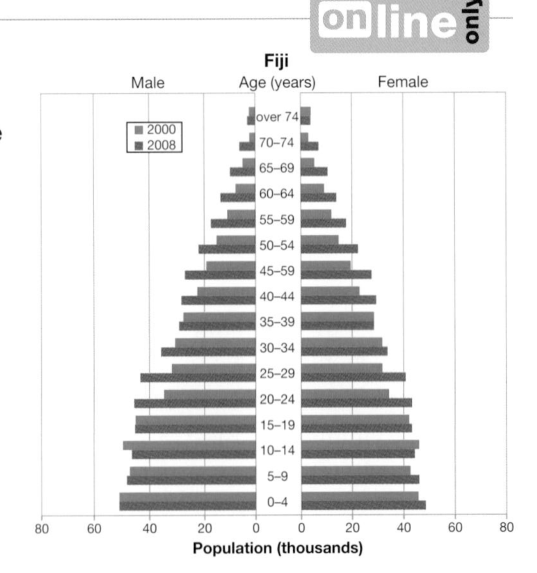

Go online to access:

- a clear step-by-step explanation to help you master the skill
- a model of what you are aiming for
- a checklist of key aspects of the skill
- a series of questions to help you apply the skill and to check your understanding.

Watch this eLesson: Using multiple data formats (eles-1761)

Try out this interactivity: Using multiple data formats (int-3379)

11.11 Review

11.11.1 Review

The Review section contains a range of different questions and activities to help you revise and recall what you have learned, especially prior to a topic test.

11.11.2 Reflect

The Reflect section provides you with an opportunity to apply and extend your learning.

Access this subtopic at **www.jacplus.com.au**

TOPIC 12 Trapped by conflict

12.1 Overview

Numerous **videos** and **interactivities** are embedded just where you need them, at the point of learning, in your learnON title at www.jacplus.com.au. They will help you to learn the content and concepts covered in this topic.

12.1.1 Introduction

People's wellbeing is put under stress when a country's development is threatened by pressures on society, economies, politics and the environment. Sometimes the tension results in outbreaks of conflict.

Societies pressure governments for change. Improvements are sought in living conditions, and freedoms may be demanded. Tension can spill over into conflict. People can find themselves forced to fight or flee.

Over time, countries change, but always somewhere in the world there are people trapped by conflict.

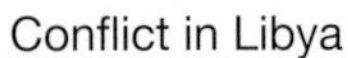

Conflict in Libya

Starter questions

1. Make a list of the best things in your life — those things that make you content.
2. List some conflicts that affect you. Are these personal conflicts that affect your wellbeing, or related to the wellbeing of your community?
3. In the media there is often reference to people whose wellbeing has been affected. In the past month, recall those ***places*** and people that you have heard about or seen mentioned.
4. Conflicts in ***places*** on the other side of the world can affect you. How do some of these conflicts connect with your life?

INQUIRY SEQUENCE

12.2 Where is wellbeing affected by conflict?

12.2.1 Global conflicts

Figure 1 shows an uneven distribution of conflict affecting wellbeing across the world in the early twenty-first century. By 2016 there were 31 active armed and numerous other conflicts of varying degrees of intensity. Most conflicts are **civil wars**, where the victims are mostly the residents of the country. In World War I, less than 5 per cent of the casualties were civilians; however, in today's conflicts the figure is nearer 80 per cent.

12.2.2 Three ways to identify conflicts

Cause of conflict

- Religious and cultural conflicts are based predominantly on characteristics of people or society. The breakup of Yugoslavia (1992–95) into Serbia, Bosnia-Herzegovina, Montenegro, Slovenia, Croatia and Macedonia saw the mass movement of ethnic groups to areas of safety.
- Economic conflicts involve monetary value. Securing the supply of oil from the Middle East, including shipping routes, has been important to the wellbeing of Americans. Large-scale deforestation and mining across Asia has destroyed the environment, forced people off the land and brought conflict between users of the land.
- Resource conflicts are those where resource distribution and use are the issue. A river crossing national borders is prone to manipulation of river flows, such as along the Nile River.
- Political conflicts can arise where people speak their opinions. The search for democracy across the Arab world (ongoing) signifies a break from dictatorships.

- Land conflicts, or territorial disputes, are often ongoing issues or revivals of past situations; for example, the consequence of colonialism. Conflict in the Middle East between Palestine and Israel, for instance, is an ongoing issue with British and French colonialism a key factor in its development.

FIGURE 1 World conflicts, 2015

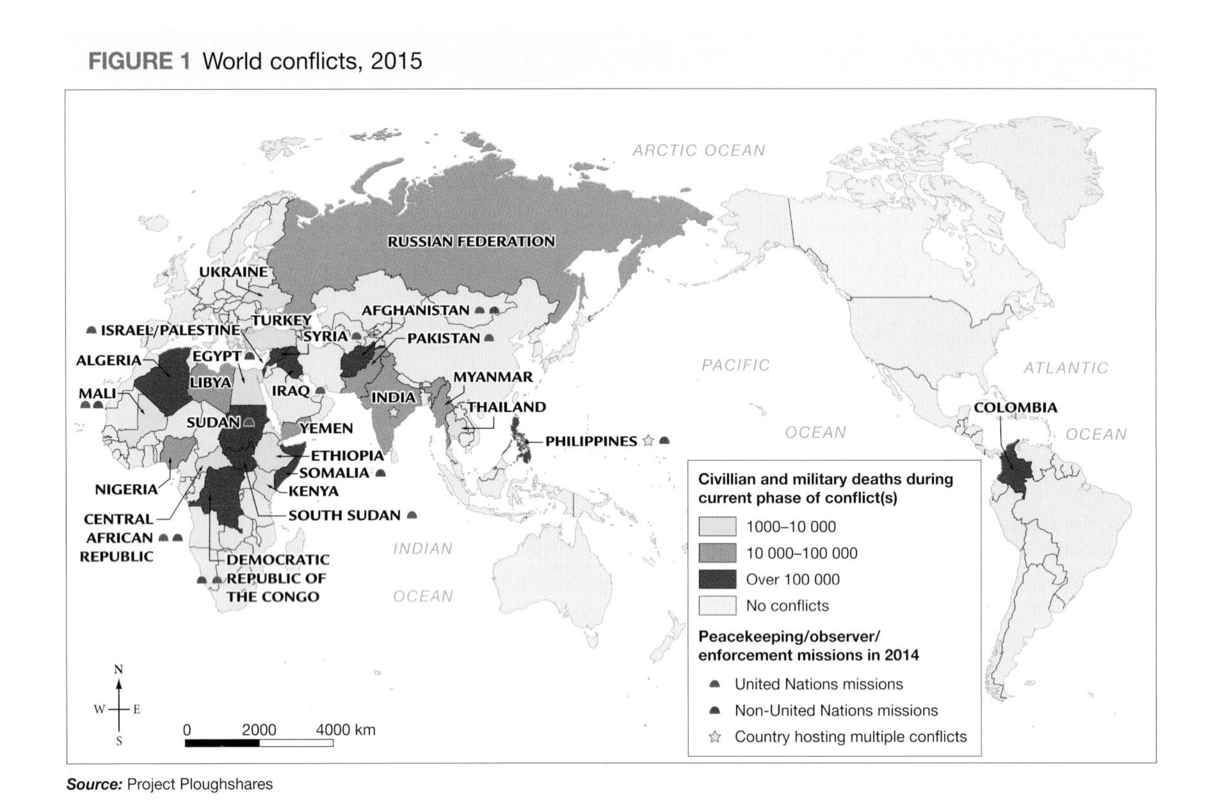

Source: Project Ploughshares

Length of conflict

Short-term conflicts are those that last a limited time and have reduced ongoing impact on people. Long-term conflicts are those in which months or years are taken for a resolution to be achieved, and where even then there are ongoing tensions.

Scale of conflict

International conflicts about the power to control land and civil conflict can destroy a nation. Conflict at this scale may become war. Small-scale or local conflicts are disagreements, generally over planning issues, which enter a dispute phase. The establishment of wind farms across southern Australia triggered conflict in local communities. Coal seam gas explorations in New South Wales and Queensland have had protest groups on the march.

Conflicts are very expensive for both the countries where conflict occurs (see figure 2) and for the countries supplying military equipment for the conflict (see figure 3). Money is often drawn away from basic essential services, such as education and health, affecting wellbeing across the world.

12.2.3 Diplomacy and conflict

Diplomats across the world strive to prevent conflict. They endeavour to negotiate successful outcomes between opposing groups to deter conflict from breaking out. Everyone would prefer stability. Soldiers in the field, as in figure 4, work with civilians to return to cooperation and peace, and improve the wellbeing of a country's population.

FIGURE 2 Percentage of countries by **Human Development Index (HDI)** ranking hosting armed conflict

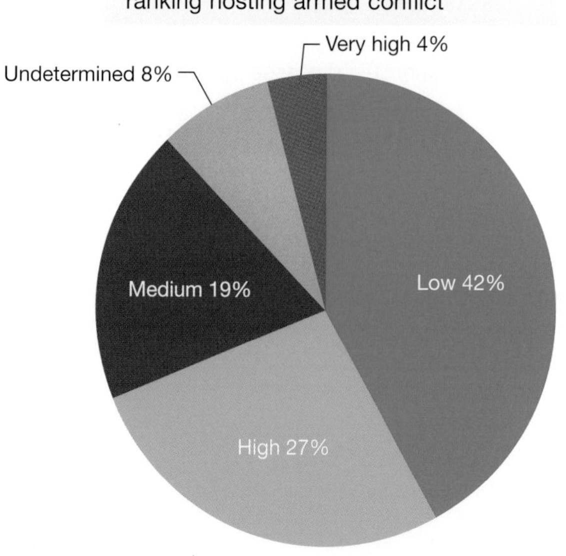

FIGURE 3 World military spending

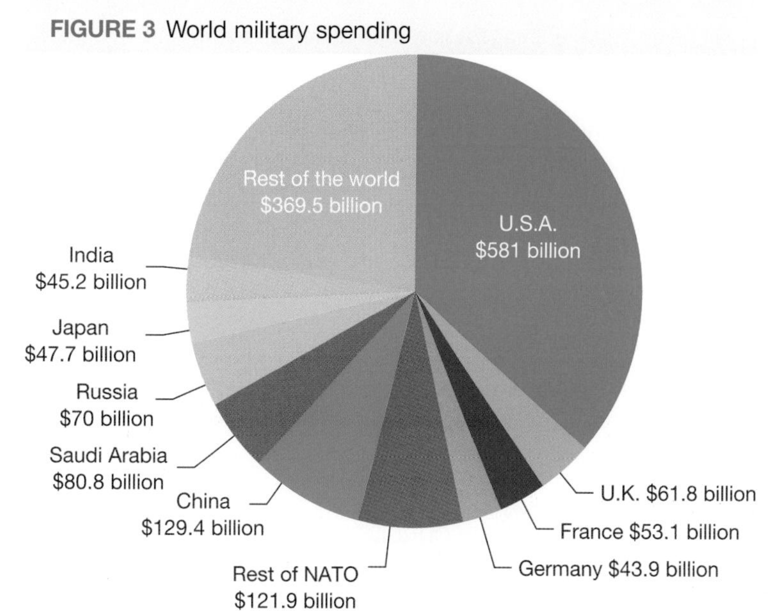

Source: IISS, *The Military Balance 2015.* All figures in US dollars

FIGURE 4 Working to avoid conflict

12.2 Activities

To answer questions online and to receive **immediate feedback** and **sample responses** for every question, go to your learnON title at www.jacplus.com.au. *Note*: Question numbers may vary slightly.

Remember

1. Outline three **classification** systems (or ways) used to identify conflicts.

Explain

2. (a) What classification is used on the figure 1 world map of conflicts in 2015?
   (b) Which ***places*** have experienced most deaths by armed conflict?
   (c) Which continent has the greatest number of deaths from armed conflict?
   (d) Which countries have fought in wars only outside their own borders in the twenty-first century?
   (e) Describe the distribution pattern of conflict in the early twenty-first century.
   (f) Is there a relationship between the number of deaths from conflict and the involvement of diplomatic missions? Provide evidence for your response.

3. Using figure 2, suggest how the HDI ranking of countries affects the likelihood of conflict.

Discover

4. What events in late 2010 would ***change*** the world map of conflict in the twenty-first century?
5. Why do developed countries of the world contribute vast quantities of money to military spending (see figure 3)?

Predict

6. On a blank world map, colour and label what you consider might be the areas of armed conflict in 2025. Justify your map.

Think

7. Will the world ever be without conflict? Consider the viewpoints of both an optimist (a person who sees hopefulness) and a pessimist (a person who takes the worst view).

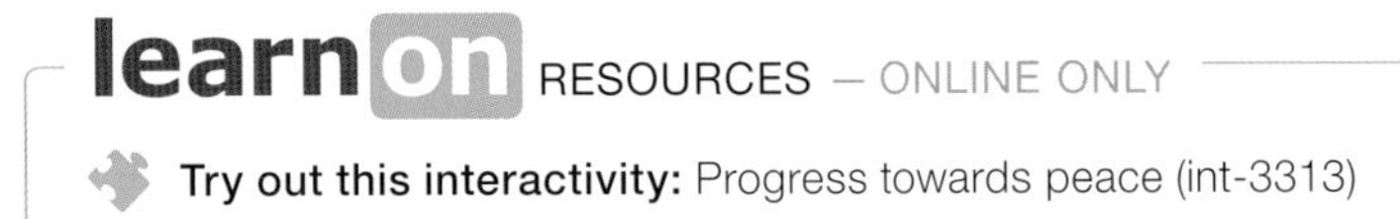

RESOURCES — ONLINE ONLY

Try out this interactivity: Progress towards peace (int-3313)

12.3 Why is there conflict over land?

12.3.1 The Gaza Strip

Families have a strong bond with their homelands. Conflicts over land are about who the land belongs to. These conflicts stretch over long time periods, from time to time erupting into hostilities. The question is: who has the right to the land?

FIGURE 1 The Gaza Strip

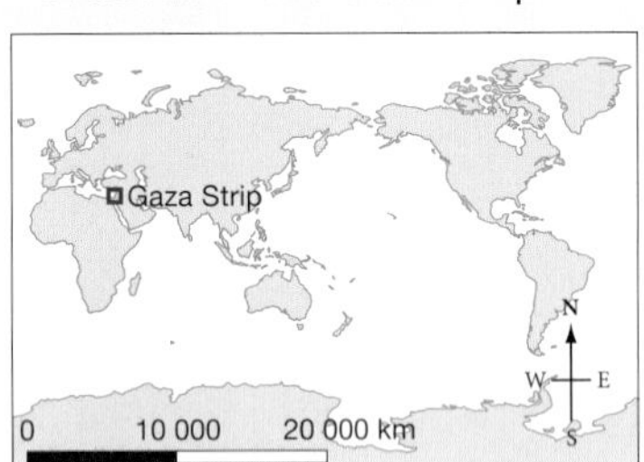

Source: Spatial Vision

12.3.2 How long has there been conflict in the Gaza Strip?

The Gaza Strip is one of the most densely populated places in the world — more than 5000 people per square kilometre in an area 40 kilometres long and eight kilometres wide. This strip of land on the south-eastern end of the Mediterranean Sea lies between the borders of Egypt to the south-west and Israel, defined by the ceasefire lines following the 1948 Arab–Israeli War. Palestinians came to this area as **refugees** when part of their traditional homeland was incorporated into the new state of Israel. Over 50 years later hostilities between the Palestinians and Israelis are ongoing.

12.3.3 Conflict has fragmented lives

Figure 2 shows how people's lives have been affected by the hostilities in the Gaza Strip.

- Since 2007 GDP per capita has been reduced by 50 per cent.
- Forty-three per cent of Gaza's workforce, including more than 60 per cent of its youth, is unemployed.
- Ninety per cent of schools in Gaza run on double shifts, with class sizes averaging 38 students.
- Forced labour is estimated at about 104 000 children, some as young as six years, working collecting building rubble, in factories, in garages or street selling.
- Only 45 per cent of the electricity demand is met; there are blackouts of 12 to 16 hours per day.
- Medical services relying on electricity for surgical equipment are greatly affected; it can take 18 months to receive elective surgery.

FIGURE 2 Impacts of the Israeli blockade of the Gaza Strip, which has been in place since 2007

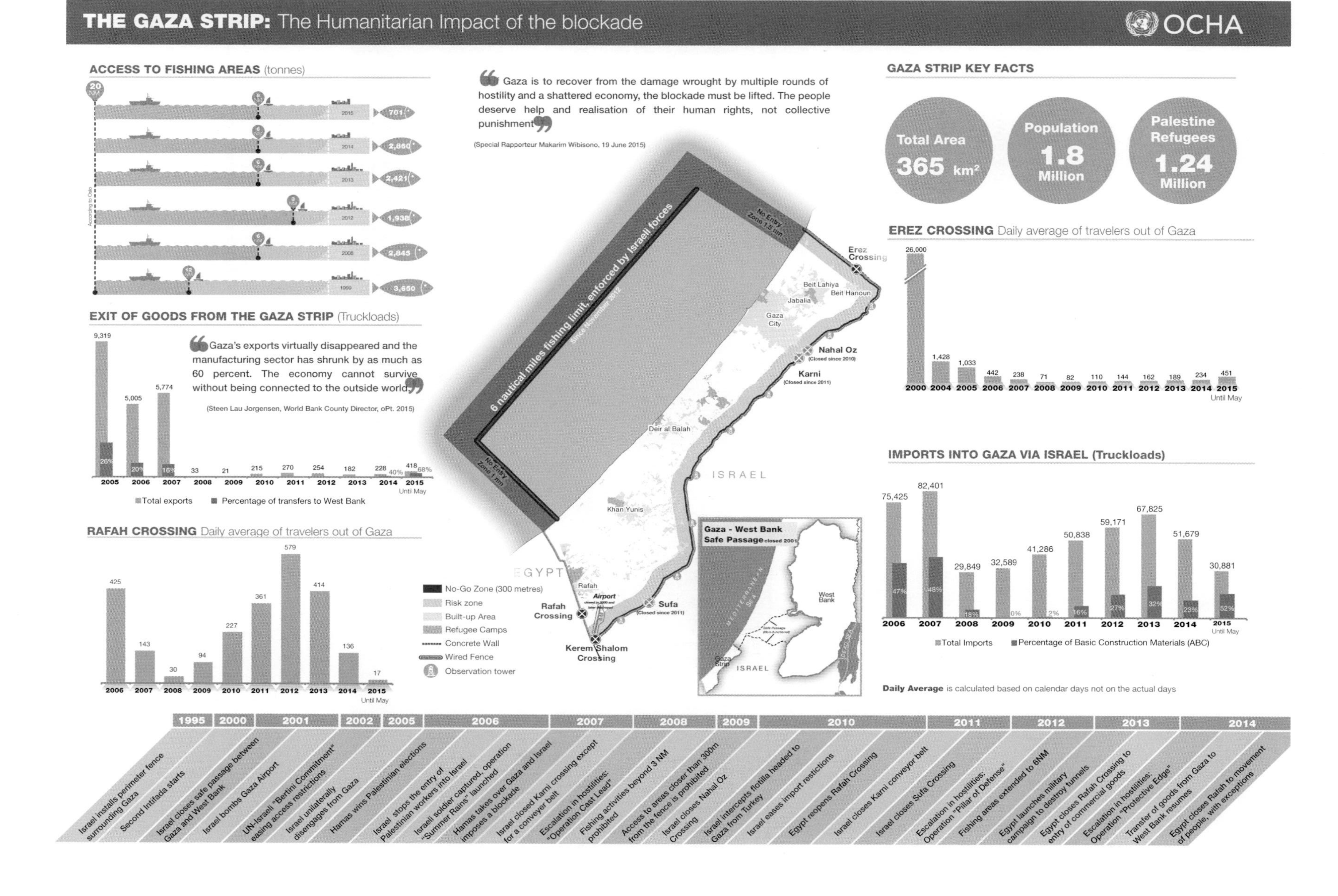

- More than 60 per cent of Gazan homes are supplied with piped water for six to eight hours once every two to four days.
- Up to 90 million litres of partially treated sewage is pumped into the Mediterranean Sea every day.
- Farmland within the No-Go Zone is inaccessible, and within the Risk Zone (1.5 kilometres from the barrier) farmers can be fired on by the Israeli army.
- It is estimated that 75 000 metric tonnes of food cannot be produced due to the limited access to land.
- Forty-four per cent of Gazans are food insecure, and about 80 per cent are aid recipients.
- Less than one per cent of materials required for housing repairs after conflict, and for population growth, have entered Gaza.

12.3.4 What does the future hold?

During 2012, the Rafah Crossing from Gaza into Egypt was opened by Egypt's new government. No longer did the people of Gaza need the cross-border tunnels as their supply line for goods. But in 2014 a change in the Egyptian Government saw the crossing closed again. Community spirit remains high as tensions continue (see figure 3).

FIGURE 3 Families visit the beach on weekends in the Gaza Strip.

12.3 Activities

To answer questions online and to receive **immediate feedback** and **sample responses** for every question, go to your learnON title at www.jacplus.com.au. *Note*: Question numbers may vary slightly.

Remember

1. Claiming territory (land) as a resource is at the heart of the conflict in the Gaza Strip. Why do you think this strip of land that is 40 kilometres long and 8 kilometres wide is so important?
2. Outline the significance of border crossings for the Gazans.

Explain

3. Use figure 2 to answer the following:
   (a) How does the denial of access to fishing grounds affect the Gazans?
   (b) Food security is affected by a lack of access to farmland. What activities prevent use of the ***environment?***
   (c) In 2011, Israel closed the border crossings to workers and supplies. How did this affect employment, exports and daily life? Use the **Israel** weblink in the Resources tab to find out more.
4. Figure 3 seems unusual in a region of conflict. How ***sustainable*** is life in the Gaza Strip?

Discover

5. Use the internet to research the impact that either electricity cuts of 12 to 16 hours per day or an irregular water supply have on Gazans' wellbeing.
6. Use the **UDHR** weblink in the Resources tab to access an illustrated book version of the Universal Declaration of Human Rights. Identify those rights that you believe may be violated in this region (in both Israel and the Gaza Strip).

Predict

7. What might be the 'stumbling blocks' to a peaceful resolution in this region?

Think

8. How are the lives of the Israeli people affected by the long-term conflict with the Gaza Strip?

Explore more with these weblinks: Israel, UDHR

12.4 How can conflict change regions?

12.4.1 The Arab Awakening on the move

The year 2011 saw an increased level of global conflict, particularly across the Arab region, where life had changed little for over 40 years. Internet access made Arabs realise that people across the world were more interconnected than them. Arabs wanted work, a future for their children and to be able to live with freedom.

Tunisia was the starting place for change (see figure 1). Across the region tensions grew. Modern technology, particularly the internet, had shown Arabs how the rest of the world lived; in particular, that people could have freedom of speech, choice with their lives, and expectations of the government to provide basic services. In Egypt, Cairo youth gathered at the beginning of 2011 (see figure 2). This massing of youth included not just the poor, but also the politically connected, educated middle class. Mobile phones and social media spread the word. Egypt and Tunisia experienced mass civil revolts. Libya watched and clamped down heavily on its people. Violence broke out in the cities along its Mediterranean coastline and Libya was thrust into civil war (see figure 1).

In many of the Arabian Peninsula countries, wealth from oil exports was used by the governments to bring a halt to protests. Salaries were increased, public services such as health care were provided, education became accessible and reform was promised. However, in Syria the government resisted the rebels, playing on time to wear down the internal revolt. Syria does not export oil, and oil forms only 20 per cent of its GDP. Civil war broke out and is ongoing today.

FIGURE 1 Changing times in the Arab world

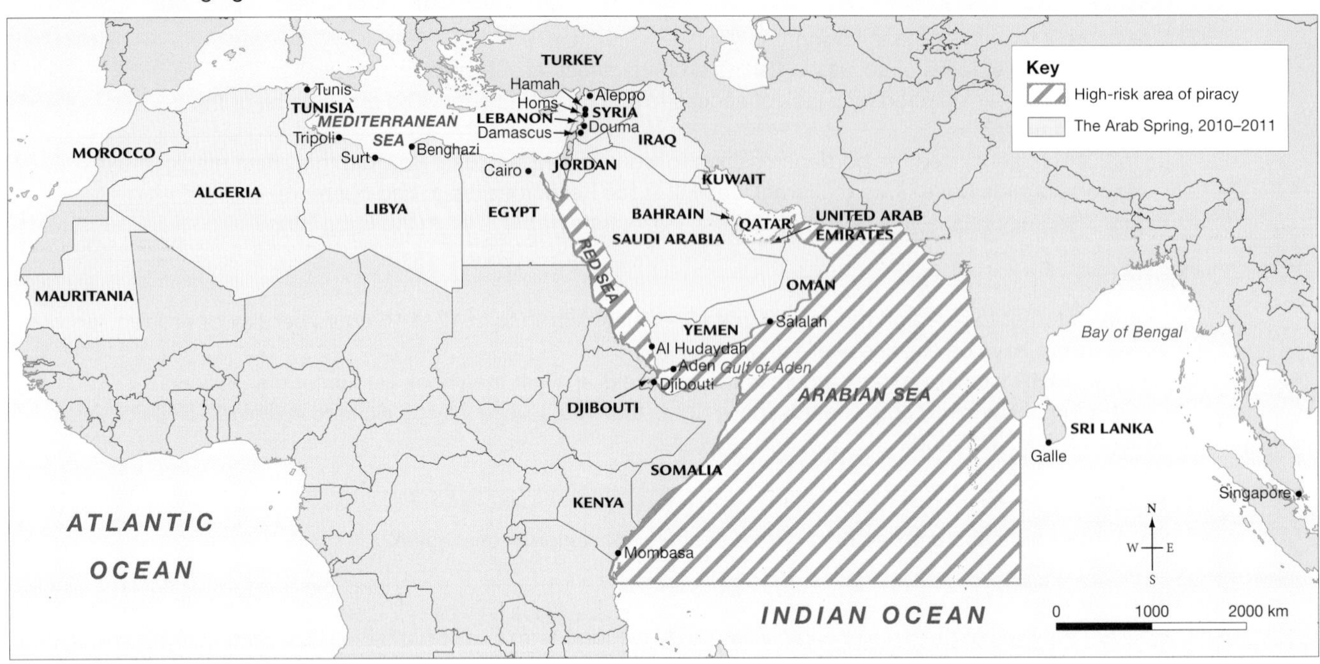

Source: EPOS - International Mediating and Negotiating Operational Agency, http://eposweb.org/; Lowy Institute, http://lowyinstitute.org/

1. Somali pirates

- Piracy peaked in 2011 — 176 attacks
- No ships hijacked since 2013
- Shipping routes and travel times are returning to normal

2. Tunisia

17 December 2010–14 January 2011

- Tourism collapsed
- Food prices increased
- Unemployment increased
- Influx of refugees from Libya (almost 2 million)
- Economic reforms brought some stability

3. Egypt

25 January 2011–11 February 2011

- Tourism collapsed
- Food prices increased
- Unemployment increased

4. Libya

17 February 2011–23 August 2011

- Oil infrastructure damaged by war, reducing production by 90 per cent
- Food prices increased
- About 40 000 dead; 434 000 displaced; 1 million refugees (most fled to Tunisia)
- Militia control large areas and government forces control the remainder, resulting in chaos
- Not prepared to work with Western efforts for peace deals

5. Syria

16 March 2011–ongoing in 2017

- 400 000+ dead and 6.3 million internally displaced persons
- Urban areas destroyed and besieged across the country
- Northern forests burnt from shelling
- World Heritage sites destroyed and looted
- Children psychologically affected
- Cross border clashes with Turkey
- In early 2017 there were about 5 000 000 refugees.

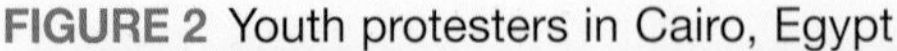

FIGURE 2 Youth protesters in Cairo, Egypt

12.4.2 Impacts on places

Figure 1 shows the negative impacts of conflict on places in the region. There are positives too. In Tunisia people are proud to have stood up and made a change. The leadership has changed and political debate is heard in the streets, cafes and in the media. Egypt has new political players and elections have been held, but unrest continues. The shops of Libya are well-stocked and media outlets have appeared, but there are many opposing viewpoints on the way forward. For Syria, rebel-controlled communities are sustaining themselves.

How was tourism affected?

In Tunisia, tourism fell 31 per cent over a few months (see figure 3). Tour companies evacuated tourists from Cairo and along the Nile River, and dropped Egypt from tour programs. The companies feared lawlessness and confrontations at major tourist sites. Jordan, free of upheaval, lost 22.2 per cent of its tourists. Tour packages of which it was a part were dropped. Other stable economies, such as Dubai (11 per cent) gained tourists. Tunisia is working with European tourism companies to win back its share of the market. Egypt is keen to increase tourism to environmental sites outside Cairo. The GDP of countries reliant on tourism was sharply affected.

FIGURE 3 Trends in tourism across the Arab world

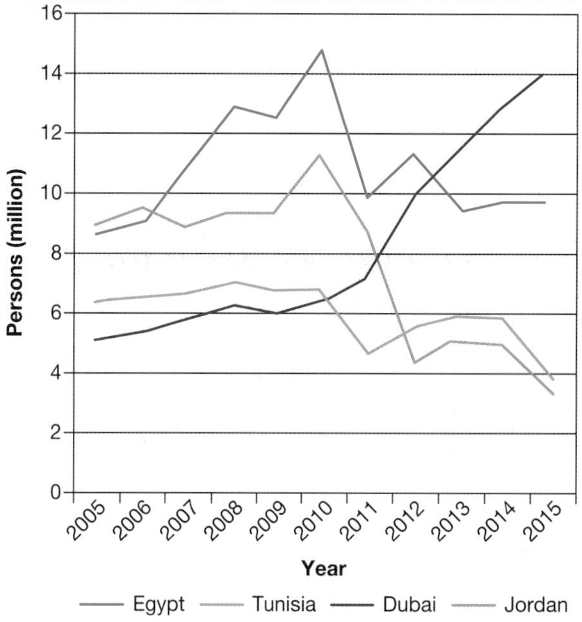

Pirates ahoy!

The oil reserves of this region supported change. Rebels, however, threatened the oil shipping routes, especially through the Persian Gulf and the Gulf of Aden. Unemployed youth of Somalia, hard hit by economic conditions, civil war and drought, turned to piracy off their country's long coastline (see figure 1). International governments sent naval forces to the region to ensure an ongoing oil supply to Europe and the United States. Ships travelled in convoy with armed guards (even the tourist ships). Shipping is now returning to normal.

12.4 Activities

To answer questions online and to receive **immediate feedback** and **sample responses** for every question, go to your learnON title at www.jacplus.com.au. *Note*: Question numbers may vary slightly.

Remember

1. Over what time periods did conflict occur in each of the countries involved in the Arab Awakening?
2. Use the **Timeline** weblink in the Resources tab to find out in what ways the type and ***scale*** of conflict ***changed*** across the region.

Explain

3. Using figure 1 and the text in section 12.4.2 'Impacts on places', rank from highest to lowest the level of impact of conflict on countries in the region. Justify your ranking.
4. Using the data in figure 3, describe the trend in tourism seen with the ***change*** in conflicts across the region.

Discover

5. Using the internet, explore in greater detail piracy on the seas in the Middle East. Why is Somalia a suitable ***place*** to launch pirate attacks from?

Predict

6. Tourism plays a significant role in the region's economy. What part will tourism play in the region's economy in the future?

Think

7. For a region so rich in oil, conflict seems unusual. Choose two countries from the region and research their living condition indicators for wellbeing. Did these living conditions lead to conflict? Support your answer with evidence.

12.5 How does conflict affect people?

12.5.1 People on the move

Global conflict saw more than 60 million people flee their homes in 2015 (see figure 1), according to the United Nations High Commission for Refugees (UNHCR) report; many have fled more than once. These people feel they have no choice but to move. Each year the number of people on the move is different. Each year different places are in conflict. In 2015 one in every 122 people were forced to flee their home.

FIGURE 1 Who is hosting the world's refugees?

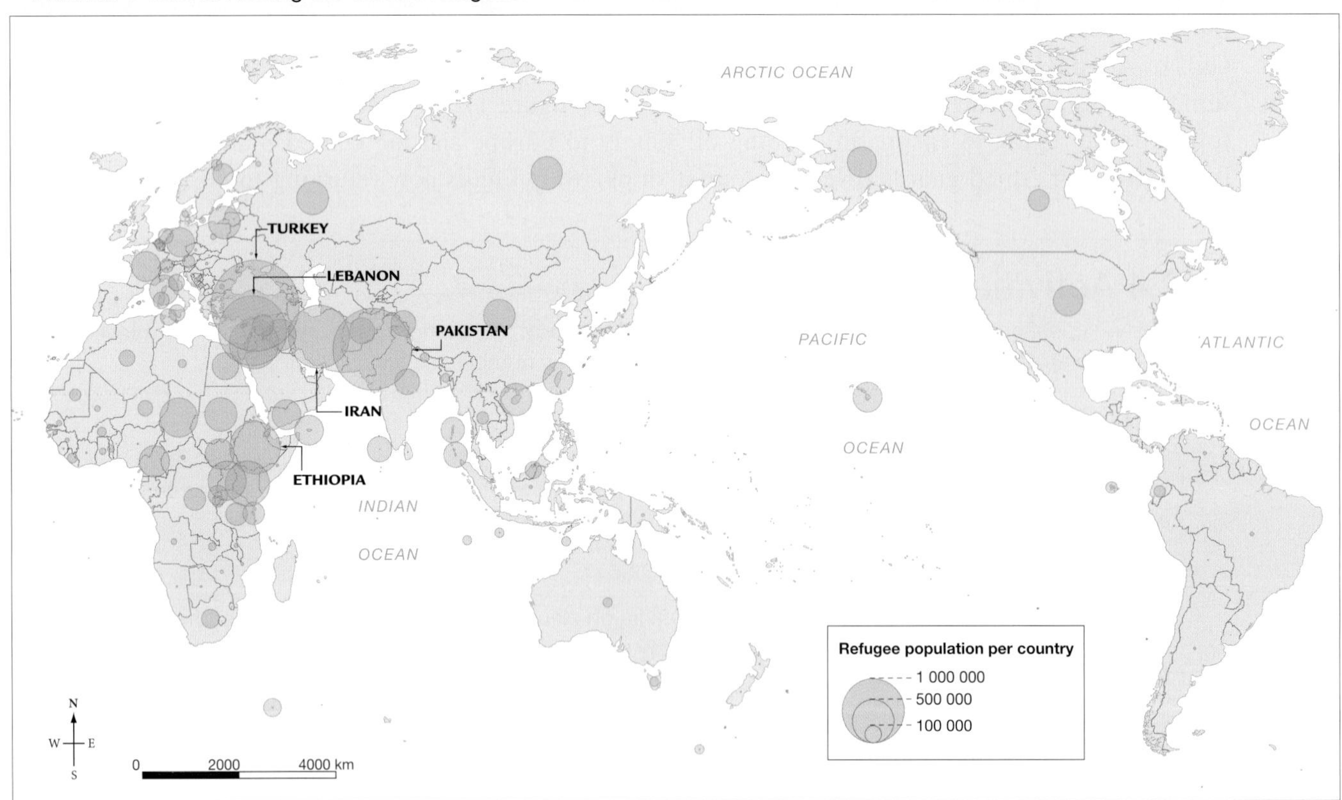

Source: United Nations High Commissioner for Refugees

12.5.2 Who moves?

In 2015, the largest group of people on the move was the 38.2 million **internally displaced persons (IDP)**. In addition, some 20 million refugees left their country of origin, and 80 per cent of these arrived in a neighbouring country. Afghans make up one in every four of the world's refugees, 95 per cent of whom are in Pakistan and Iran. More than 10 million remain **stateless people** for long periods of time.

Most of those who flee have experienced conflict, although some are 'environmental refugees', especially those escaping prolonged drought. Others flee as 'economic refugees', finding the living conditions of their country unacceptable and choosing to seek a better lifestyle.

FIGURE 2 Major refugee groups and their host countries, 2013–2015

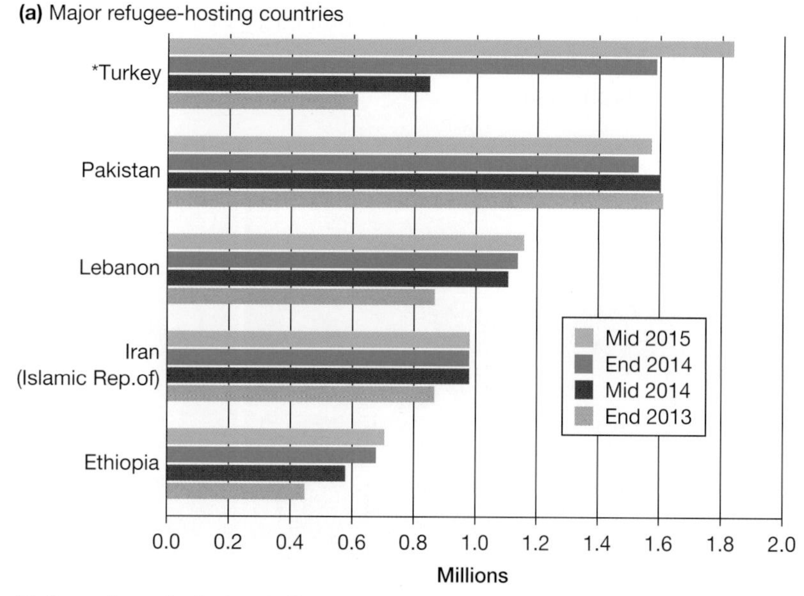

*Refugee figure for Syrians in Turkey is a government estimate.

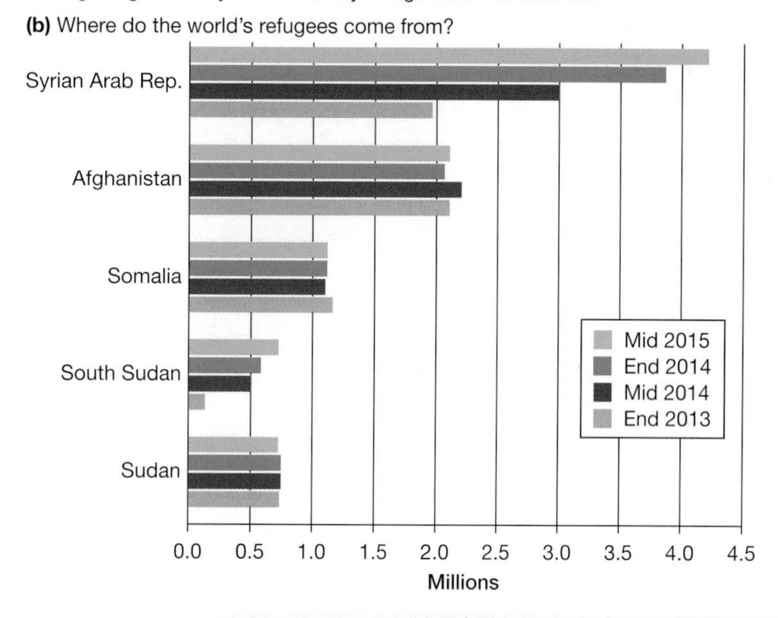

12.5.3 Life as a refugee

People who flee often are forced to make the decision quickly. These people are distressed by the situation that they find themselves in and simply take with them possessions that can be carried — every family member carries something (see figure 3). People walk to safety or cram into vehicles. Families and friends are torn apart.

The UNHCR is the international organisation charged with leading and coordinating international action for the worldwide protection of refugees. The UNHCR monitors the movement of refugees. As the need arises, the UNHCR, NGOs and governments respond and establish camps across the borders from the conflict to accommodate people in the short term.

People of concern to the UNHCR are predominantly women (48 per cent of refugees) and children (46 per cent are under the age of 18). Camp life is basic. Women and children are at risk.

FIGURE 3 Refugees arrive at Dadaab refugee camp, Kenya

The Dadaab refugee camp (see figure 4) in north-east Kenya is the largest complex in the world, and is home to more than 330 000 mostly Somali refugees. Life is no longer 'normal' for them. In 2015, fewer than 100 000 refugees returned home voluntarily (less than 1 per cent of the refugees), the lowest number in 30 years. Many refugees spend more than 4–5 years in camps awaiting resettlement or for peace to return in their homelands.

FIGURE 4 Dadaab refugee camp, Kenya

- Space is restricted.
- Privacy is lacking.
- Essential food items only are provided.
- Cooking facilities are basic.
- Water is provided at a central location.
- Sanitation can be limited.
- Medical support is stretched to its limits.
- Education facilities are lacking.
- Time is on everyone's hands; there is no work.
- Family values need to be maintained.
- Violence against women and children can spread in the camp.

12.5 Activities

To answer questions online and to receive **immediate feedback** and **sample responses** for every question, go to your learnON title at www.jacplus.com.au. *Note*: Question numbers may vary slightly.

Remember

1. What is the difference between a refugee and an asylum seeker?
2. What is the role of the UNHCR?

Explain

3. Figure 1 shows the global distribution of people on the move. Summarise the ***places*** in the world generating most refugees.
4. Using figure 2 (a) and (b) answer the following questions:
   (a) Which region of the world provides most refugees?
   (b) Will the number of refugees from Africa increase? Explain your answer.
   (c) How have these graphs ***changed*** over the years 2013–2015?

Discover

5. Research the Za'atari refugee camp established in 2012 in the desert of Jordan. Write an extended paragraph on how it has evolved into a 'new city', and include two images to illustrate your comments.

Think

6. A mother in a refugee camp speaks to the media about the plight of her family. Working with a partner, create the interview questions and responses. Present the ideas using technology.

learnON RESOURCES — ONLINE ONLY

Try out this interactivity: Leaving home (int-3314)

12.6 A case study: How is wellbeing affected in the Syrian Arab Republic?

12.6.1 Introduction

The civil war in the Syrian Arab Republic is becoming a long-term event. It began in 2011 as part of the uprising of its people against the government in the Arab Awakening. Civil war does not mean that everyone living in the country is involved in the war, but everyone living in the country is affected by the war. Life and wellbeing is changed.

FIGURE 1 The movement of IDPs in the Syrian Arab Republic

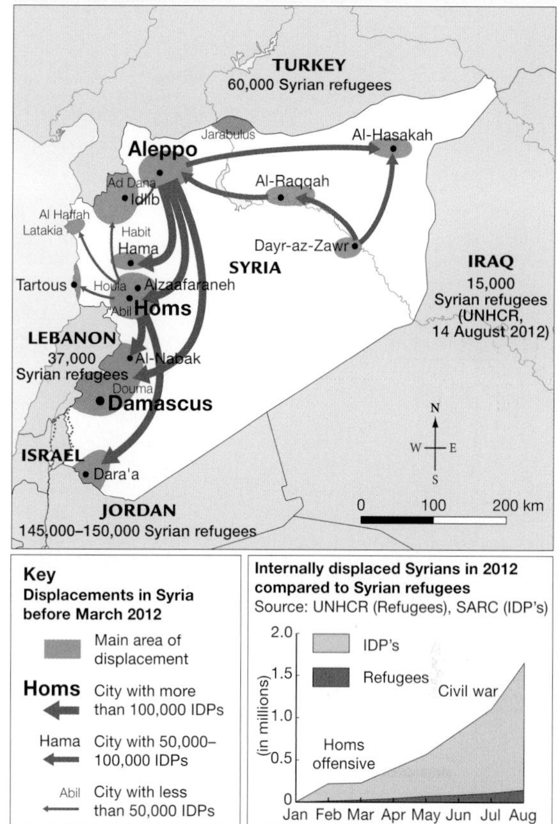

Source: © Internal Displacement Monitoring Centre

12.6.2 How has the Syrian population been affected?

The Syrian people had four choices when government hostilities broke out against their protests in the Arab Awakening: join the Syrian Arab Republic's army, join the rebels, leave the fighting zones, or stay in their homes.

By 2017, 6.3 million Syrian people — especially women, children and young men — had fled areas of conflict to somewhere else within the Syrian Arab Republic as a first option, becoming internally displaced persons (IDPs) in their own country (see figure 1). These people make up one in five of all IDPs globally — this is the largest displaced population worldwide.

In 2013, the Assad-led government declared 'surrender or starve' to its people and began sieges on key cities, particularly the capital city, Damascus, and large populations to the north of the country. Sieges 'lock' people within a city's boundaries, preventing easy movement out and denying entry to the city. In early 2016 it was estimated that between 390 000 and 1.9 million people were trapped in cities.

Multiple opposition groups formed in a wider context throughout the region and began to have a presence in the Syrian Arab Republic. Some of these groups have a religious base and others are terrorist cells. Since then the pressures of conflict in different areas have seen many IDPs flee again, often at night to avoid detection. Some of these people achieved a border crossing into surrounding countries to become refugees (see figure 2), massing in 'tent cities' on the border with Turkey. The level of liveability for the Syrian people declined.

FIGURE 2 Numbers and locations of people fleeing conflict in the Syrian Arab Republic, February 2016

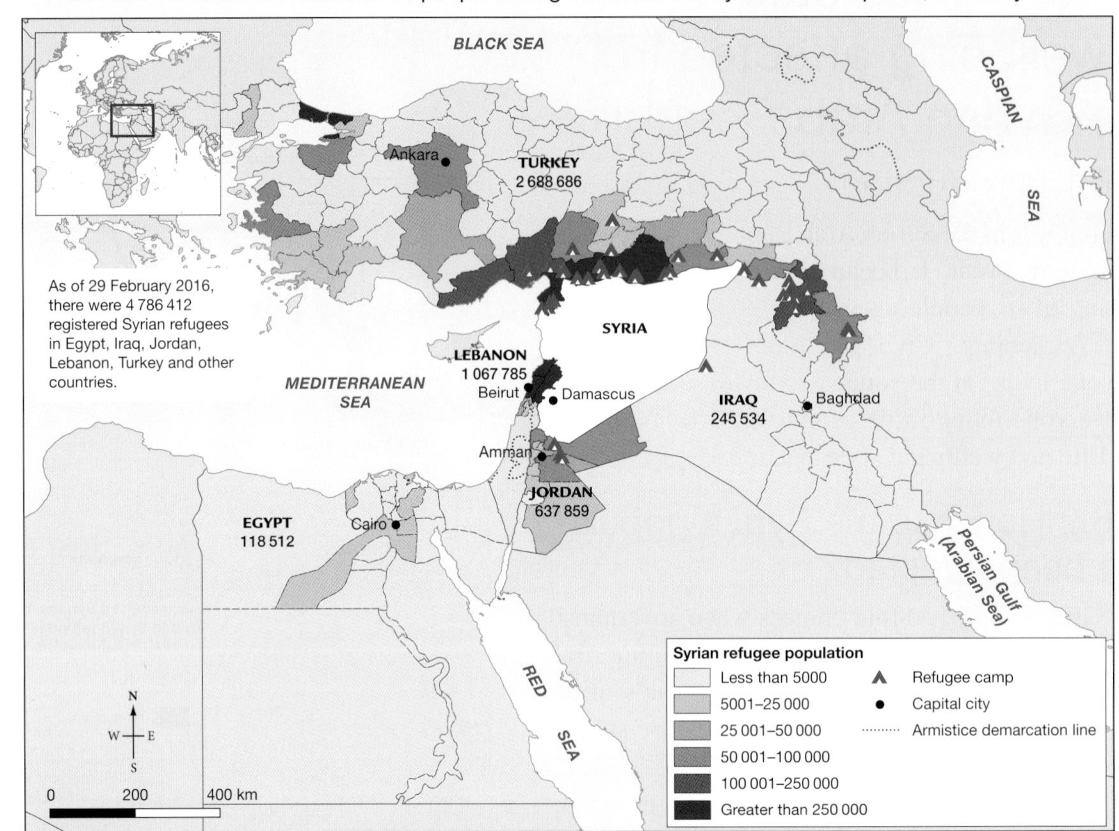

Source: UN High Commissioner for Refugees (UNHCR), Syria Situation Map (as of 29 February 2016) available at: http://www.refworld.org/docid/56de7fdd4.html [accessed 22 June 2016]

12.6.3 Are homes safe places in the Syrian Arab Republic?

In figure 3, Damascus shows the greatest change in liveability from 2010 to 2015. Living conditions have changed: safety in homes is at risk, there is food insecurity and children are traumatised. Global relief organisations estimate that 13.5 million Syrians need humanitarian aid.

The street-to-street fighting that is a key element of civil war has destroyed buildings, including houses, in major cities of the Syrian Arab Republic such as Aleppo and Homs. Public services such as electricity, running water and gas supplies no longer operate. There is no transport system. Without oil, people rely on wood fires for heating and cooking, but this has brought about local deforestation. War continues to injure and kill more than 320 000 local people who have remained in their homes. In late 2014 the United States, UK and France began airborne bombing of cities to reduce the threat of rebel groups; Russia began air strikes in late 2015. These bombing raids further destroyed buildings.

FIGURE 3 Change in liveability score 2010–2015, showing the change in Syria

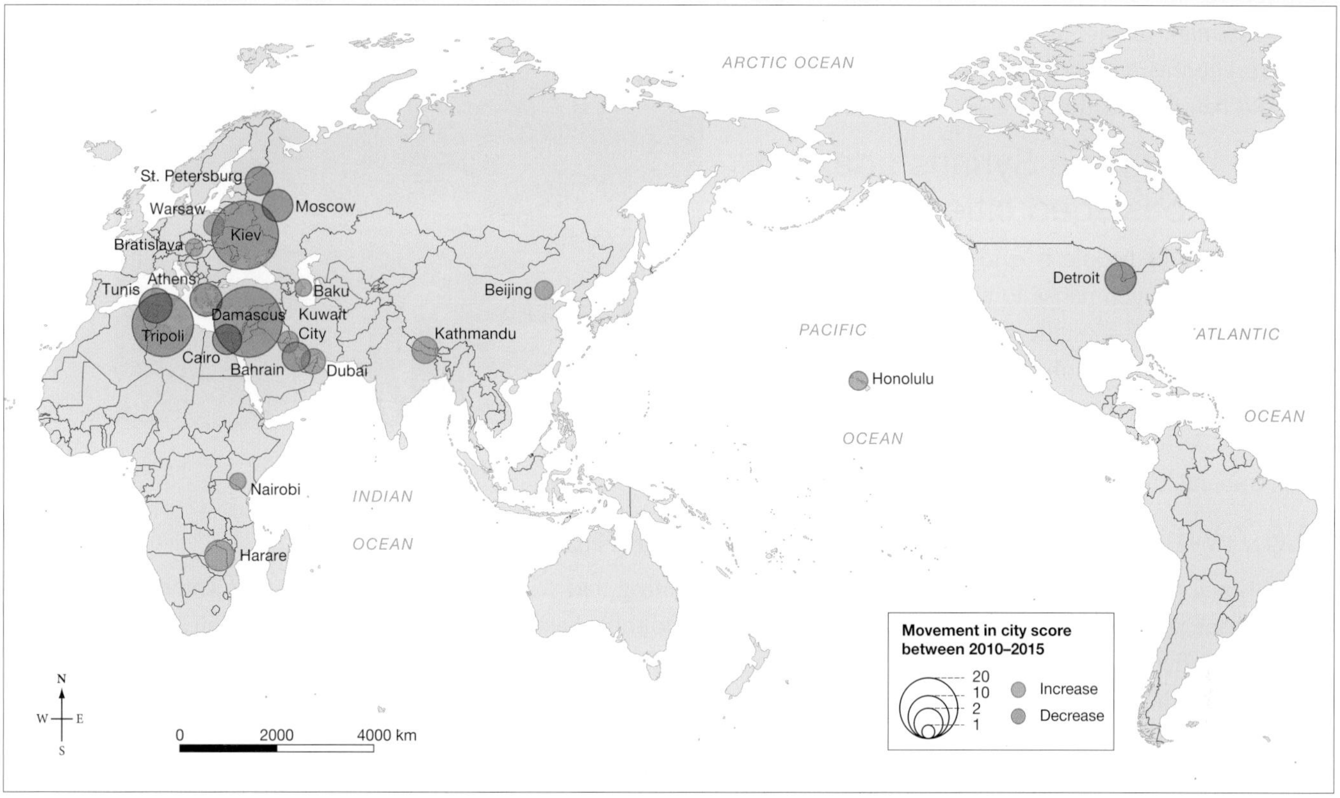

Source: Economist Intelligence Unit

12.6.4 Is there enough food?

Food insecurity is a daily issue for the war-torn areas of the Syrian Arab Republic. It is not safe to be outside for too long tending plants. Transport cannot get to the besieged cities with tinned and fresh foods. Only about one per cent of the people requiring food aid received food in 2015. Reports of malnourishment surfaced in 2016 at besieged Madaya (where 40 000 people are trapped) when social media began reporting that families were stripping the trees of leaves and boiling these to provide one meal a day. Aid organisations negotiated with the Assad government to be allowed to enter the city — and be protected from attack — with a convoy of trucks bringing food, but this is only a short-term solution.

FIGURE 4 A Syrian refugee makes a cooking fire.

12.6.5 How are children affected?

Children in any war-torn zone have their lives dramatically changed. The streets are no longer playgrounds. Education is disrupted or abandoned for months or years. Fear enters their lives — the sounds of aircraft, bombing and shooting punctuate their days and nights. Deafness in children becomes a problem. Families are torn apart, with some people fleeing and others staying.

FIGURE 5 The destruction of their homes is just one of the many significant effects the war has had on the children of Syria.

Children miss their friends. Young males and men are recruited for the fight by both sides of the conflict with blackmail, threats, fear and propaganda. Life is insecure, confusing and scary; children 'literally' grow old before their time.

12.6.6 How have Syrians adapted to life in besieged cities?

The resilience of people is evident in these besieged cities as people acclimatise to a basic lifestyle. Innovation is required — static bicycles are pedalled to generate power for mobile phones; medicines are produced from home remedies; plastic is burnt to extract oil derivatives; and rooftop gardens produce small amounts of vegetables.

FIGURE 6 Residents of Homs go about daily life among the destruction of homes.

12.6.7 What are the costs to the Syrian Arab Republic?

International peace talks have brought ceasefires in the fighting, but will peace ever be achieved? The costs to the Syrian Arab Republic are immense. The soul of the country has been changed forever. So many of its people have fled — more than 4.6 million are refugees and 6.6 million are IDPs. Some of those who fled will return to the Syrian Arab Republic, but they too have changed as a result of the experiences they have been through. And how will those who remained perceive the returnees and those who stay away? Families have been forever changed. The cities will take years to rebuild; more than 50 per cent of some cities has been destroyed. Services and food supplies will need to be re-established. Children will have years of schooling to catch up on. The Syrian Arab Republic has been changed.

12.6 Activities

To answer questions online and to receive **immediate feedback** and **sample responses** for every question, go to your learnON title at www.jacplus.com.au. *Note*: Question numbers may vary slightly.

Remember

1. Who are the sides in the civil war in the Syrian Arab Republic?
2. What was the government policy that forced great hardship on the Syrians? Explain its implications.

Explain

3. Using figure 1:
   (a) Describe the movement of IDPs within the Syrian Arab Republic.
   (b) Explain why the flows are both to and from some cities.
4. Figure 2 shows the cross-border movement of Syrian refugees.
   (a) Rank the neighbouring countries from highest to lowest in the number of Syrian refugees registered in each country in February 2016.
   (b) Is the distribution of Syrian refugees even across the neighbouring countries? In particular, refer to the situation in Turkey.
   (c) Suggest why the refugee camps are found along the borders.

Discover

5. (a) Copy and complete the following table using data from the current Human Development Report.
   (b) Using the indicators in the completed table, describe life in Syria.
   (c) From the table, is there an indication of why refugees opt to go to Turkey and Lebanon in preference to other cross-border countries? Support your answer using statistics.

Country Indicator	Syrian Arab Republic	Turkey	Jordan	Lebanon	Iraq	Egypt
HDI ranking						
GNI per capita						
Public expenditure on education (% of GDP)						
Local labour market (% answering good)						
Public health expenditure (% GDP)						
Internet users (% of population)						

6. Use the **Children's stories** weblink in the Resources tab to listen to some children describe their life in Syria during conflict. Which parts of their secure, pre-conflict life do the children miss?

Think

7. As a small group activity, use the internet to create a photographic essay of at least six images of life in Syria during conflict. Choose the images carefully and include some qualitative and quantitative data by adding text to each image to support your choice of images and to show your understanding of the situation and the impact on the wellbeing of Syrians.

Predict

8. Is the wellbeing of the Syrian refugees likely to improve in a cross-border country? Give reasons for your answer.

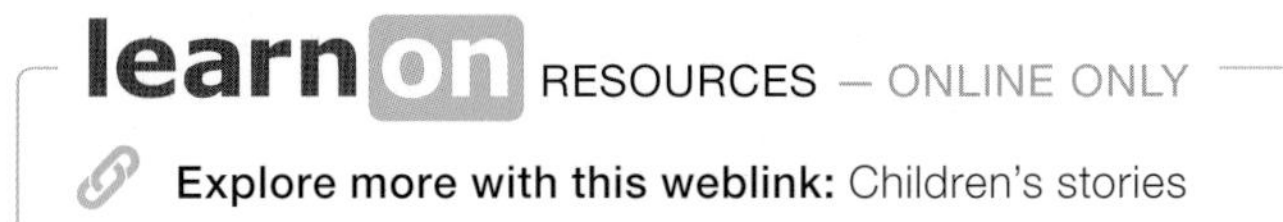

RESOURCES — ONLINE ONLY

Explore more with this weblink: Children's stories

12.7 What is the impact of people moving through Europe?

12.7.1 Who are asylum seekers?

Refugees flee conflict and cross a border into another country to seek relief from the trauma of war and make a home elsewhere. All refugees (those who cannot return home due to a fear of persecution because of race, religion, nationality or membership of a social group) have been **asylum seekers**, but not all asylum seekers are found to be refugees. Asylum seekers have either not satisfied the UNHCR criteria to be a refugee or have gone outside of the process to seek a place to live.

12.7.2 Where are these people coming from to Europe?

Most of those arriving in Europe are fleeing the civil war in the Syrian Arab Republic, with other significant numbers arriving from the ongoing chaos in northern Africa since the Arab Awakening. The Syrians have fled through Turkey to reach the shores of the Aegean Sea, from which on a clear day the Greek islands of Lesbos and Kos can be seen a mere 4 kilometres away (figure 1). The Africans come across the Mediterranean Sea, particularly from Libya. However, movement across these waters is treacherous in small boats and dinghies; loss of life by drowning is high (there were more than 3770 deaths in 2015).

Greece's islands are the first point of arrival, where the refugees are fingerprinted, photographed and given a document allowing legal residency for 30 days in Greece. Greece does not accommodate the mass of people arriving on its shores. It is costly for the already poor country to rescue people from the seas and process their movement.

FIGURE 1 Refugees board a dinghy with all their possessions to make the short crossing from Turkey to Greece.

FIGURE 2 Tents of migrants and refugees in the port of Piraeus, Athens, Greece

12.7.3 Who helps the people move?

It is estimated that 90 per cent of refugees have their journeys organised by criminal gangs, including individual people smugglers and migrant smuggling networks across Europe. Money is extorted for the risky sea crossings (it probably costs more than A$3000 to cross from Turkey to Greece) and on trains within Europe. High prices are paid for accommodation and fake documents such as passports that allow refugees to apply for asylum elsewhere in Europe, especially in Germany and Sweden. People smugglers often instruct that when a coast guard ship is in sight the boat or dinghy should be destroyed to ensure the refugees' rescue, a meal and health checks before arriving on European soil. Figure 3 shows the movement of refugees throughout Europe.

FIGURE 3 The flow of people from northern Africa and west Asia across Europe

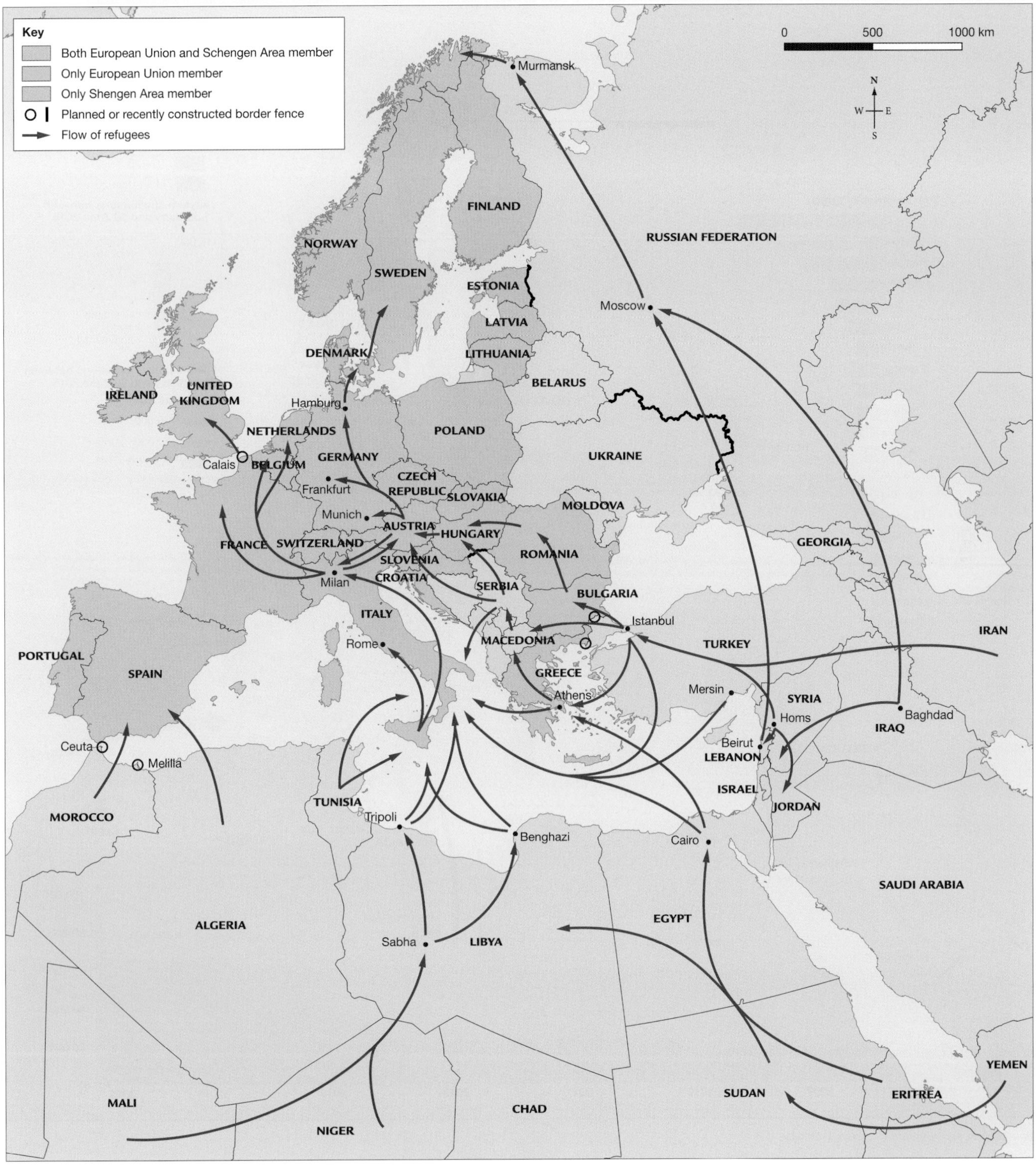

Source: Business Insider Inc

12.7.4 Where are the places that refugees and asylum seekers go?

Figure 4 shows that in 2015 Europe hosted 3.1 million refugees (22 per cent of the world's total), predominantly from the Syrian Arab Republic (1.7 million refugees). Of these numbers, more than one million refugees arrived in 2015.

FIGURE 4 Refugee and asylum-seeker numbers in Europe, 2015

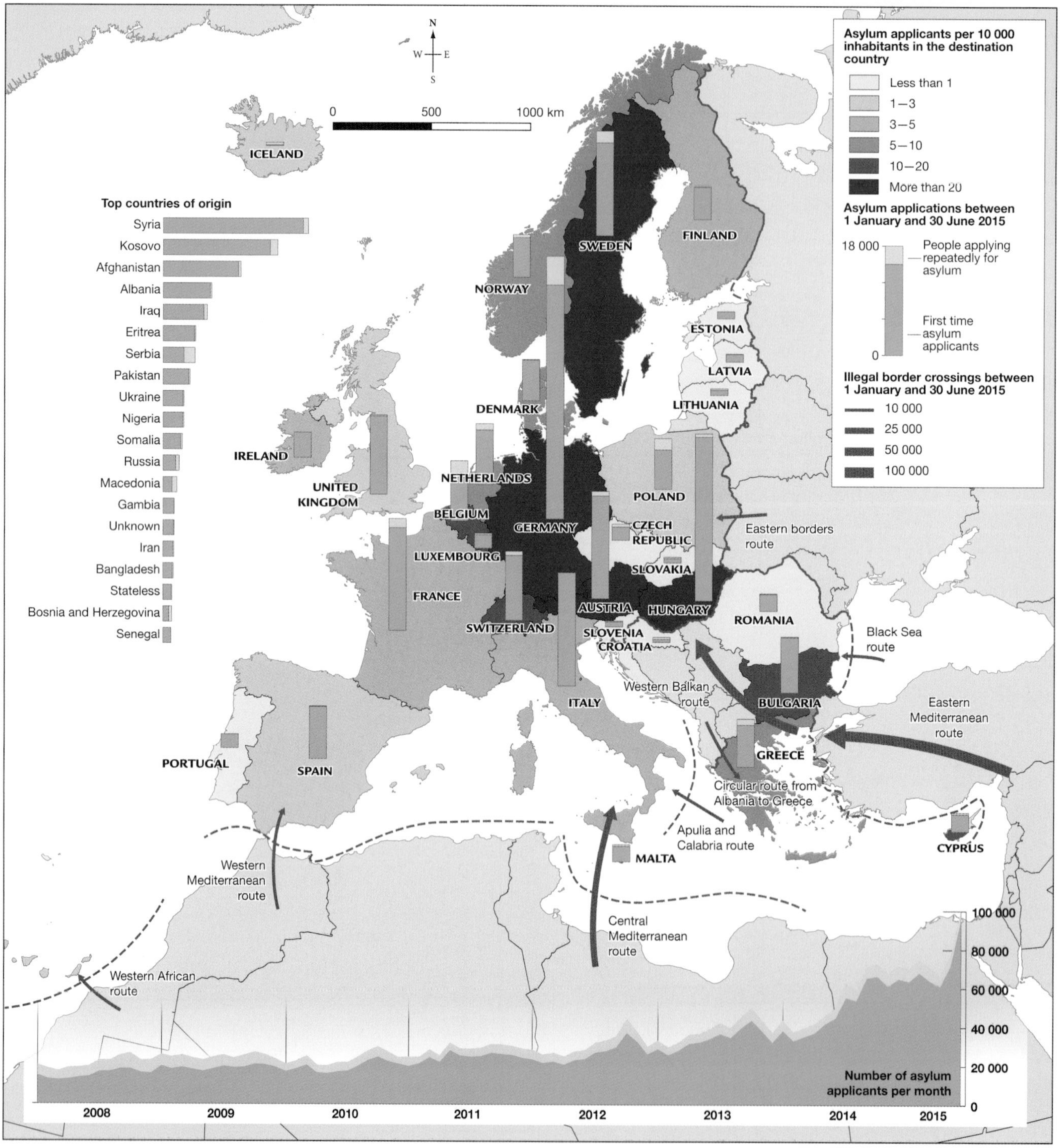

Source: Eurostat, Worldometers.info, Frontex.

12.7.5 How has the European region been affected?

In 2015, four times as many refugees arrived in Europe as in 2014. Germany, with its developed economy, high living standards and political compassion has been targeted as a place to go. The German community initially showed open-minded goodwill and generosity (see figure 5a), but in 2016 attitudes began to change; the numbers of migrants became overwhelming. By early 2016, 1.1 million people had sought asylum in Germany. Concerns about the impact on the German way of life began to grow, with issues such as housing availability and infrastructure pressure, as well as how people with different languages and cultures would live together being raised.

Sweden had a very open approach to asylum seekers, providing safety for people in need of protection (see figure 5b). Permanent residency permits were offered to those with appropriate documents. Accommodation, a small daily allowance, health care and schooling were provided. Early in 2016, Sweden announced tougher rules as it felt that it had reached its limit regarding the numbers of asylum seekers that it could take. Some scenes of violence and criminal activity had changed Swedish attitudes and expulsion of asylum seekers began.

Hungary saw itself as a stepping stone for those moving north, but the sheer number of people moving through the country along disused railway lines, on roads and across paddocks struck fear within the government. In late 2015, a 4-metre-high wire fence was erected along the border with Serbia and patrolled by police with tear gas and water cannons (see figure 5c), but refugees found gaps and cut holes to continue their movement north-west, or changed their path to go through Croatia.

FIGURE 5 (a) Welcome to Germany (b) Volunteers providing supplies in Sweden (c) The Hungarian fence (d) Sleeping at an Italian shelter (e) Tent city near Calais

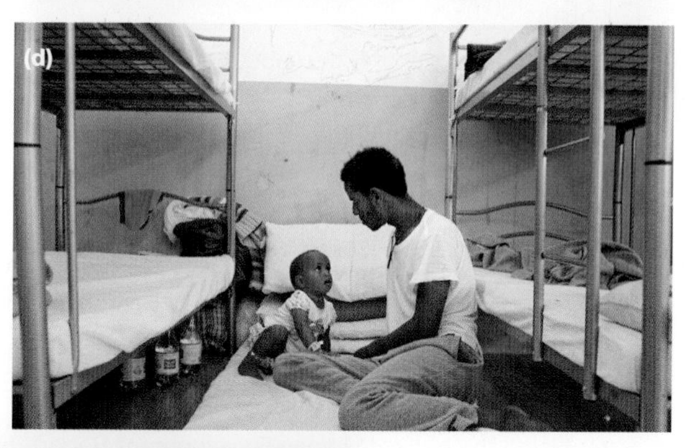

Italy, with its influx of refugees from northern Africa, has given the task of caring for the refugees to charities, companies, cooperatives and individuals; the government pays for food, health checks and psychology appointments and provides approximately A$3.85 per day as living expenses. Shelters are often substandard and overcrowded (see figure 5d). Italy hopes that the people will move on from the southern regions, through Milan and on to other European countries.

France has settled many of the northern African refugees within its cities. Most are French speaking from France's colonial dominance of northern Africa in the nineteenth century. Some refugees aim to reach Britain and strive to board ferries via the trucks transporting goods through the tunnel under the English Channel. Refugees established a tent camp city near Calais (see figure 5e) while they waited to attempt a crossing; authorities did not approve and in October 2016, amid protests and clashes, the camp was closed and dismantled.

12.7.6 Will the asylum seekers find a better life?

The European countries have tried to find a regional solution to the flood of migrants and now find themselves bickering with each other over decisions made within one country that affect a neighbouring country — a domino effect. Greece and Italy, as major entry points, feel the pressure as countries close their borders and restrict the on-flow of migrants. Some countries feel they have taken their 'fair share' of the numbers of asylum seekers and are turning away those that can't prove their status. Economies are stretched with the level of support required to provide basic needs for the refugees. In 2016 Europol set up a European Migrant Smuggling Centre to try to stem the tide of refugees. And yet the refugees keep coming.

12.7 Activities

To answer questions online and to receive **immediate feedback** and **sample responses** for every question, go to your learnON title at www.jacplus.com.au. *Note*: Question numbers may vary slightly.

Remember

1. What is the difference between a refugee and an asylum seeker?
2. What is the role of people smugglers in the mass movement of people across Europe?

Explain

3. Using figures 3 and 4, describe the scale of the journeys and direction of movement along the major routes taken by the refugees. Use country names in your answer.
4. Figure 4 shows the numbers seeking asylum (blue bars) and the number of asylum seekers per 10 000 residents of the host country (pink tones) in Europe for a six month period in 2015.
   (a) Which country had the highest number of applicants for asylum in that six month period?
   (b) Name five other countries that had a considerable number of asylum applicants.
   (c) Which countries have the highest ratio of asylum seekers to the population of their country?
   (d) Using the line graph, describe the change in trend of the number of asylum seekers per month from 2008–2015.

Discover

5. (a) Using the current Human Development Report, find the HDI ranking for each country.

Country	Greece	Italy	Hungary	Germany	France	Sweden
HDI Ranking						

   (b) How does the HDI ranking for the 'first ports-of-call' of the refugees compare to the other countries listed?
   (c) How might the HDI rankings help to explain the movement of the asylum seekers through Europe?
   (d) According to the HDI rankings, which countries might be best placed to cater for the wellbeing of large numbers of people on the move?
6. Research the meaning of the European Union and the Schengen areas as shown in figure 3. In what ways may these organisations have contributed to the mass movement of people?

Think

7. Imagine you are a refugee moving from place to place on your journey across Europe seeking asylum. In small groups, write a series of 'tweets' for the social media site Twitter that describe your wellbeing in a number of countries.

12.8 How does Australia contribute to global human wellbeing?

12.8.1 Australia's international assistance to global wellbeing

After World War II Australia welcomed many European migrants, especially from Italy and Greece, who were hoping to improve their wellbeing after the traumas of war and the decline of their economic and social structures. In the 1970s, refugees and 'boat people' (asylum seekers), fearing conflict and persecution within Vietnam, fled to Australia. With the break-up of Yugoslavia in the 1990s, eastern European nationalities sought visas for Australia. Over time, Australia has become a multicultural society.

12.8.2 Who comes to Australia seeking asylum in the twenty-first century?

Conflict in Afghanistan, Iran, the Islamic Republic of Iraq, Pakistan, Sri Lanka, Myanmar and many parts of Africa has seen deterioration in the living standards and human wellbeing in those countries. Many people have left via approved migration channels, while others have gone outside the legal system and become asylum seekers, either travelling by air or by boat.

The number of boats bringing asylum seekers from across Asia (see figure 1) to Australia's territorial waters, in particular Christmas Island, the Cocos Islands and Ashmore and Cartier Islands, became pronounced from 2009 to 2013 (see figure 3).

FIGURE 1 The movement of asylum seekers across 'stepping stones' and through hubs en route to Australian territorial waters

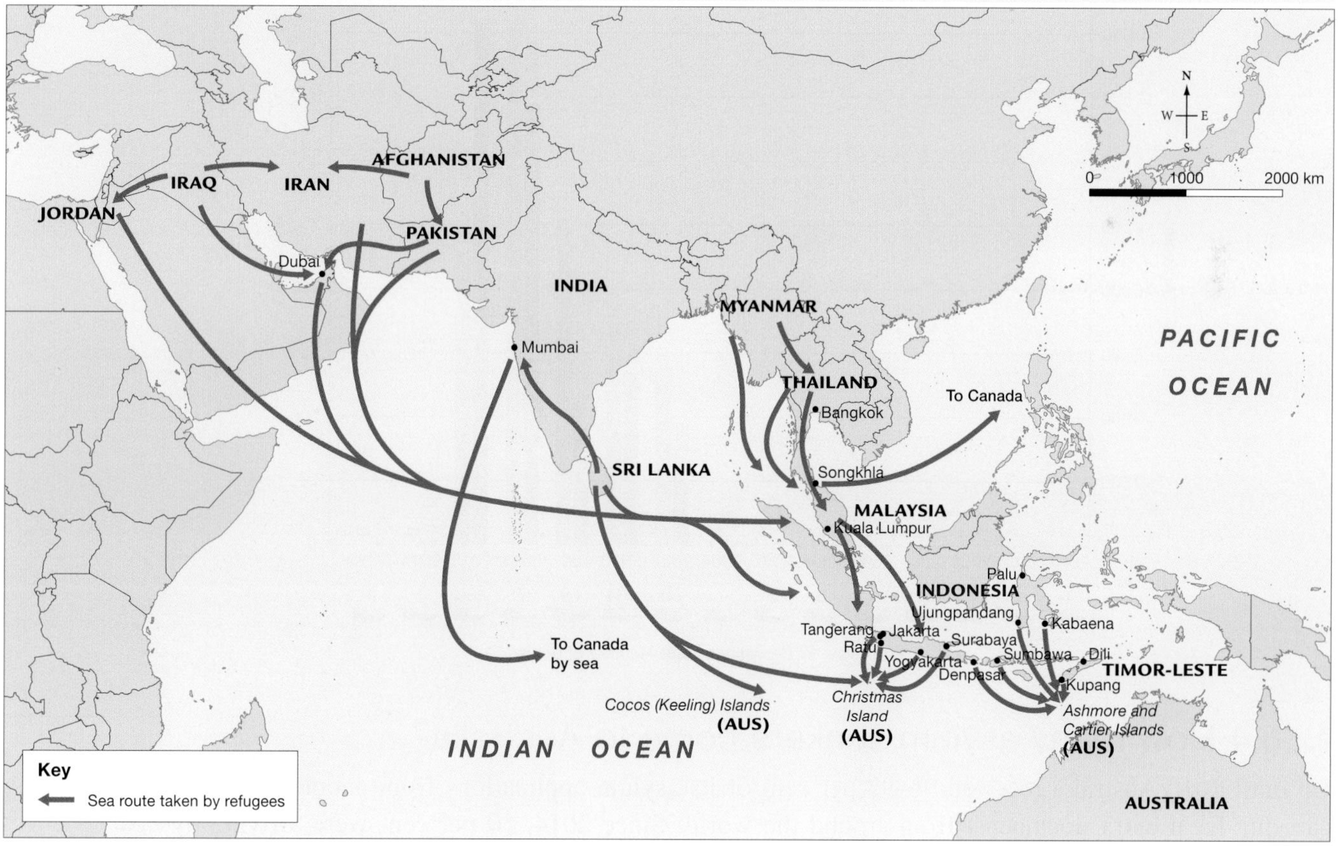

Source: World Vision

12.8.3 Why do these people come by boat?

A range of 'push and pull' factors see people becoming refugees and asylum seekers. Some of these factors occur in the home country; some occur in the host country (see figure 2).

FIGURE 2 The 'push and pull' factors that influence people movement across the seas

Push
• Instability
• Violence
• Need for physical safety
Pull
• Stability
• Empowerment
• Economic prospects
• Education
• Existing diaspora (dominant factor)
• Perception of prospects of remaining permanently
• Available support —health, housing, welfare
Home country
Host country

FIGURE 3 Australian asylum seeker arrivals, boat numbers and drownings, 2001–2015

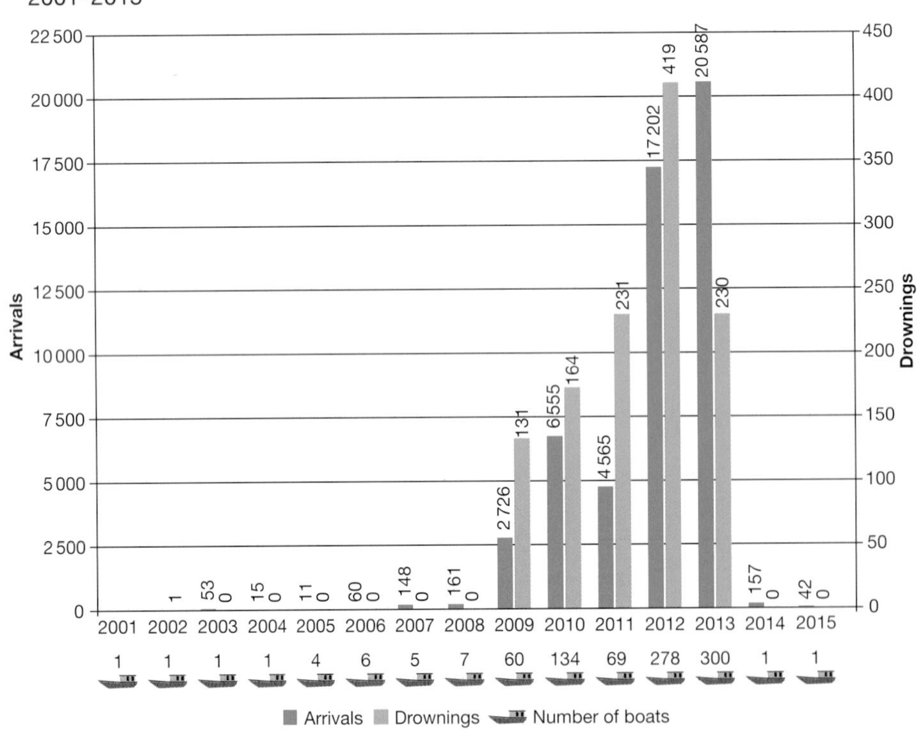

12.8.4 How many asylum seekers come to Australia?

Up until 2010 Australia received 96–98 per cent of its asylum applications from people arriving by air carrying legal entry documents from around the world. Since 2014, 50 per cent were arriving by boat without legal entry documents. It is these arrivals that have drawn the attention of the government and the media.

12.8.5 Australia's role in its region

A timeline of the Australian Government's policy change on asylum seekers:

- 2001–2008: Australia's Pacific Solution policy saw few asylum seekers reach Australia (see figure 3). Navy ships stopped the boats and transferred these asylum seekers to offshore detention centres for processing on the islands of Nauru and Papua New Guinea.
- Late 2007: a change of government dismantled this policy.
- By 2013 asylum seeker boat arrivals in Australian maritime territories, and deaths at sea, had increased considerably (see figure 3). Government policy reverted to detention and processing on the offshore islands after the High Court ruled that detaining asylum seekers in Malaysia was illegal.
- Late 2013 saw another change in government and the introduction of a 'turn-back the boats' policy.
- In 2014 Indonesia expressed its displeasure at this policy and suspended diplomatic relations with Australia.
- By 2016 more than 630 asylum seekers on 20 boats had been turned back.

Developing countries host more than 86 per cent of all refugees; Australia hosts less than one per cent of all asylum applications. Nauru receives one-third to one-half of its revenue from Australian aid and payment of $1000 a month in visa fees for each asylum seeker. The centre employs hundreds of Nauruans and is the biggest source of revenue for the country. Papua New Guinea receives similar aid from Australia to detain and process young male asylum seekers on Manus Island and resettle the people who gain refugee status there.

Diplomatic discussion continues with third party countries, such as Cambodia and the Philippines, to accept people from Nauru who have been declared refugees. Other diplomatic discussions take place with country-of-origin countries, such as Iran, about the return of refugees to their homeland. Australia continues to declare that these asylum seekers will never live in Australia.

In response to the Syrian refugee crisis, in 2015 Australia announced it would take an additional 12 000 humanitarian refugees from the refugee camps of Lebanon, Jordan and Turkey. Australia would also provide A$44 million in assistance for food, water, health and education within the Syrian region.

12.8.6 What role do NGOs play in assisting asylum seeker wellbeing in Australia?

There is a range of non-government organisations (NGOs) working to aid asylum seekers after arriving in Australian territory. Some of these are part of global organisations such as the UNHCR and Amnesty International; others are national organisations such as the Refugee Council of Australia, the Asylum Seeker Resource Centre and ACOSS (Australian Council of Social Services); while some are specifically state orientated, such as St Vincent de Paul Society in Victoria, Starts in Western Australia and the NSW Human Rights Law Centre. These groups are concerned for the wellbeing of asylum seekers.

12.8 Activities

To answer questions online and to receive **immediate feedback** and **sample responses** for every question, go to your learnON title at www.jacplus.com.au. *Note*: Question numbers may vary slightly.

Remember

1. How do we know these people are asylum seekers rather than refugees?
2. Where in Australia have asylum seekers arrived to seek residency?

Explain

3. Using figure 1:
   (a) Describe the probable route taken by an Iraqi, Sri Lankan, Afghan and Myanmar asylum seeker to reach Australian territory. In your answer include a sense of scale or distance travelled.
   (b) Suggest why journeys go through multiple countries and hubs.
   (c) Why do people go outside the UNHCR assessment programs? Figure 2 will be useful to discuss 'push' and 'pull' factors.

Discover

4. Look at the graph in figure 3.
   (a) Outline the change in boat numbers, number of arrivals and deaths at sea from 2010 to 2015.
   (b) Suggest how these statistics reflected Australia's approach to asylum seekers during two phases in this time period.
   (c) How would you assess Australia's attitude to the wellbeing of the asylum seekers from 2010 to 2015?
5. Using the Department of Immigration and Border Security site find Australia's migration levels for 2015 to 2019. In what way might these quotas affect the decision of people to become asylum seekers?
6. Choose two NGOs working on behalf of asylum seekers and research their aims and undertakings to assist human wellbeing.

Think

7. Australia's policies have affected its neighbours. Outline the impact on wellbeing in the developing countries of Indonesia, Nauru and Papua New Guinea.
8. Is Australia's multicultural society affected by twenty-first century political policy developments?
9. Hold a class debate. Should Australia take more asylum seekers? The SkillBuilder, 12.9 'Debating like a geographer' may be of use.

Predict

10. Suggest how asylum seekers in a detention centre, expecting to find a home in Australia, would feel about their future on a developing Pacific Island with no hope of ever living in Australia.

12.9 SkillBuilder: Debating like a geographer

WHAT DOES DEBATING LIKE A GEOGRAPHER MEAN?

Debating like a geographer is being able to give the points for and against any issue that has a geographical basis, and supporting the ideas with arguments and evidence of a geographical nature.

Go online to access:

- a clear step-by-step explanation to help you master the skill
- a model of what you are aiming for
- a checklist of key aspects of the skill
- a series of questions to help you apply the skill and to check your understanding.

FIGURE 1 Palm cards for a debate on whaling in the Southern Ocean

Affirmative speaker 1
(Introduces key ideas)
- Where is the Southern Ocean?
- Who is whaling?
- Which countries are involved in the issue?
- How far is it from Japan?
- Whale species
- Uses of whale meat
- The role of tradition
- Scientific research

Negative speaker 1
(Negates affirmative speaker 1 and introduces key ideas)
- Southern Ocean is a whale sanctuary
- Why don't the trawlers work closer to home?
- What is so important about the whale hunting that the benefits outweigh the costs?
- Global food chains affected
- Animal cruelty

Affirmative speaker 2
(Negates negative speaker 1 and expands on key ideas—provides the facts, statistics, emotional argument)
- Whale numbers
- Scientific research: what is research achieving?
- Importance of tradition

Negative speaker 2
(Negates affirmative speaker 2 and expands on key ideas—provides the facts, statistics, emotional argument)
- Global food chains: facts
- How are whales caught? Is it humane?
- The work of Greenpeace, its actions, the conflict
- International Whaling Commission, its work, the global ban

Affirmative speaker 3
(Negates negative speaker 2 and sums up key ideas)
- Emphasises that resource is well managed: whaling is not the only threat to species

Negative speaker 3
(Negates affirmative speaker 3 and expands on key ideas)
- Emphasises the resource is being degraded and conflict is rife

Watch this eLesson: Debating like a geographer (eles-1762)

Try out this interactivity: Debating like a geographer (int-3380)

12.10 Is all well in paradise?

online only

Access this subtopic at **www.jacplus.com.au**

12.11 How can wellbeing be addressed during conflict?

12.11.1 Caring for wellbeing

The countries in conflict vary from year to year. Conflict disrupts life. Sometimes conflicts are short term and a country moves towards peace. In other countries, conflicts linger and the level of development of a country is affected — health, education, wealth and population structure.

International NGOs assist the well-being of civilians caught up in conflicts. Three significant organisations are as follows:

FIGURE 1 Médecins Sans Frontières personnel assist children in South Sudan.

1. Médecins Sans Frontières provides emergency medical care. During conflict, local health systems often fail and hospitals close. In refugee camps, waterways may become contaminated, waste abounds and there is a lack of sanitation, all of which can lead to an outbreak of disease (see figure 1).
2. The International Red Cross and Red Crescent Movement is the largest humanitarian network. It aims to alleviate human suffering, protect life and health, and uphold human dignity, especially when the population structure is imbalanced by the predominantly male involvement in conflict.
3. World Food Programme (WFP) steps in when the distribution of food and other resources for the population is disrupted. WFP saves lives and protects livelihoods, reduces chronic hunger, and restores and rebuilds lives, especially for women and children.

12.11.2 Moving towards peace

The United Nations (UN) peacekeeping operation started in the Middle East in 1948. The UN implements peace agreements; monitors ceasefires; assists in **disarmament**, **demobilisation** and **reintegration** of former combatants; protects civilians; protects and promotes human rights; assists in restoring the rule of law; and assists in the redevelopment of a country. UN peacekeeping missions (2015) are found in 16 countries and are made up of personnel from the UN countries.

The Regional Assistance Mission to the Solomon Islands (RAMSI) is a partnership between the people and government of the Solomon Islands and 15 contributing countries of the Pacific region. RAMSI commenced in 2003 at the request of the Solomon Islands in an attempt to regain peace and stability in the region. The Solomon Islands had seen simmering civil unrest since 1999. RAMSI helped the Solomon Islands to return to long-term stability, security and prosperity. RAMSI also assisted with the development of improved laws, justice and security. The democratic government has become more effective, with an improved economic base and greater development of services for the people (see figure 2).

12.11.3 The Global Peace Index

The Global Peace Index (see figure 3) uses 23 indicators and 30 other factors of wellbeing to assess a country's 'peacefulness'. Among the criteria used are elements of peace at home (government stability, democratic processes, community relations, security and trust between people) and peace in foreign relations (military spending levels, commitment to the United Nations and avoidance of war).

FIGURE 2 RAMSI personnel assisted with improving the wellbeing of residents and are now in a policing role.

FIGURE 3 Global Peace Index, 2015

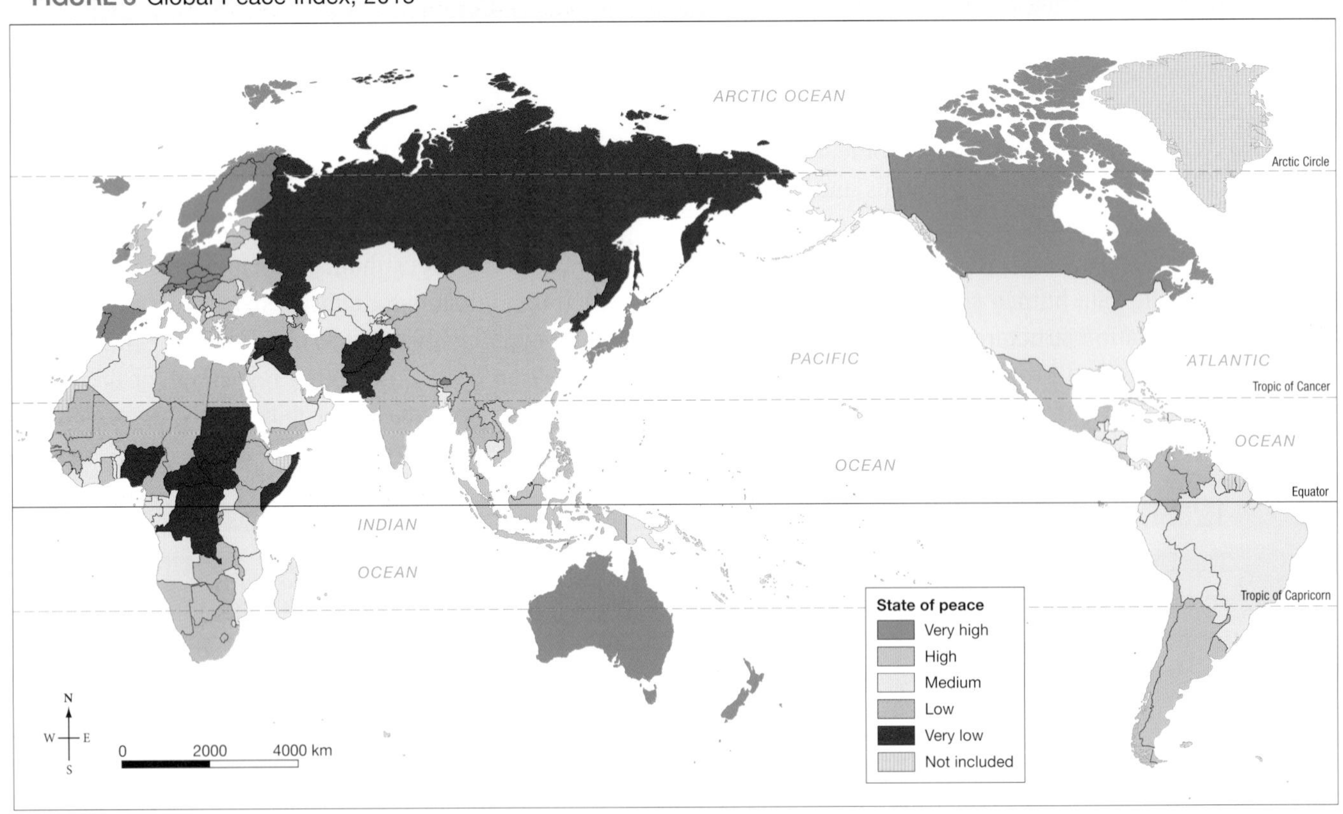

Source: Institute for Economics and Peace

12.11 Activities

To answer questions online and to receive **immediate feedback** and **sample responses** for every question, go to your learnON title at www.jacplus.com.au. *Note*: Question numbers may vary slightly.

Remember

1. What role do NGOs play in restoring wellbeing to a country?
2. What roles does the UN provide for a country to move towards peace?

Explain

3. The Global Peace Index (2015) is mapped in figure 3. Describe the distribution of ***places*** with a very low level of peacefulness and those with a high level of peacefulness.
4. Explain why criteria for assessing peace levels of a country use indicators at home and in foreign relations.

Discover

5. Using the internet, research an international NGO and show how it is working towards peace in areas of very low peacefulness. Focus on a country not studied in this topic.
6. Use the **Peacekeeping** weblink in the Resources tab to describe how peacekeeping has ***changed*** over time.
7. Research the work done by RAMSI. Outline two tasks that have been completed to ensure a move towards peace.

Predict

8. From the conflicts discussed in this topic, which of the countries would you expect to be rated differently on the Global Peace Index in 2020?
9. When RAMSI pulls out of the Solomon Islands, do you expect peace to be maintained?

Think

10. Write an essay to show how a country's HDI ranking and Peace Index levels indicate that people's wellbeing may change. Use examples from three countries that are in conflict.
11. Is power gained through conflict?

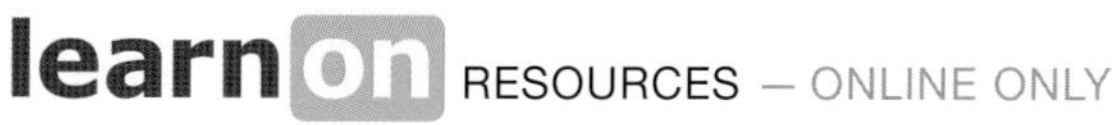

Explore more with this weblink: Peacekeeping

Deepen your understanding of this topic with related case studies and questions.
Defence and peacekeeping

12.12 SkillBuilder: Writing a geographical essay

WHAT IS A GEOGRAPHICAL ESSAY?

A geographical essay is an extended response structured like any essay, but it focuses on geographical facts and data, particularly data that can be mapped.

Go online to access:

- a clear step-by-step explanation to help you master the skill
- a model of what you are aiming for
- a checklist of key aspects of the skill
- a series of questions to help you apply the skill and to check your understanding.

FIGURE 1 An essay plan

Introduction: A freeway should not go through the urban parkland. Three reasons, or themes, are listed.

Theme 1: Noise levels from traffic. Currently peaceful environment. Sound barriers don't work.

Theme 2: House and land prices will decrease. People will not buy property because of the noise. Lifestyle is changed; roads and pathways are divided by the freeway; many people can't get to the parkland.

Theme 3: Animals will lose habitat and movement routes. Currently the area is home to kangaroos, and the habitat will be diminished. Vegetation may not support the kangaroos, animals will suffer.

Conclusion: If a road has to go through this area, it must be a tunnel under the parkland.

learnon RESOURCES — ONLINE ONLY

Watch this eLesson: Writing a geographical essay (eles-1763)

Try out this interactivity: Writing a geographical essay (int-3381)

12.13 Review

12.13.1 Review

The Review section contains a range of different questions and activities to help you revise and recall what you have learned, especially prior to a topic test.

12.13.2 Reflect

The Reflect section provides you with an opportunity to apply and extend your learning.

Access this subtopic at **www.jacplus.com.au**

TOPIC 13 Fieldwork inquiry: Comparing wellbeing in the local area

13.1 Overview

Numerous **videos** and **interactivities** are embedded just where you need them, at the point of learning, in your learnON title at www.jacplus.com.au. They will help you to learn the content and concepts covered in this topic.

13.1.1 Scenario and your task

You may have noticed that there are distinct variations across space in any city, suburb or regional community in terms of human wellbeing. Your council has asked for locals to inform them about differences in wellbeing they notice within their local areas, and what could or should be done about these in the future. Investigation of the topic will require you to undertake some fieldwork in order to make first-hand observations in the field, collect, process and analyse data.

Your task

Your task is to produce a fieldwork report you could present to your local council that outlines variations within your local area, reasons for the differences and strategies to improve the situation in the future. The aim of the fieldwork is for you to explore some of these variations by comparing two places at the local scale. The key inquiry questions the council wants to know answers to are:

- How does wellbeing vary between area X and area Y in the local area?
- What factors might explain the variations in wellbeing?
- How can wellbeing be improved in the local area?

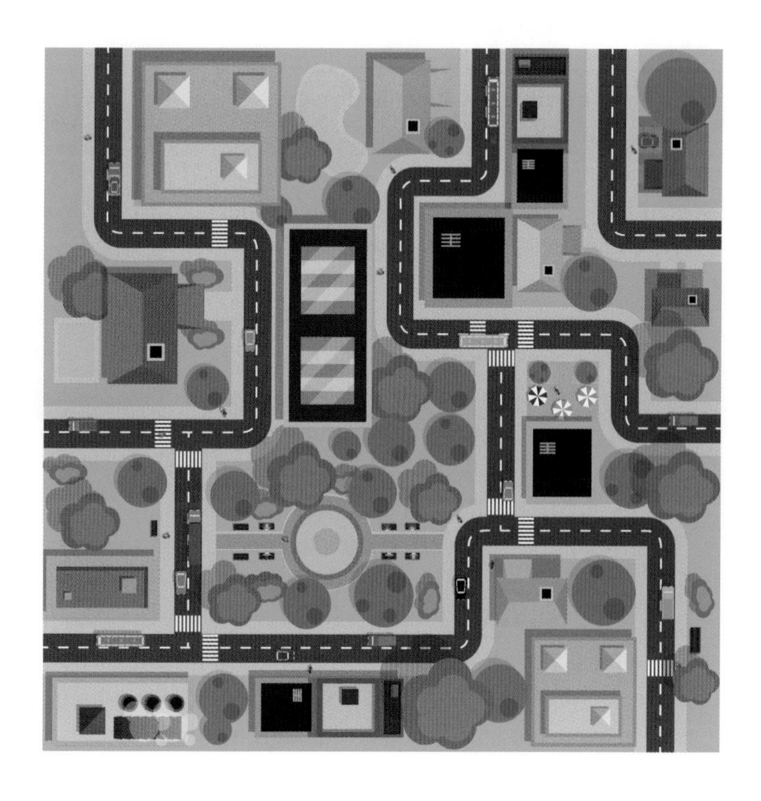

13.2 Process

13.2.1 Process

- Go to **www.jacplus.com.au** to access and watch the introductory video lesson for this fieldwork enquiry.
- **Planning:**
  1. As a class, discuss the types of indicators you would use as a basis for comparing wellbeing in your local area; for example, surveys. Decide on teams and allocate tasks, or different streets, to each team member. Download the fieldwork planning document from the Resources tab. Use the task list to help you plan your fieldwork.
  2. You will need to determine the features of the houses and streets that you wish to gain data about. How will you record this data on the day (per house block or per street block)? Think carefully and plan your data record sheet so it is easy to use and also easy to summarise. Download the sample street analysis template from the Resources tab to help you plan and to record your housing data.
  3. If you are planning to survey people, you will need to plan and prepare survey questions. Download the community sample survey template from the Resources tab to help you plan and to record your data.

13.2.2 Collecting and recording data

1. Prior to going on your field trip, prepare a simple map to show the location of your fieldwork site(s) relative to key features such as your school or city centre. You will need a separate location map if your second site is not in the same area.
2. Prepare a more detailed map of your fieldwork site(s). Use a street directory, Google Earth or local council map as a guide. Include streets, street names, schools, preschools, shops and shopping centres, parks, public transport and other community facilities. Complete your map with BOLTSS.
3. During the field trip you may be required to survey houses whereby you record key features, take photographs (*Hint:* Keep a record of the location of photographs taken) and survey local residents in public places such as parks and shopping centres.
4. Download the fieldwork report document from the Resources tab to help you prepare your report.

13.2.3 Analysing your information and data

- An important skill is the ability to analyse the information you have collected on your field trip and any other supplementary data, in order to write the findings of your inquiry into a fieldwork report. A key part of your report is to determine any patterns or trends revealed in the data. At the same time, try to identify any anomalies (variations) from the patterns or trends. Download the analysis document from the Resources tab to help you further analyse the data you have collected.

13.2.4 Communicating your findings

- Formally write your observations as a fieldwork report using these suggested sub-headings:
  - Background and key inquiry question (include location descriptions and map(s))
  - Conducting the fieldwork (planning and collection of data)
  - Findings (results of data analysis)
  - Future (How might wellbeing in the local community be improved upon? What could local councils and other community-based organisations do to improve living conditions? You might like to put forward a proposal to local council outlining your suggestions.)

  You may wish to add your own headings.

13.3 Review

13.3.1 Reflecting on your work

- Think about how you approached this fieldwork project and how you, personally, were able to organise yourself and contribute to the working of the team. Access and complete the reflection document in the Resources tab. Be honest in assessing your strengths and areas where you think you could do better next time.
- Submit your reflection document along with your completed fieldwork report.

GLOSSARY

ageing population: an increase in the number and percentage of people in the older age groups (usually 60 years and over)
algal blooms: rapid growth of algae caused by high levels of nutrients (particularly phosphates and nitrates) in water
alluvial plain: an area where rich sediments are deposited by flooding
alpha world city: a city generally considered to be an important node in the global economic system
antenatal care: the branch of medicine that deals with the care of women during pregnancy, childbirth and recovery after childbirth
aquaculture: the farming of aquatic animals and plants; also called fish farming
aquifers: layers of rock which can hold large quantities of water in the pore spaces
arable: refers to land that is suitable for growing crops
asylum seekers: people who are awaiting confirmation of their refugee status
atoll: a coral island that encircles a lagoon
base flow: water entering a stream from groundwater seepage, usually through the banks and bed of the stream
biocapacity: the capacity of a biome or ecosystem to generate a renewable and ongoing supply of resources and to process or absorb its wastes
biodegradable: capable of being decomposed through the actions of microorganisms
biodiversity: the variety of plant and animal life within an area
biophysical environment: all elements or features of the natural or physical and the human or urban environment including the interaction of these elements
bioremediation: the use of biological agents, such as bacteria, to remove or neutralise pollutants
booms: floating devices to trap and contain oil
carbon credits: term for a tradable certificate representing the right of a company to emit one metric tonne of carbon dioxide into the atmosphere
child: any person below 18 years of age
child soldiers: child who is, or who has been, recruited or used by an armed force or armed group in any capacity, including but not limited to children, boys and girls, used as fighters, cooks, porters, messengers, spies, or for sexual purposes. This term does not refer only to child who is taking, or has taken a direct part in hostilities.
Chlamydia: a sexually transmitted disease infecting koalas
civil wars: a war between citizens of the same country
classification: the categorisation of characteristics, changes, factors into distinctive groups
climate change: any change in climate over time, whether due to natural processes or human activities
continental shelf: a shallow area of sea which is a gentle continuation of the neighbouring landmass. It is usually less than 300 metres deep. After the shelf, the continental slope drops into deep ocean.
conurbation: an urban area formed when two or more towns or cities (e.g. Tokyo and Yokohama) spread into and merge with each other
Coriolis force: force that results from the Earth's rotation. Moving bodies, such as wind and ocean currents, are deflected to the left in the Southern Hemisphere and to the right in the Northern Hemisphere.
cottage industry: an industry where the creation of products and services is home-based rather than factory-based
cull: selective reduction of a species by killing a number of animals
deltaic plain: flat area where a river(s) empties into a basin
demobilisation: disbanding or discharging troops in a move towards peace

demographic transition model: a graph attempting to explain how a country's population characteristics change as the level of wellbeing in a country improves over time

dependent population: those in the under 15 years and over 60 years age groups. People in these age groups are dependent on those in the working age groups, either directly or indirectly for support.

desertification: the transformation of land once suitable for agriculture into desert by processes such as climate change or human practices such as deforestation and overgrazing

developing nation: a country whose economy is not well developed or diversified, although it may be showing growth in key areas such as agriculture, industries, tourism or telecommunications

development: According to the United Nations, development is defined as 'to lead long and healthy lives, to be knowledgeable, to have access to the resources needed for a decent standard of living and to be able to participate in the life of the community'.

diplomats: people who manage international relations

disarmament: reducing, limiting or abolishing weapons

diversion: (water) man-made project to divert the water from a river

downturn of the global economy: a recession or downturn in a nation's economic activity which includes increased unemployment and also decreased consumer spending

dyke: an embankment constructed to prevent flooding by the sea or a river

ecological footprint: the amount of productive land used on average by each person (in the world, a country, etc.) for food, water, transport, housing, waste management, and other purposes

ecological service: the benefits to humanity from the resources and processes that are supplied by natural ecosystems

ecosystems: systems formed by the interactions between the living organisms (plants, animals, humans) and the physical elements of an environment

emissions trading scheme: a market-based, government-controlled system used to control greenhouse gas as a cap on emissions. Firms are allocated a set permit or carbon credit and they cannot exceed that cap. If they require extra credits, they must buy permits from other firms that have lesser needs or a surplus.

enhanced greenhouse effect: the observable trend of rising world atmospheric temperatures over the past century, particularly during the last couple of decades

enhanced greenhouse gas emissions: describes the observable trend of rising world temperatures over the past century, particularly during the last couple of decades, due to the burning of fossil fuels

environmental worldview: varying viewpoints, such as environment-centred as opposed to human-centred, in managing ecological services

ephemeral: describes a stream or river that flows only occasionally, usually after heavy rain (e.g. Todd River, Alice Springs)

eutrophication: a process where water bodies receive excess nutrients that stimulate excessive plant growth

exotic species: species introduced from a foreign country

female infanticide: the killing of female babies, either via abortion or after birth

fertility rate: the average number of children born per woman

floating settlements: anchored buildings that float on water and are able to move up and down with the tides

flood mitigation: managing the effects of floods rather than trying to prevent them altogether

fossil fuels: carbon-based fuels formed over millions of years, which include coal, petroleum and natural gas. They are called non-renewable fuels as reserves are being depleted at a faster rate than the process of formation.

GDP: (gross domestic product) market value of all goods and services produced within a country in a given period

geothermal: (power) describes power that is generated from molten magma at the Earth's core and stored in hot rocks under the surface. It is cost-effective, reliable, sustainable and environmentally friendly.

ghost nets: rafts of plastic fishing nets that have been lost at sea, abandoned or deliberately discarded. They can continue to float in ocean currents and trap fish, birds and anything else that crosses their path.

global warming: increased ability of the Earth's atmosphere to trap heat

gross domestic product: (GDP); a measurement of the annual value of all the goods and services bought and sold within a country's borders. We normally discuss GDP per capita (total GDP divided by the population of the country).

groundwater: water held underground within water-bearing rocks or aquifers

groundwater salinity: presence of salty water that has replaced fresh water in the subsurface layers of soil

gyre: swirling circular ocean current (similar to water swirling around a plug hole)

high tide line: the line on land where the water's surface is at maximum height reached by a rising tide

hinterland: the land behind a coast or shoreline extending a few kilometres inland

historical architecture: urban environment that has significant value due to its unique form and history of development

housing affordability: relates to a person's ability to pay for their housing. In Australia, those spending more than 30 per cent of their income on housing, while earning in the bottom 40 per cent of the income range, are considered to be in housing affordability stress.

Human Development Index (HDI): measures the standard of living and wellbeing by measuring life expectancy, education, literacy and income

human–environment systems thinking: using thinking skills such as analysis and evaluation to understand the interaction of the human and biophysical or natural parts of the Earth's environment

humanitarianism: concern for the welfare of other human beings

humus: decaying organic matter that is rich in nutrients needed for plant growth

impervious: a rock layer that does not allow water to move through it due to a lack of cracks and fissures

indicators: a value that informs us of a condition or progress. It can be defined as something that helps us to understand where we are, where we are going and how far we are from the goal.

Indigenous: Australia's Indigenous peoples are made up of Aboriginal people (who live all around Australia) and Torres Strait Islanders (who settled the many small islands to the north of Cape York Peninsula in Queensland).

Industrial Revolution: started in approximately 1750 in the United Kingdom, eventually spreading to the rest of the world. It brought major technological changes in agriculture, manufacturing, mining and transportation. A major invention of the era was the application of coal and eventually oil and gas power to industry.

industrialised: having developed a wide range of industries or having highly developed industries

infrastructure: the basic physical and organisational structures and facilities (e.g. buildings, roads, power supplies) needed for the operation of a society

infrastructure: the structural framework, features or systems of the urban environment that provide services to a country city or area

internally displaced persons (IDP): people travelling within their country to 'safer' places, legally remaining under the control of their government

International Bill of Human Rights: the informal name given to the Universal Declaration of Human Rights and the two International Covenants

International Covenants: a multilateral treaty adopted by the United Nations General Assembly in force from 1976. It commits those who have signed the Covenant to respect the civil and political rights of individuals and their economic, social and cultural rights.

invasive plant species: commonly referred to as weeds; any plant species that dominates an area outside its normal region and requires action to control its spread

Kyoto Protocol: an internationally agreed set of rules developed by the United Nations aiming to reduce climate change through the stabilisation of greenhouse gas emissions into the atmosphere

lagoon: a shallow body of water separated from the sea by a sand barrier or coral reef

life expectancy: the number of years a person can expect to live, usually when they are born, based on the average living conditions within a country

manufacturing and industrial base: to make or process materials into finished products by means of large- and medium-scale industrial operation

maternal mortality: the death of a woman while pregnant or within 42 days of termination of pregnancy

medium-density housing: a form of residential development such as detached, semi-attached and multiunit housing that can range from about 25 to 80 dwellings per hectare

micro hydro-dams: produce hydro-electric power on a scale serving a small community (less than 10 MW). They usually require minimal construction and have very little environmental impact.

monsoon: a wind system that brings heavy rainfall over large climatic regions and reverses direction seasonally

mulch: organic matter such as grass clippings

national park: an area set aside for the purpose of conservation

natural increase: the difference between the birth rate (births per thousand) and the death rate (deaths per thousand). This does not include changes due to migration.

nautical mile: a unit of measurement used by sailors and/or navigators in shipping and aviation. It is the average length of one minute of one degree along a great circle of the Earth. One nautical mile corresponds to one minute of latitude. Thus, degrees of latitude are approximately 60 nautical miles apart.

neonatal intensive care: the specialised nursing practice of caring for newborn infants.

non-government organisation (NGO): a citizen-based association that operates independently of government, usually to deliver resources or serve some social or political purpose

perennial: describes a stream or river that flows permanently

photodegradation: action of sunlight breaking down plastic into minute particles

population density: the number of people within a given area, usually per square kilometre

population distribution: the spread of people across the globe

population structure: the number or percentage of males and females in a particular age group

qualitative indicators: usually consists of a complex set of indices that measure a particular aspect of quality of life or describe living conditions. Useful in analysing features that are not easily calculated or measured, such as freedom or security.

quantitative indicators: easily measured and can be stated numerically, such as annual income or how many doctors there are in a country

rainwater harvesting: the accumulating and storing of rainwater for re-use before it soaks into underground aquifers

ratified: to sign or give formal consent to a treaty, contract, or agreement, making it officially valid

recharge: the process by which groundwater is replenished by the slow movement of water down through soil and rock layers

refugees: people who flee through fear of persecution — for reasons of race, religion, nationality, membership of a social group, or because of a political opinion — and cross outside their home borders

regional and remote areas: areas classified by their distance and accessibility from major population centres.

reintegration: repairing the fractures (breaks) in society after conflict

remediation: act or process of correcting a fault

replacement rate: the number of children each woman would need to have in order to ensure a stable population level — that is, to 'replace' the children's parents. This fertility rate is 2.1 children.

reservoirs: large natural or artificial lakes used to store water, created behind a barrier or dam wall

ringbark: remove the bark from a tree in a ring that goes all the way around the trunk. The tree usually dies because the nutrient-carrying layer is destroyed in the process.

river delta: a landform composed of deposited sediments at the mouth of a river where it flows into the sea

river fragmentation: the interruption of a river's natural flow by dams, withdrawals or transfers
salinity: an excess of salt in soil or water, making it less useful for agriculture
salt scald: the visible presence of salt crystals on the surface of the land, giving it a crust-like appearance
sanitation: access to facilities that safely dispose of human waste (urine, faeces and menstrual waste)
sex ratio: the number of females per 1000 males
slum: rundown area of a city with substandard housing
socioeconomic: of, relating to or involving a combination of social and economic factors
standard of living: a level of material comfort in terms of goods and services available to someone or some group. This is often measured on a continuum; for example, a 'high' or 'excellent' standard of living compared to a 'low' or 'poor' standard of living.
stateless people: frequently lack identification documents, live on the edges of society and are subjected to discrimination
stewardship: the caring and ethical approach to sustainable management of habitats for the benefit of all life on Earth
storm surge: a temporary increase in sea level from storm activity
subsidence: the gradual sinking of landforms to a lower level as a result of earth movements, mining operations or over-withdrawal of water
teenage pregnancy rate: number of girls aged 15–19 years who fall pregnant per thousand
terminal lake: a lake where the water does not drain into a river or sea. Water can leave only through evaporation, which can increase salt levels in arid regions. Also known as an endorheic lake.
thermoline circulation: refers to the flow of ocean water caused by changes in water density. Salt and temperature levels can change; for example, fresh water added by rain, snow melt or river run-off. Temperatures can change from contact with the atmosphere.
topsoil: the top layers of soil that contain the nutrients necessary for healthy plant growth
tsunami: a powerful ocean wave triggered by an earthquake or volcanic activity under the sea
Universal Declaration of Human Rights: the first specific global expression of rights to which all human beings are inherently entitled
urban environment: the human-made or built structures and spaces in which people live, work and recreate on a day-to-day basis
urban infilling: the division of larger house sites into multiple sites for new homes
urban renewal: redevelopment of old urban areas including the modernisation of household interiors
urban sprawl: the spreading of urban developments (as houses and shopping centres) into areas on the city boundary
urbanisation: the social and economic processes whereby cities increase in size largely due to population growth
water rights: refer to the right through ownership to use water from a water source such as a river, stream, pond or groundwater source
watertable: upper level of groundwater, or level below which the earth is saturated with water
weathering: the breaking down of rocks
weeds: any plant species that dominates an area outside its normal region and requires action to control its spread
weirs: walls or dams built across a river channel to raise the level of water behind. This can then be used for gravity-fed irrigation.
wellbeing: a good or satisfactory condition of existence; a state characterised by health, happiness, prosperity and welfare
wetland: an area covered by water permanently, seasonally or ephemerally. They include fresh, salt and brackish waters such as rivers, lakes, rice paddies and areas of marine water, the depth of which at low tide does not exceed 6 metres.

INDEX

Y